Brickwork Level 1

Brickwork Level 1 has been adapted from John Hodges's classic *Brickwork for Apprentices* – the established textbook on brickwork for bricklayers. Designed to meet new requirements of the City and Guilds bricklaying programmes, this book has been written to match the latest industry-based requirements and technical developments in the field, including recent changes to the Building Regulations. Each chapter follows the syllabus and contains a section of multiple-choice questions to provide trainees with vital practice for the job knowledge and multiple-choice tests.

Highly illustrated throughout and now in full colour, this is the essential reference for qualified bricklayers and other professionals working in the construction industry, as well as students wishing to embark on a career in bricklaying.

There is also the facility to access the Support Material on the Routledge website that includes:

- PowerPoint slides for each chapter
- Lesson plans and schemes of work
- Multiple-choice questions and answers
- Job knowledge questions and answers
- Practical drawings and mark sheets

Malcolm Thorpe is a Life Member of the Guild of Bricklayers and former Head of Construction at West Nottinghamshire College, UK. He has acted as a CITB adviser and was involved in the drafting of bricklaying programmes. He was a verifier for CITB and spent several years producing examination material for examining bodies. He also spent over thirty years designing and organising Guild of Bricklayers local and national bricklaying competitions.

Brickwork Level 1

For Construction Diploma and Apprenticeship Programmes

Second Edition

Malcolm Thorpe

Routledge
Taylor & Francis Group
LONDON AND NEW YORK

Second edition published 2021
by Routledge
2 Park Square, Milton Park, Abingdon, Oxon, OX14 4RN

and by Routledge
52 Vanderbilt Avenue, New York, NY 10017

Routledge is an imprint of the Taylor & Francis Group, an informa business

First edition published by Routledge 2010

British Library Cataloguing-in-Publication Data
A catalogue record for this book is available from the British Library

Libary of Congress Cataloging-in-Publication Data
A catalog record has been requested for this book

ISBN: 978-0-367-62535-1 (hbk)
ISBN: 978-0-367-62534-4 (pbk)
ISBN: 978-1-003-10960-0 (ebk)

Typeset in Utopia
by MPS Limited, Dehradun

Visit the Support Material: www.routledge.com/9780367625344

Contents

Preface . . . *vii*

Chapter 1 The Construction Industry (Residential) . . . 1

Chapter 2 Health and Safety in the Construction Industry . . . 35

Chapter 3 Communication . . . 97

Chapter 4 Construction Technology . . . 139

Chapter 5 Tools . . . 173

Chapter 6 Preparing and Mixing Concrete and Mortar . . . 189

Chapter 7 Setting Out Basic Masonry Structures . . . 245

Chapter 8 Laying Bricks to Line . . . 277

Chapter 9 Laying Blockwork to Line . . . 309

Chapter 10 Cavity Walling . . . 329

Chapter 11 Joint Finishes . . . 351

Chapter 12 Answers to Multiple-Choice Questions . . . 367

Index . . . 371

Preface

Changes in construction training have led to the need to update this series of books which incorporate the new Occupational Standards.

The content of each book follows the On-Programme Training Specification.

The On-Programme specification is divided into Knowledge Requirements (K) and Skill Requirements (S).

After the initial chapter, which gives the apprentice an insight into the construction industry they are entering, each chapter follows very closely the Training Specification for Apprenticeship and Diploma units.

The aim of each book is to provide an informative resource and workbook for all building craft apprentices. It can be used to provide teaching and assessment material, or used simply to reinforce college lectures.

Each chapter has a set of multiple-choice questions designed to test your level of knowledge before you move on to the next chapter.

There is also the facility to access the Support Material on the Routledge website, which includes lesson plans, schemes of work, completed units tracking sheets, mapping sheets, PowerPoint displays, activity packs, job knowledge questions and answers, multiple-choice questions and answers, practical training units and practical competency units with marking sheets.

It can be accessed here: www.routledge.com/9780367625351

Malcolm Thorpe

CHAPTER 1

The Construction Industry (Residential)

This chapter will cover the following City and Guilds units:

- Apprenticeship Module 1 K1.1
- Diploma Unit 101/501 2.1, 2.2
  7.1, 7.2, 7.3, 7.4, 7.5

This chapter is about:

- The construction industry
- Building from different eras
- Types of residential buildings
- Listed buildings and their grading
- Types of construction
- The building team

The following City and Guilds performance criteria will be covered:

- Know building types
- Types of construction

The following Diploma outcomes will be covered:

- Know building types
- Types of construction

Introduction

When students are thinking of entering the building trade they may ask many questions. The three main questions are:

- What is the construction industry?
- What can the construction industry offer me?
- What type of education will I need?

The construction industry

The construction industry is one of the largest employers of labour in the country, with a labour force of just over one million, a figure that has dropped steadily over the past years.

Construction means creating, not only the houses we need to live in but many other buildings such as schools, hospitals and shopping centres.

The majority of buildings and structures are designed and constructed for a specific purpose. The use of the building will determine the size, shape, style and ultimately the cost.

Every person employed in the construction industry makes a direct contribution to the community in general but also to the nation.

The industry is made up of a large number of firms which can be classified as:

- builders
- contractors
- subcontractors, etc.

The firms range in size from one-person firms to multinational companies.

- A small company is defined as having between one and 49 employees.
- A medium company is defined as having between 50 and 249 employees.
- A large company is defined as having more than 250 employees.

There are also several different types of construction work to consider when thinking of joining the construction industry.

The whole industry can be further divided into four:

- *New work* refers to all types of building work and services which are about to start.
- *Maintenance work refers* to any work on an existing building which requires damaged or out-of-date items to be brought up to an acceptable standard.

Examples of maintenance work include new kitchen and bathroom units, or external brickwork requiring repointing, etc.

- *Refurbishment* is when an old building has been allowed to fall into a state of disrepair and it needs to be brought back to standard. Changing existing buildings for another purpose is also classed as refurbishment.

An example is when an old warehouse has been changed into a block of flats.

- *Restoration work* is when an old building is brought back to its original state.

Examples are old historic buildings bought by the National Trust and then painstakingly restored to their former glory.

The construction process

The construction process is said to be the most complex of all industries.

People employed in manufacturing industries travel each day to the same place and do the same type of work.

In the construction industry the employees move to a different place of work as soon as the particular job has been completed. The distance depends on the nature and size of the contractor and the availability of work.

No two construction sites are ever the same, and it is seldom that more than a few dwellings are the same.

The construction industry does not lend itself to production-line methods, so it is very labour intensive. The construction team is therefore comprised of people possessing a vast range of skills, from the tradespeople to the professionals.

The construction industry differs from other industries in the following ways:

- Work is carried out in the open and is subject to stoppages from the weather.
- Every day the plan of work is different.
- The labour force is not static and can change daily.
- Great distances have to be travelled by employees, so they are often many miles from the head office.
- Every job is different, so there is no repetition through which employees can produce more after gaining experience.
- Many of the contracts are completed by one person after being designed by another person.
- Safety in the industry has a very poor record.
- The industry is very labour intensive.

All the above statements can cause many problems and it is very difficult for any one person to rectify them; therefore the construction team becomes very important indeed.

Building styles through the ages

This subject is so vast that a short introduction can touch upon only a part of it. For that reason this chapter attempts in its text to give only a brief summary dealing with the buildings in Britain and mainly concentrating on England.

To give a wider background, a date chart is shown in Chart 1.1 to show the various styles through to the present day.

Change and development proceed at different paces and to a great deal depending upon political conditions and the economic prosperity.

The word 'architect' comes from the Greek word 'architon' meaning 'master and not 'architect' It was during the Middle Ages that the architect was referred to as a 'master and not 'architect'

It was arguable that be using the term 'chief craftsman' the person whom controls the building process was the architect.

In the past there have been many different architectural styles.

Date	Period		Features
420	Roman		Hadrian's Wall
420–1066	Saxon		Cruck houses
1066–1189	Norman		Stone built manor houses
1189–1307	Early English	Gothic	Gothic arches
307–1377	Decorated	Gothic	More intricate Gothic arches
1377–1485	Perpendicular	Gothic	Tall Gothic arches
1485–1558	Tudor		Red brick, palaces, decorative brick chimneys, inglenooks
1558–1602	Elizabethan	Renaissance	Great Fire of London
1603–1625	Jacobean	Renaissance	Cob walling
1625–1702	Stuart	Renaissance	Window tax
1695–1725	Baroque	Georgian	Town planning, terraces
1720–1760	Palladian	Georgian	Sliding sash windows
1760–1800	Adam	Georgian	Brick tax
1810–1837	Regency	Georgian	Buckingham Palace
1837–1901	Victorian		St. Pancras, Kings Cross
			Crystal Palace
			Industrial revolution
1901–1910	Edwardian		Gothic revival
1910 onwards	Modern		Garden Cities
			Two World Wars
			High rise construction

CHART 1.1
Historical styles of building

Roman Empire

The practice of architecture continued during the period of the Roman Empire. The Romans brought with them highly developed systems of building which they adapted to suit the local conditions and material sources.

Many walled towns were built on a regular grid pattern. These had well built houses, baths, temples; town halls in addition the Romans organized an efficient water system.

The buildings were constructed of stone, bricks and mortar, finished with plaster. The roofs were of tile and slates. The floors were hollow to allow warm air from a furnace to circulate underneath them.

Thousands of small coloured stones were set in cement to form intricate patterns and pictures on the floor (mosaic). Their windows were glazed with thick glass.

One of the most famous remains of Roman construction is Hadrian's Wall along the border of Scotland and England. This is about 117 kilometres long, 4–5 metres high and 3 metres wide. Forts were built at regular intervals along the wall itself and a deep trench was dug on its North side (Figure 1.1).

Saxon

After the departure of the Roman legions Britain was constantly raided by Angles, Saxons and Jutes.

The appearance of Britain gradually returned to what it was before the Roman invasion. New communities were developed by the Saxons and most buildings were constructed of timber and wattle and daub walls, though where stone was available it was also used for walling (Figure 1.2).

FIGURE 1.1
Hadrian's Wall

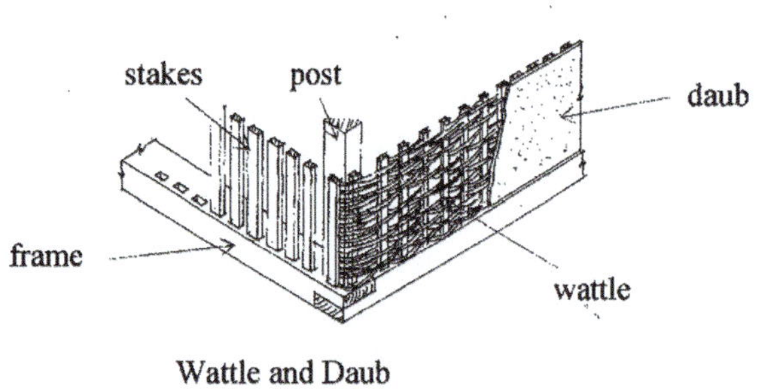

FIGURE 1.2
Wattle and daub

The common style was the Cruck form of house, built in bays, each bay was about 5 metres wide, and the number varied according to the wealth of the owner.

These crucks were massive tree trunks in pairs, meeting at the top in gable shape to support the ridge pole of the roof. Smaller timbers, called rafters, helped to hold up the roofing, which was of thatch or wood shingles.

Early entrance doors and openings were covered in animal skins, but later doors were constructed (Figure 1.3).

Later Saxon architecture, which some survives today, were of stone. Examples of Saxon towers can be seen which have vertical lines. Saxon windows were round or triangular, sometimes set in pairs producing narrow openings (Figure 1.4).

Norman

After the Norman's conquered England in 1066, Willian the conqueror developed the feudal system, all land becoming the property of the King.

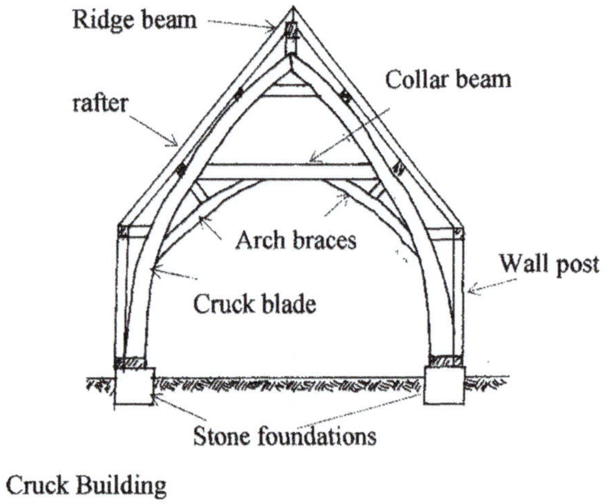

FIGURE 1.3
Cruck building

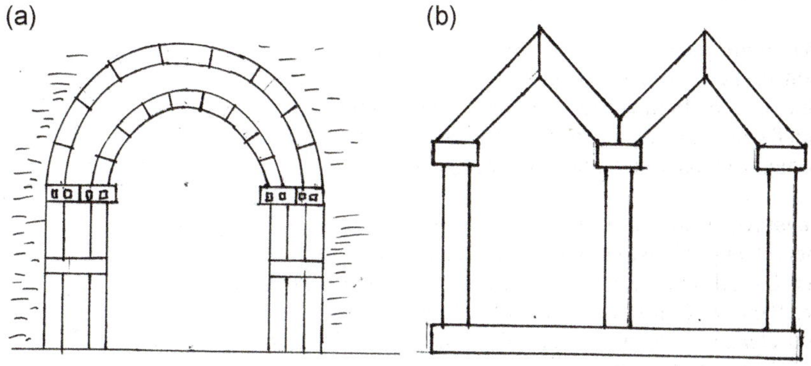

FIGURE 1.4
Saxon doorway (left) and Saxon pointed windows (right)

It was divided up between the Barons, each required a castle, and the Barons in turn gave land to their Knights who built manor houses (Figure 1.5).

Grouped around these stone buildings were the workers houses of simple wattle and daub construction. Stone built manor houses were erected with a ground floor, known as the undercroft, for storage and animals. The living accommodation was on the first floor level.

Gothic

Not many buildings survived intact from the middle ages. The greatest hazard to medieval towns was fire.

In 1189 a law was passed making it compulsory to build at least the lower part of a building in stone.

FIGURE 1.5
Manor house

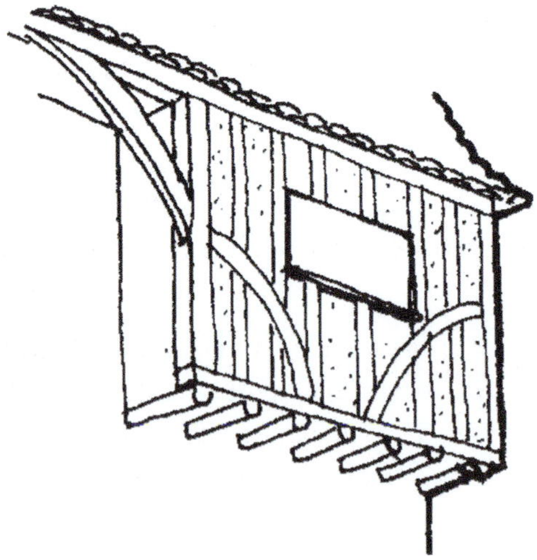

FIGURE 1.6
Overhanging upper storey

The majority of houses were built in wood and most have perished. This is because in towns where space was limited the streets were narrow and overhanging upper storey caused the loss of many houses due to the easy spread of fire (Figure 1.6).

Many buildings of this period contained arches with a pointed top. (Gothic).

Gothic arches were developed in three stages:

Early English, Decorated and Perpendicular.

Early English was a simple lancet arch. Canterbury Cathedral has many examples (Figure 1.7).

Decorated Gothic designs had larger openings and had rich coloured glass and very complex stone tracery in the head (Figure 1.8).

York Minster is an excellent example.

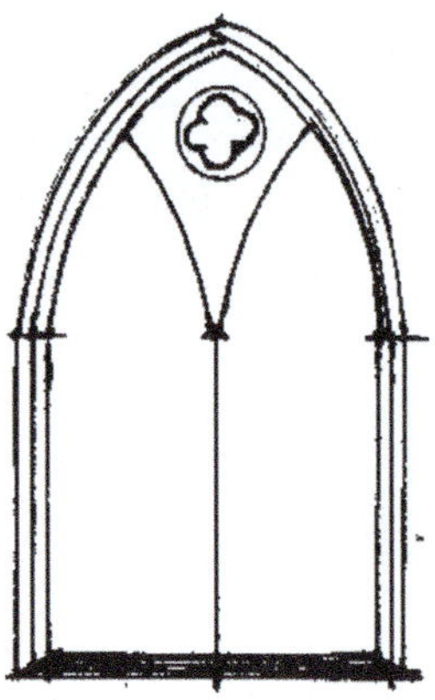

FIGURE 1.7
Early English

FIGURE 1.8
Decorated

Perpendicular as its name indicates were very tall and included panelled decoration (Figure 1.9).

Tudor

Brick was the main building material for most of the large mansions built during this period, but many had to be imported.

A prominent feature of these mansions were the gables. The roofs were very often tiled with clay or slate. The chimney stacks were very ornate (Figure 1.10).

Coal had only just started to be used as a fuel in larger mansions.

Inglenook fireplaces became a feature of this period house. A fireplace built outwards to form a small sitting area each side of the fire opening.

FIGURE 1.9
Perpendicular

FIGURE 1.10
Decorated gables

The majority of small houses were constructed from timber using a box frame principle, laid on a stone foundation. Windows were mostly frames in stone with either a flat stone head or a four centered arch. Stone mullions were used to divide the window area and leaded lights were used to glaze them.

The roofs were either of thatch, clay or slate tiles.

Renaissance

Three periods followed, Elizabethan, Jacobean and Stuart, and are all grouped together and known as the Renaissance.

The term 'renaissance' means the rebirth or rediscovery.

Elizabethan

Shakespeare's England was typical of the Elizabethan period. England was prospering and the aristocracy commissioned large houses.

Stone houses were gradually replacing timber houses, but the majority was still half timbered.

The skylines of these houses are one of their attractive and characteristic features.

They are varied and very decorative, from curving, projecting gables to decorative chimney stacks rising from mainly flat roofs.

As there was no necessity to design for defence the windows became larger.

During this period Hardwick Hall (Figure 1.11) was completed and with it the slogan 'Hardwick Hall - all glass - no wall'.

FIGURE 1.11
Hardwick Hall

Jacobean

The Jacobean period saw more elaborated features in furnishings and fittings.

Heating and lighting became a priority.

Stuart

Indigo Jones was the first important Architect in the country and he brought the Italian Renaissance to Britain which caused a revolution in designs (Figure 1.12).

He built the Queen's House in Greenwich and the Banqueting Hall in Whitehall.

FIGURE 1.12
Decorated staircase

The exteriors were of a more regular symmetric design, the roof line did not project above the walls, windows and doors were topped with pediments.

The small houses of this period continued to be mainly timber frame. In areas where clay was available for the production of bricks, the lower walls and tile hanging to upper walls became common.

In areas where chalk was available, 'knapped' flint in a brickwork frame were common (Figures 1.13(a) and (b)).

Cob walls (layers of pressed mud, gravel and straw) and thatched cottages were popular (Figure 1.14).

(a)

(b)

FIGURE 1.13
Knapped flint walling

FIGURE 1.14
Cob walling

The Great Plague spread through London in 1665 and the Great Fire of London in 1666 destroyed over 13000 buildings.

During this unfortunate period another of the Great Architects, Christopher Wren, was appointed Assistant to the Surveyor General.

He planned the new City of London and St. Paul's Cathedral was one of his greatest achievements (Figure 1.15).

He introduced Portland stone to London for church buildings.

FIGURE 1.15
St. Paul's Cathedral

FIGURE 1.16
Window filled in to avoid the window tax

Wren's designs became the inspiration for other houses to be built.

During this period a chimney tax of 2 shillings was levied for every chimney.

The tax was soon followed by the window tax of 1697 but was disbanded in 1851 (Figure 1.16).

Georgian

This Georgian period was further divided into Baroque, Palladian, Adam and Regency.

Baroque

This period is often referred to as the 'age of elegance' due to the fact that many grand and highly decorative and extravagant houses were built.

Blenheim palace was designed by Sir John Vanburgh which showed off his style.

Castle Howard was another gigantic residence which later became obsolete because of their sheer size. Chatsworth House in Derbyshire is still today used and open to the public. The design did not include attics or hipped roofs (Figure 1.17).

FIGURE 1.17
Chatsworth House, Derbyshire

Palladian

This period of architecture returned to the more sober classical designs.

During this period whole streets were developed instead of single detached houses creating the ideas for town planning.

The Circus in Bath, 1754, is a typical example were a circle of 33 houses were designed, and still there today (Figure 1.18).

This idea of designing town houses by the street, or even around a square, was initiated by Indigo Jones in Covent Garden.

FIGURE 1.18
The Circus in Bath

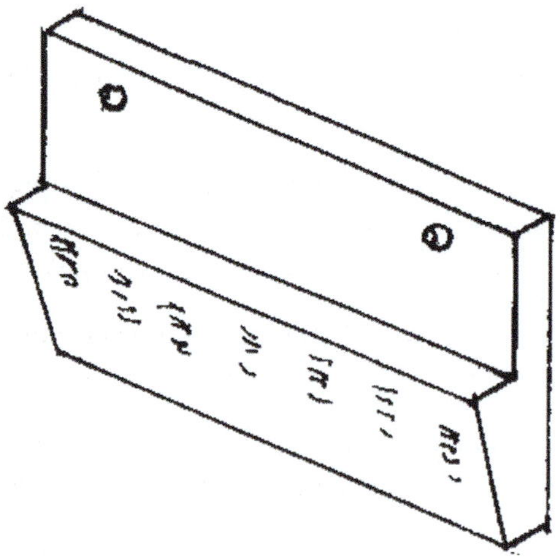

FIGURE 1.19
Mathematical tile

Adam

This period was more famous for the interior designs through Robert Adam.

The smaller houses of this period were either constructed totally in brick or timber frame covered in weather boarding.

Tile hanging was also used extensively during this period.

Mathematical hanging tiles were introduced as an alternative to bricks (Figure 1.19).

The brick tax helped increase the usage of hanging tiles and weather boarding as a way of avoiding the brick tax. Bricks were taxed per 1000 used.

The tiles were laid in a partly overlapping pattern (Figure 1.20).

Their lower section—the part intended to be visible when the tiling was complete—was thicker; the upper section would slide under the overlapping tile above and would therefore be hidden.

In the top corners there were holes for nails to be inserted. They would then be hung on a lath of wood, and the lower sections would be moulded together with an infill of lime mortar to form a flat surface.

Mathematical tiles had several advantages over brick: they were cheaper, easier to lay than bricks (skilled workmen were not needed), and were more resistant to the weathering effects of wind, rain and sea-spray, making them particularly useful at seaside locations such as Brighton.

Various colours of tile were produced: red, to resemble brick most closely; honey; cream; and black.

Regency

The Regency period is the intermediate time between the end of Georgian Britain and the beginning of the Victorian Period.

Georgian Britain was based on an agricultural economy.

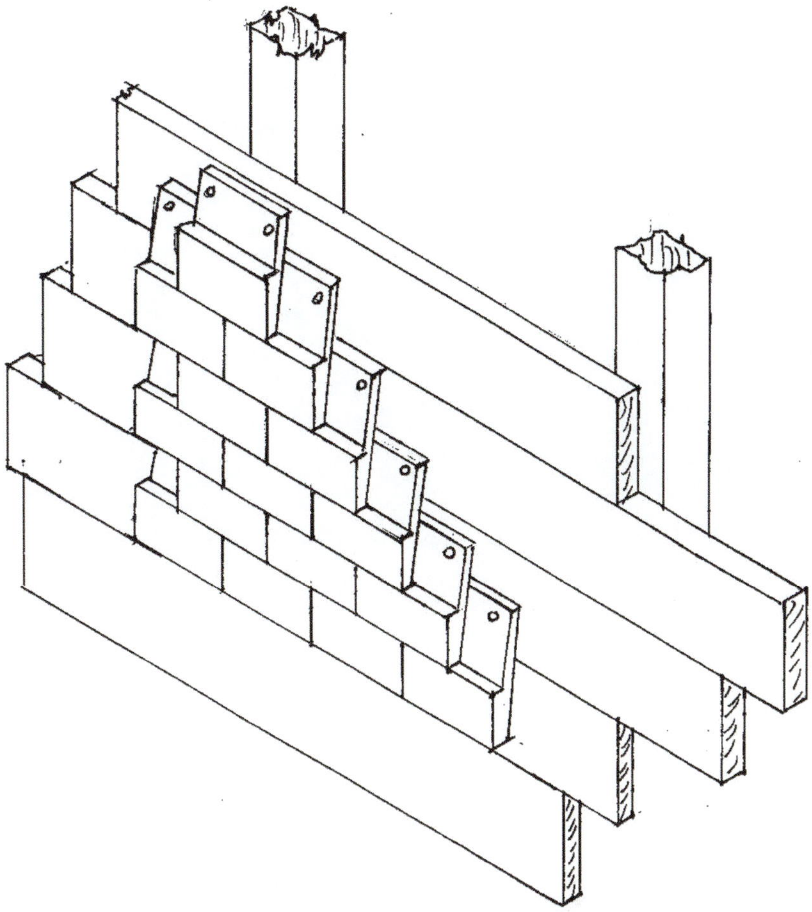

FIGURE 1.20
Mathematical tiles nailed to wooden planks, overlapped and mortared to give the appearance of a brick surface

Adams continued to influence the planning and house designs; among his most famous designs are Buckingham Palace (Figure 1.21), Marble Arch and the Brighton Royal Pavilion.

Nash introduced stucco to cover the whole building, not just the lower storey has Adam had done (Figure 1.22). Terraced housing was very popular during this period.

What are the main characteristics of a Georgian property?

- Townhouses were arranged over three or four storeys
- Sash windows with smaller panes - tall windows on the first two floors and smaller windows on the top storeys
- Symmetrical flat exterior and balanced interior layout
- Stucco-fronted exterior, meaning it is rendered in a plaster material that covers the construction material beneath. In earlier Georgian designs, the ground floor was rendered and the rest of the exterior was exposed brickwork, while in the later Regency style, houses were rendered from top to bottom.

FIGURE 1.21
Buckingham palace

FIGURE 1.22
Terraced housing with stucco

- Render painted white or cream
- Built around garden squares, as the houses did not have their own garden

Victorian

The Victorian era, which included the industrial revolution in Britain, introduced many changes to society and the way buildings were constructed. As such there are even differences to be seen between early, middle and later Victorian architecture.

Whilst early Victorian properties typically incorporated more complex design features; porches, bay windows and overly elaborate interior decoration. As the period went on the style became simpler and the industrial revolution made possible the use of new building materials such as iron and glass.

This period saw a vast amount of building take place and a great deal still remains to this day.

The population increased from 8.9 million in 1801 to 32.5 million in 1901 so there was an urgency for huge amounts of housing. Many more houses were built along with the required infrastructure such as roads, railways and canals.

By the turn of the 19th century electric lighting had been introduced into towns and most houses had a cold water tap and an inside toilet.

Engineering and architecture advanced also and many large structures were designed and built.

The great railway stations of King's Cross (1852) (Figure 1.23) and St. Pancras (1865–1875) along with Crystal Palace (1850–1851) were some of the great landmarks of this period.

Most of the housing was either detached or semi-detached being built of brick or stone. Many Victorian houses are tall rather than wide and have steep gables and chimney stacks. Inner city housing often included basements, two or three storeys high and had attics and dormers in the roof (Figure 1.24).

In general the exterior of a Victorian period property can be identified by the inclusion of: Bay sash windows, Terracotta tiles, Ornamental stonework and multi-coloured brickwork often in red.

FIGURE 1.23
King's Cross station

FIGURE 1.24
Victorian architecture

The factory workers homes were very often back-to-back houses. Many had external communal earth toilets at the bottom of a communal yard (Figure 1.25).

These buildings were soon to be labelled slums.

Bathrooms in these houses were a rarity.

What are the main characteristics of a Victorian property?

- Coloured brickwork
- High pitched roof
- Ornate gable trim

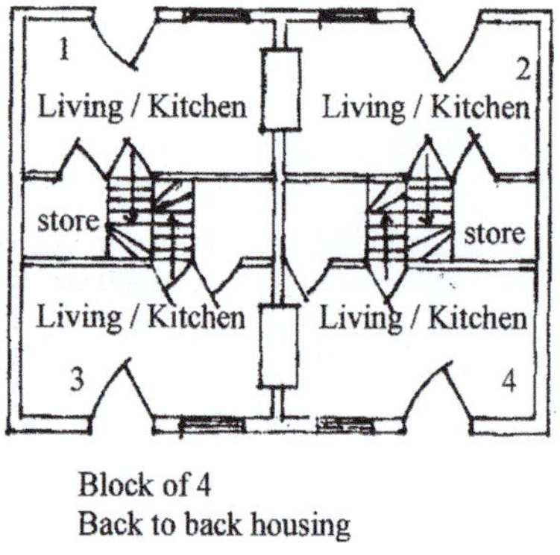

FIGURE 1.25
Block of four: back-to-back housing

- Tall rather than wide
- Geometric tiled hallways
- A brickwork porch
- Front door to the side of the façade
- Narrow hallway
- Stained glass windows
- Electricity and water with inside toilet
- Bay windows to sit in, for reading and writing
- Dark furniture and wood floors
- Fireplace in every room
- All wall surfaces became decorated
- Patterned wallpaper - typically heavy floral designs
- Elaborate design details that reflect the wealth of the owner and those coming into 'new' money

Edwardian

Styles in the 20th century did not suddenly change and the Victorian style continued for a while.

British architecture had until this period been either Gothic or Classical.

Extending beyond the reign of King Edward VII, the Edwardian era of architecture was a period of revivalism influenced especially by the Baroque, Georgian and the Arts and Crafts movements. The context for the period was a rapidly increasing population and completion of new railway lines, which gave rise to the 'suburbs'.

By choosing to live in the suburbs, the Edwardians were able to build their homes often on larger plots of land. So, unlike the smaller, darker Victorian homes, Edwardian houses were lower, wider and more roomy, with bigger hallways and more windows.

Edwardian homes tend to be shorter than equivalent Victorian residences, partly because the middle classes who lived in these homes had less of a need for servants, unlike the Victorian generations before them. Gone were the cellars and the second floors, but in came larger halls and spacious gardens.

Due to the mix of influences meant the exterior Edwardian homes might be identified by: Dutch gables, deep bay windows, sash windows and pilasters (Figure 1.26).

Also it was in fashion to half-clad the property exterior in timber or sub-divide windows into smaller square panes in order to create an aesthetic appeal.

The Edwardian period from 1901 to 1910 was very short but heavily influenced by the Arts and Crafts Movement.

FIGURE 1.26
Edwardian housing

There was a tendency to have simple designs and an appreciation for the handmade as against the mass production in the Victorian Period.

It's common for an Edwardian property to have a front garden and be set back from the pavement, as there was an ever-increasing desire for privacy at that time. Living rooms would sometimes have windows at both ends, covered by a small sloping roof on the outside.

What are the main characteristics of an Edwardian property?

- Houses built in a straight line
- Red brickwork
- Porch with wooden frames
- Mock-Tudor cladding and timbers at the top of the house
- Wide hallway
- Parquet wood floors
- Wider, brighter rooms
- Simple internal decorative features

Modern

Modern Architecture is the term given to the style of building which evolved after the First World War, and which has culminated in the current designs of glass, concrete and steel, based on modular construction.

There were four primary causes for the emergence of modern architecture:

(a) The 19th century population explosion, together with the Industrial revolution, caused by the urgency for buildings.

(b) The development of new building material

(c) A desire by the architects to change style

(d) The destruction by the two wars.

Unfortunately the building boom encouraged people with very little knowledge of house design which led to devastating consequences.

These people used shoddy materials, short cut methods and had very poor tradespeople producing what became known as 'Jerry Builders'.

Another type of construction was created after the Wars by using redundant factory workers, who had been working for the war effort, to produce pre-fabricated units in the factory for the erection on site.

These became known as 'Prefabs' and were supposed to be a temporary accommodation for those requiring to be re-housed (Figure 1.27).

Some of these are still being used today.

Because of the shortage of land coupled with the urgency to replace bomb damaged houses another style was created known as high rise buildings (skyscrapers) (Figure 1.28).

This was a fundamental different design from earlier designs.

New methods of construction with reinforced concrete and steel allowed the higher structures.

The previous tall buildings only had stairs. The invention of lifts allowed for much higher buildings.

As the years progressed traditional building methods of building were used less and less.

During this period mathematical and scientific theory was used.

FIGURE 1.27
Prefab

FIGURE 1.28
High rise buildings

Problems were tackled by means of calculations, combined with scientific tests of materials to prove their strength.

The 1946 New Towns Act and 1947 Town and Country Planning Act were introduced to ease the housing shortage and prevent the return to 'Jerry Building'.

The New Towns Act enabled the establishment of New towns financed by the government. Milton Keynes, Corby and Stevenage are some examples.

They have all been designed to contain mixed developments with planned expansions built into the scheme.

The 'modern look' means simplicity in form and design. Modern architecture is based on abstraction, which is created by clean lines, basic shapes, and forms. Thus, simple, plain, geometric forms, rectangular shapes, and linear elements make the characteristics of modern architecture.

The form of a modern building is also determined by its functionality. Functionality means performing for intended use, fitness for purpose.

In modern architecture, the emphasis is best defined with the concept of open space. Rather than traditional separated spaces (or rooms), the modern style creates undivided living and working spaces.

The initial post-war era was characterised by a break from the past with little inspiration being taken from older building styles. However, by the mid-1960s, enthusiasm for older styles had been reignited and interest in preserving older buildings was on the rise.

A growing number of old buildings were being renovated and modernised without their most charming features being dispensed with.

The mid-to-late sixties was the era of Postmodern Architecture.

This style was seen as an attempt to improve upon Modernism or the International Style, which was often seen as cold, ugly and merely functional, rather than pleasant and welcoming.

Brutalism also became popular in the post-war era. This movement was based on exposed concrete and retains a large number of admirers, but many buildings created in this style have since been demolished.

Today's houses are full of labour-saving equipment.

Most modern housing has either attached garages or separate garages with street parking reduced to a minimum.

Windows gradually became much larger with the introduction of cheap float glass, then in the 1980s, the size was reduced to push towards double glazing.

During this time the insulation of houses was very much improved in an attempt to save energy.

This is the theme for the future.

Types of building

Many different types of construction are required to fulfil the needs of today's ever-demanding society.

These consist of the following and are shown in Figure 1.29:

- dwelling units – for people to live in
- communal buildings – for all people to share
- industrial units – for people to work in
- recreational units – for people to relax in
- communications – roads, rail, sea and air networks – to allow people to move from one place to another.

Residential buildings

For the purpose of Level 1, only buildings for residential purposes will be dealt with.

One of the main functions of a building is to provide shelter for families.

Residential buildings can be found in various shapes and sizes, from a small bungalow to high-rise flats.

Classification of dwellings

A dwelling is a place of residence, a place where people live. A wide range of dwellings is required to meet the various needs of the occupants.

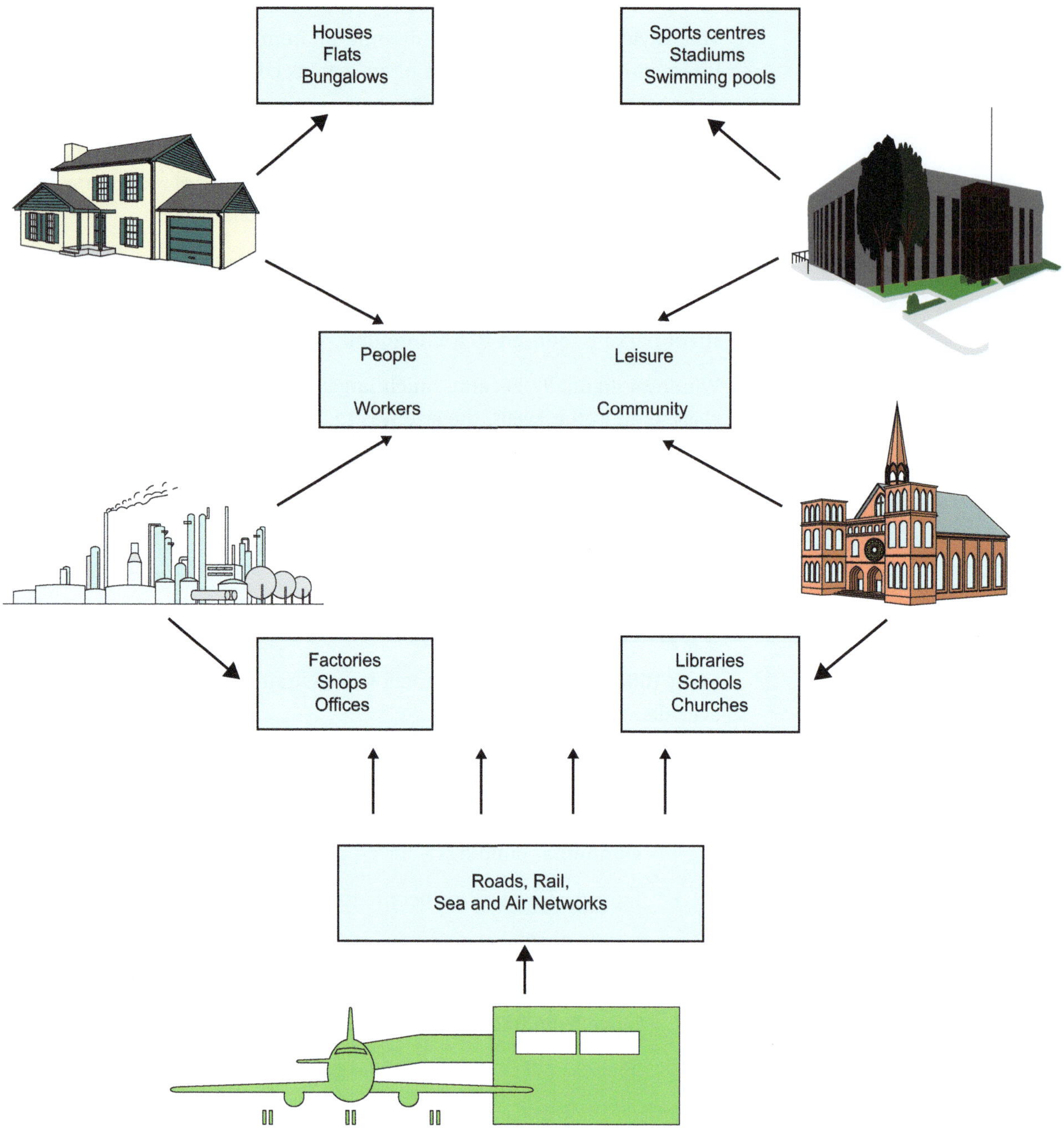

FIGURE 1.29
Types of construction

Dwellings can be constructed as individual units or grouped together to form one building. Various terms are used to describe the way in which dwellings, or groups of dwellings, are constructed.

Detached

When a dwelling is detached, it is not connected to any other dwelling unit. It stands alone on its own plot of land. It can be a house or a bungalow (Figure 1.30).

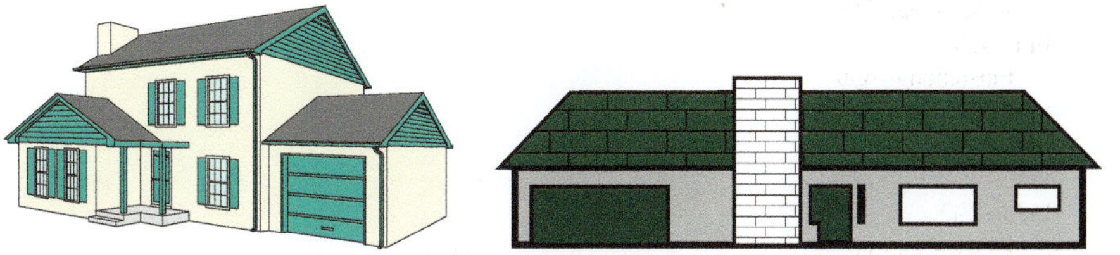

FIGURE 1.30
Detached house and detached bungalow

Semi-detached

A semi-detached dwelling is one of a pair built side by side across the boundary of adjacent plots of land and sharing a common dividing wall. The dwellings can be either houses or bungalows, as shown in Figure 1.31.

Terraced

When three or more dwellings are connected together in a row they are terraced, the middle building having two common or party walls. They can also be houses or bungalows.

Link-detached

Link-detached dwelling units are linked to the next unit by a garage or an outbuilding. They form a continuous row of buildings. Each one is detached, but there is no break in the line of building.

Low-rise flats

Blocks of flats built up to three storeys high.

Medium-rise flats

Blocks of flats built up to seven storeys high.

High-rise flats

Blocks of flats built over seven storeys high (Figure 1.32).

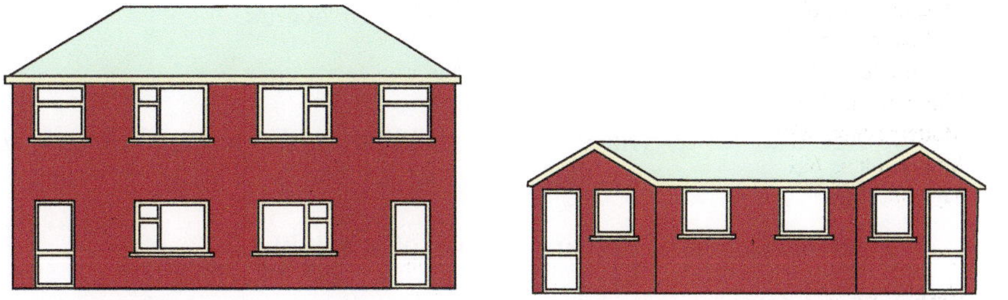

FIGURE 1.31
Semi-detached house and bungalow

FIGURE 1.32
Low-, medium- and high-rise flats

Restoration and heritage

Listed buildings

A listed building is one that has been placed on one of the four statutory lists of maintained buildings by:

- Historic England
- Historic Environment Scotland
- Cadw, in Wales
- Northern Ireland Environmental Agency in Northern Ireland

There are THREE types of listed status for buildings in England and Wales.

- Grade I: Buildings of exceptional interest.
- Grade II^{X}: Particularly important buildings of more than special interest.
- Grade II: Buildings that are of special interest, warranting every effort to preserve them.

A listed building may not be demolished, extended or altered without special permission from the local planning authority, which in turn consults with the relevant central government agency.

Almost anything can be listed; it does not have to be a building.

The criteria for listing include architectural interest, Historic interest and close historical associations with significant people or events.

The specific criteria include the following:

1. Age and rarity: The older the building is, the more likely it is to be listed.
2. Aesthetic merits: The appearance of a building.
3. Selectivity: Where a large number of buildings of a similar type survive.
4. National interest: Significant or distinctive regional buildings.

FIGURE 1.33
Example of a Grade I listed building: Buckingham Palace, the official residence of the British monarch

FIGURE 1.34
Manchester Town Hall

Owners of listed buildings are, in some circumstances, ordered to repair and maintain them and could face criminal prosecution if they fail to do so, or if they perform any unauthorized alterations.

When permission has been granted to carry out alterations or repairs, the owners have to use specific materials or techniques.

Examples of Grade I listed buildings are:

- Buckingham Palace (Figure 1.33)
- Clifton Suspension Bridge, Bristol
- Humber Bridge, East Riding of Yorkshire/North Lincolnshire
- Blackpool Tower
- Warwick Castle

Examples of Grade II^{x} listed buildings are:

- Manchester Town Hall (Figure 1.34)
- Battersea Power Station
- Coliseum Theatre, London
- Shibden Hall, Calderdale

Examples of Grade II listed buildings are:

- BT Tower, London (Figure 1.35)

FIGURE 1.35
BT Tower, London

- India House, London
- Birmingham Back to Back Houses, Birmingham
- Abbey Road Studios, London

There are currently 500,000 actual buildings listed.

Multiple-choice questions

Self-assessment

This section of the book is designed to allow you to check your level of knowledge. The section consists of revision questions for this chapter. The questions are all multiple choice and have four possible answers. The answers are to be found at the end of the book.

The main type of multiple-choice question will be the four-option multiple-choice question. This will consist of a question or statement, known as the stem, followed by a choice of four different answers, called the responses. Only one of these responses is the correct answer; the others are incorrect and are known as distracters.

You should attempt to answer the questions by choosing either (a), (b), (c) or (d).

Example

The person employed by the local authority to ensure that the Building Regulations are observed is called the:

(a) clerk of works
(b) building control officer
(c) council inspector
(d) safety officer

The correct answer is the building control officer, and therefore (b) would be the correct response.

The construction industry

Question 1 One of the specific criteria for a building to be listed is its:

(a) position
(b) architectural value
(c) low age
(d) high value

Question 2 State the type of dwelling unit shown.

(a) house
(b) bungalow
(c) flat
(d) bed sitter

Question 3 Blocks of flats up to three storeys high are known as:

(a) high-rise
(b) medium-rise
(c) low-rise
(d) skyscrapers

Question 4 In which historical period was Hadrian's Wall constructed?:

(a) Saxon
(b) Norman
(c) Tudor
(d) Roman

Question 5 The type of arch shown is known as:

(a) early English
(b) perpendicular
(c) pointed
(d) decorated

Question 6 Dwelling units are designed to fulfil which of the following main needs?

(a) for people to work in
(b) for people to live in
(c) for people to be entertained in
(d) for people to relax in

Question 7 Why was the window in the drawing below filled in with brickwork?

(a) to save on brick tax
(b) to save on glass
(c) to make a decorative feature
(d) to save on heat loss

Question 8 A series of dwelling units built in a long line all joined together is known as:

(a) attached units
(b) detached units
(c) terraced units
(d) semi-detached units

CHAPTER 2

Health and Safety in the Construction Industry

This chapter will cover the following City and Guilds units:

• Apprenticeship	Module 2	K1.1, K1.2, K2.1, K3.1	
• Diploma	Unit 201/601	1.1, 1.2, 1.3, 1.4, 1.5	2.1, 2.2, 2.3, 2.4, 2.5, 2.6
		3.1, 3.2, 3.3, 3.4, 3.5	4.1, 4.2, 4.3
		5.1, 5.2, 5.3	6.1, 6.2, 6.3, 6.4
		7.1, 7.2, 7.3, 7.4	8.1, 8.2, 8.3, 8.4, 8.5
		9.1, 9.2, 9.3, 9.4	
	Unit 104	2.6	

This chapter is about:

- Awareness of relevant current statutory requirements and official guidance
- Employee and employer responsibilities to workplace safety, wearing appropriate personal protective equipment and compliance with warning signs
- Risk assessments
- Accidents and emergencies
- Health and hygiene on site
- Security in the workplace

The following City and Guilds performance criteria will be covered:

- Know the health and safety regulations affecting bricklaying work
- Understand how site procedures contribute to maintain health and safety
- Understand specific health and safety considerations relating to bricklaying work

The following Diploma outcomes will be covered:

- Know the health and safety regulations, roles and responsibilities
- Accident, first aid and emergency procedures
- Identify hazards
- Health and hygiene
- Safe handling of materials
- Working platforms
- Electricity
- Personal protective equipment
- Signs and notices

Safety legislation

The construction industry is often involved in very difficult and often hazardous sites. It is therefore very important that the new recruit is aware of these dangers and that there are various regulations in place to control and reduce these possible hazards.

Prevention of hazards in the workplace

Hazards within a workplace can occur because of several circumstances. There may be faults in equipment, tools, stored substances, dangerously stacked materials, materials obstructing safe access, or simply a lack of site safety.

The health and safety of employees at their workplace and any other persons at risk through work activities are covered through various Acts of legislation and regulations.

These include the following:

- The Health and Safety at Work Act 1974
- The Control of Substances Hazardous to Health Regulations 2002 (COSHH)
- The Noise at Work Regulations 2005
- Work at Height Regulations 2005
- Reporting of Injuries, Diseases and Dangerous Occurrences Regulations 2013 (RIDDOR)
- The Personal Protective Equipment at Work Regulations 1992, amended 2018
- The Fire Precautions (Workplace) Regulations 1997, amended 1999
- Provision of the Use of Work Equipment Regulations 1998 (PUWER)
- The Electricity at Work Regulations 1989
- The Manual Handling Operations Regulations 1992
- Control of Vibration at Work Regulations
- Lifting Operations and Lifting Equipment Regulations 1998 (LOLER)
- The Construction (Design and Management) Regulations 2015.

Note

As Regulations are updated regularly is it always advisable to check with the latest version or amendments.

The main health and safety legislation applicable to building sites and workshops is covered by the Health and Safety at Work Act 1974.

Health and Safety at Work Act 1974

The four main objectives of the HASAWA are:

- To secure the health, safety and welfare of all persons at work.
- To protect the general public from risks to health and safety arising from work activities.

- To control the use, handling, storage and transportation of explosives and highly flammable substances.
- To control the release of noxious or offensive substances into the atmosphere.

These objectives can only be achieved by involving everyone in health and safety matters.

This Act applies to all work activities. It requires employers to ensure so far as is reasonably practicable the health and safety of their employees, other people at work and members of the public who may be affected by their work.

Employers should have a health and safety policy. If they employ five or more people, the policy should be in writing.

The self-employed should ensure so far as reasonably practicable their own health and safety and make sure that their work does not put other workers, or members of the public at risk.

Employees have to co-operate with their employer on health and safety matters and not do anything that puts them or others at risk.

Employees should be trained and clearly instructed in their duties. The main purpose of this Act is to cover all aspects of safety.

The framework promotes, stimulates and encourages high standards of health and safety in the workplace.

The Act involves everyone, management, employees, the self-employed, the employees' representatives, the controllers of premises, and the manufacturers of plant, equipment and materials, in matters of health and safety.

The Act also deals with the protection of the public, where they may be affected by the activities of people at work.

Outline of the Act

The Act itself is very complex and is an extensive document with numerous parts and sections. There are four main parts.

Part 1 of the Act describes: Employers' and management duties:

1. Provide and maintain a safe working environment.
2. Ensure safe access to and from the workplace.
3. Provide and maintain safe machines, equipment and methods of work.
4. Ensure the safe handling, transport and storage of all machinery, equipment and materials.
5. Provide their employees with the necessary information, instruction, training and supervision to ensure safe working.
6. Prepare, issue to employees and update as required a written statement of the firm's safety policy.

7. Involve trade union safety representatives (where appointed) with all matters concerning the development, promotion and maintenance of health and safety requirements.

Employees' duties:

1. Take care at all times and ensure that they do not put themselves, their workmates or any other person at risk by their actions.
2. Co-operate with their employers to enable them to fulfil the employer's health and safety duties.
3. Use the equipment and safeguards provided by the employers.
4. Never misuse or interfere with anything provided for health and safety.

General legal requirements

All construction and demolition work is subject to the Health and Safety at Work Act and to certain provisions of the Factories Act.

The following information sets out some of the basic legal requirements that may apply to you.

Notification

Work on construction sites. You must notify the local Health and Safety Executive (HSE) office of your site if you are starting a 'building operation' (including demolition or maintenance work) or 'work of engineering construction' (including road works), the work will take six weeks or more and notice has not already been given (e.g. by a main contractor).

Accidents, dangerous occurrences and ill-health. The local HSE office must be notified if:

(a) a person dies as a result of an accident caused by or connected with work on the site (whether or not that person was at work);

(b) a person suffers a listed 'major injury' accident (this includes accidents where a person is admitted to hospital for more than 24 hours) or a health condition as a result of an accident caused by or connected with work on the site (whether or not that person was at work);

(c) a listed 'dangerous occurrence' takes place because of or in connection with the work (e.g. collapse/partial collapse of a scaffold more than 5 m high and certain buildings and structures);

(d) a person at work is prevented from working for three or more days as a result of an injury or illness caused by an accident at work;

(e) a person at work is affected by a 'specified disease' (e.g. lead poisoning, pneumoconiosis, vibration white finger), diagnosed by a doctor, and that person had been doing particular types of work.

If the person affected is your employee, you must notify the HSE. In all other cases it is usually the responsibility of the main contractor.

Safety supervisor

Every contractor who carries out 'building operations' or 'works of engineering construction' and normally employs a total of 21 or more workers must specifically appoint, in writing, one or more people to act as safety supervisor(s).

Small contractors usually pay for the part-time services of an outside safety supervisor who may work for a safety group or specialist safety company.

Duties

The safety supervisor will advise the contractor/employer on how to comply with the law and will regularly supervise site work to see that it is carried out safely.

The safety supervisor must be sufficiently experienced in the type of work being done, be suitably qualified, and have enough time and authority to carry out his or her duties properly.

Safety policy

Every employer who has five or more employees must prepare a written safety policy. It must be revised as appropriate and brought to the attention of all employees.

Welfare

Every contractor on site must ensure that adequate welfare facilities are available for their employees. These should include, as a minimum:

(a) adequate washing facilities

(b) adequate toilets

(c) drying sheds, huts, rooms or other accommodation for sheltering during bad weather, storing clothes and taking meals (including tables and chairs, facilities for boiling water and a supply of wholesome drinking water).

First aid

Every employer and self-employed person on site must ensure that adequate first aid is available.

It is sensible for all contractors to make arrangements with the main contractor to provide first aid (if possible).

First aid arrangements will vary with the degree of risk on the site, but should usually include, as a minimum:

(a) an adequately stocked first aid box (or boxes)

(b) a trained first aider(s); although for small sites it is sufficient to appoint a person to take charge of the first aid box and any situation where serious injury or major illness occurs (responsibilities should include phoning for an ambulance)

(c) information for workers on site about first aid arrangements, including the location of the nearest telephone.

Exclusion of children and others from work areas

Where possible, a fence should be erected to enclose the site or construction work. The fence should be at least 2 m high and difficult to climb up. Where this is not possible, e.g. on a partly occupied new housing site, special precautions should be taken, particularly in the case of children, to:

a protect them from the dangers of excavations (including shallow ones filled with water), holes or openings and badly stacked materials

b prevent tampering with vehicles and plant, electricity supplies, gas cylinders and hazardous chemicals, e.g. by providing secure compounds

c prevent access to higher levels.

The safety officer's role

Safety officers should be properly trained and experienced in the construction industry, and preferably be members of the Institution of Occupational Safety and Health.

Their job is to advise management on the following:

- preventing injury - personnel
- preventing damage - plant and machinery
- hazards causing ill-health
- legal requirements - regulations
- provision and use of protective clothing and equipment
- suitability of new or hired plant and equipment
- hazards on new contracts before work starts

They also:

- carry out site surveys
- determine the cause of accidents
- assist in training for all levels
- foster an understanding within the firm.

The site safety office will use a similar sheet to the one shown in Figure 2.1.

SITE SAFETY SURVEY			
THE SAFETY OFFICER WOULD CHECK THE FOLLOWING AREAS ON SITE			
1. Scaffolding		7. Portable Tools	
2. Ladders		8. Fire Precautions	
3. Lifting appliances		9. Welfare Facilities	
4. Hoists		10. First Aid	
5. Excavation / Earthworks		11. Registers / Records	
6. Plant and Machinery		12. Site Tidiness	

FIGURE 2.1
Site safety officer's survey sheet

The Health and Safety Inspector

The HSE has a number of branches controlling different sections of industrial activity.

The Health and Safety Executive Inspectorate (or what used to be called the Factory Inspectorate) ensures that the regulations covering safety, health and welfare are complied with in factories and on sites.

The Health and Safety Inspector has the power to enter any workplace and has legal powers to enforce the regulations if necessary.

Outline of the Health and Safety Executive

Inspectors from the HSE have the responsibility to enforce the health and safety laws set out in the Act.

The inspector has the power:

- To enter premises in order to carry out investigations.
- To take statements. An inspector can ask anyone questions relevant to the investigation and also require them to sign a declaration as to the truth of the answers.
- To check records. All books, records and documents required by legislation must be made available for inspection and copying.
- To give information. An inspector has a duty to give employees or their safety representative information about the safety of their workplace and details of any action he or she proposes to take. This information must also be given to the employer.
- To demand the seizure, dismantling, neutralizing or destruction of any machinery, equipment, material or substance that is likely to cause immediate serious personal injury.
- To issue a Prohibition Notice. This requires the responsible person to stop immediately any activities that are likely to result in serious personal injury.
- To prosecute all persons, including employers, employees, self-employed, designers, manufacturers and suppliers, who fail to comply with their safety duty.

Safety procedures and documentation

In order to comply with the various safety legislation, an employer is required:

- To display notices and certificates – e.g. a copy of a valid fire certificate.
- To notify relevant authorities – e.g. commencement of any building works likely to last in excess of six weeks has to be notified to the relevant authority.
- To keep relevant records – e.g. the accident book in which details of all accidents are recorded.

The Control of Substances Hazardous to Health (COSHH)

The Control of Substances Hazardous to Health 2002 (COSHH) must be consulted when dealing with the handling, moving, storage and finally disposal of potentially hazardous materials or products.

Alongside these regulations it is important to use any detailed codes of practice and manufacturer's advice which is often found on the packaging of the material.

A material or product that is hazardous to health could be anything that could affect your health.

There will normally be a sign stating the possible dangers, as shown in Figure 2.2.

What the regulations are about

Whether you are an employer, a contractor, a subcontractor or self-employed, COSHH requires you to protect people who may be exposed to health risks arising from hazardous substances with which you work.

The construction industry has always involved risks to the health and safety of its operatives, some of which are peculiarly its own, e.g. masons subject to silicosis from working with siliceous stones and painters being exposed to lead poisoning.

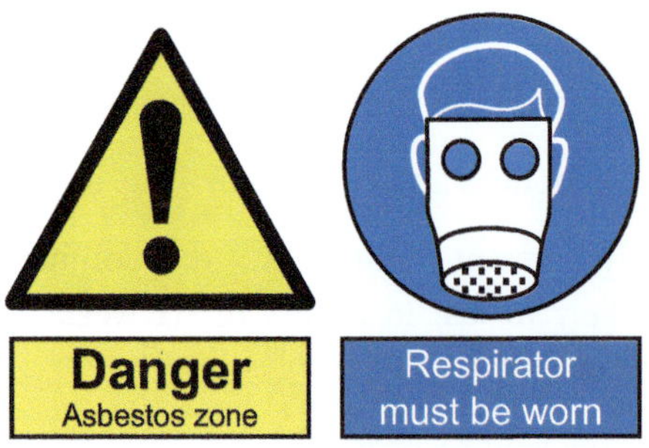

FIGURE 2.2
Asbestos warning

In more recent years, additional risks have been added by the use of new materials and processes being used, e.g. the use of epoxy resins.

The control of health hazards on construction sites is difficult owing to the continually changing working environments with the process of work in hand.

Risks common to all construction workers

Dangers to health can arise from:

- Ingestion of poisonous materials – care should be taken to avoid hazardous substances entering the mouth. Personal cleanliness is essential before food is eaten.
- Absorption through the skin by dangerous substances – continuous contact with the skin of certain materials can cause dermatitis and skin cancers.
- Exposure to certain physical conditions – without adequate protection, excessive noise can cause permanent damage to the ears.
- Inhalation of dangerous materials – breathing in any dangerous substance. Pollutants in the atmosphere that can be breathed into the body can be grouped into:
    - dust – small particles produced when emptying bags of powder such as cement and lime
    - fumes – paints and adhesives give off fumes and vapours
    - gases – these constitute a risk, particularly carbon monoxide.

How COSHH affects you

Whatever your particular business, hazardous substances will be used or met to varying degrees.

You will need to set up a system for dealing with hazardous substances on site. Some examples of the substances, jobs and precautions relevant to construction work are shown below. The list is not exhaustive.

Summary of COSHH requirements

You should consider the following requirements more fully by reading published guidance on COSHH. The six steps needed to comply with the regulations are:

1. Know what products or substances you are using.
2. Assess the hazards to health they can cause.
3. Eliminate or control the hazard if one exists.
4. Give information, instruction and training to employees.
5. Monitor the effectiveness of any controls.
6. Keep records.

The ‘substances’ covered by COSHH include:

- chemicals
- poisons
- solids
- timber
- liquids
- pesticides
- acids/alkalis
- dust – e.g. cement, gypsum, wood dust
- gas or fumes – e.g. welding fumes, hydrogen sulphide, carbon monoxide, nitrous oxide.

There are three exceptions to the regulations:

- work with asbestos and lead, as these are covered by specific regulations
- substances that are a hazard solely because they are radioactive, explosive or flammable
- substances that are a hazard solely because they are at a high or low temperature or at high pressure.

The removal of the above waste materials and substances is normally carried out by specialist contractors.

The Control of Asbestos Regulations

The Control of Asbestos Regulations 2012 came into force on 6 April 2012, updating previous asbestos regulations.

In practice the changes are fairly limited. They mean that some types of non-licensed work with asbestos now have additional requirements – notification of work, medical surveillance and record keeping. All other requirements remain unchanged.

What is the duty?

The duty to manage asbestos is contained in regulation 4 of the Control of Asbestos Regulations 2012.

It requires the person who has the duty (i.e. the 'duty holder') to:

- take reasonable steps to find out if there are materials containing asbestos in non-domestic premises, and if so, its amount, where it is and what condition it is in
- presume materials contain asbestos unless there is strong evidence that they do not
- make, and keep up to date, a record of the location and condition of

the asbestos-containing materials - or materials which are presumed to contain asbestos

- assess the risk of anyone being exposed to fibres from the materials identified
- prepare a plan that sets out in detail how the risks from these materials will be managed
- take the necessary steps to put the plan into action
- periodically review and monitor the plan and the arrangements to act on it so that the plan remains relevant and up to date
- provide information on the location and condition of the materials to anyone who is liable to work on or disturb them

There is also a requirement on others to co-operate as far as is necessary to allow the duty holder to comply with the above requirements.

The Noise at Work Regulations 2005

Building sites can be noisy places to work in. They should display signs as shown in Figure 2.3.

Excessive noise can be harmful, either by causing hearing damage or by creating nuisance, which may lead to stress.

Loud noise can cause a temporary partial loss of hearing, with recovery time varying from around 15 minutes after the noise stops to a few days, depending on the level of the noise. This temporary loss may be accompanied by ringing in the ears, or tinnitus, and this should be regarded as a warning: temporary partial hearing loss may become permanent with repeated exposure.

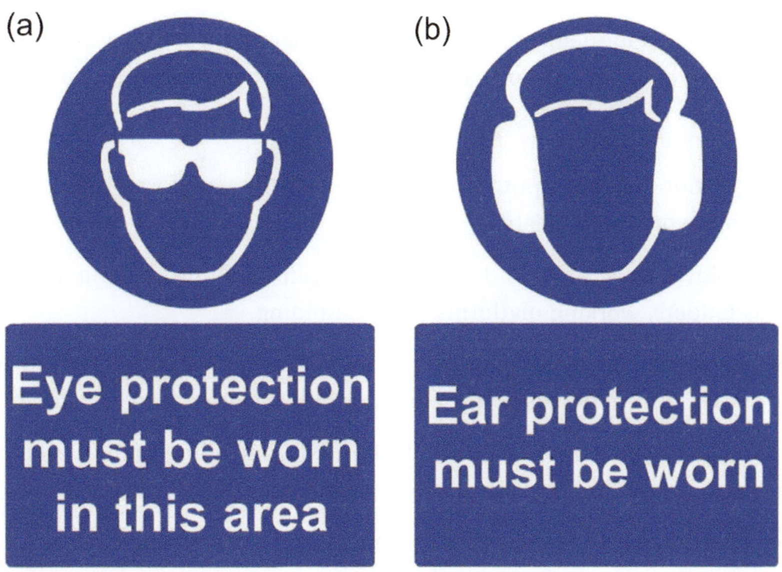

FIGURE 2.3
Eye and ear protection signs

People usually first notice permanent hearing damage when ordinary conversation starts to become difficult to understand as they permanently lose part of their hearing range. This can gradually worsen if exposure to harmful noise continues.

Even a few minutes' exposure every day to very noisy machines in the construction industry can be enough to initiate permanent hearing damage.

Employers should do as much as possible to reduce noise. However, it may not be possible to quieten all machines enough to ensure that there is no hazard, and therefore proper ear protection should also be available.

What you can do

- Play your part.
  - Keep compressor covers closed when they are running.
  - Ensure that breaker mufflers are securely fitted.
  - See that machinery panels are secured and do not rattle.
  - Do not keep machinery running unnecessarily.
  - Report any damage to your supervisor.
- Ask your supervisor for appropriate ear muffs or plugs if you are working with or near a noisy machine.
- As a general rule, if a machine is so noisy that you have to shout to carry out a conversation, you need ear protectors.
- Noise can be measured properly to assess the risk.
- Wear the ear muffs or plugs all the time when you are on a noisy part of the site. Make sure that they fit properly and are comfortable.
- Keep the muffs or plugs clean and keep them in a safe place when not in use.
- Look out for damage: if the muffs no longer fit properly or the ear seals have become hard or damaged, your supervisor must replace them.

Work at Height Regulations 2005

This regulation has been put in place to protect the workforce from injury or death from working at heights.

Your employer must provide the necessary equipment for working at heights, such as ladders, working platforms and scaffolding.

As an employee you must follow any instructions and training received while using any equipment provided and report any possible hazards to your supervisor.

A place is at 'height' if a person could be injured from falling from it, even if it is at or below ground level.

In construction work, the threshold height is 6 feet above a lower level.

Fall protection on a scaffolding is required at 10 feet.

More detailed information will be given in the section on working platforms.

Reporting of Injuries, Diseases and Dangerous Occurrences Regulations 2013 (RIDDOR)

All employers have a duty under RIDDOR to report accidents, diseases and dangerous occurrences. Reporting accidents and ill-health is a legal requirement.

This information helps the HSE and local authorities to identify where and how risks arise, and to investigate serious accidents. They can then help with providing advice on how to reduce injury and ill-health in the workplace.

Personal Protective Equipment at Work Regulations 1992, amended 2018

Personal protective equipment (PPE) is defined as 'all equipment which is intended to be worn or held by the person at work and which protects him/her against one or more risks to health or safety'.

The main requirement of the regulation is that PPE is to be supplied and used at work whenever there are risks to health and safety that cannot be adequately controlled in other ways.

To allow the correct type of PPE to be chosen the employer must carefully consider the different hazards in the workplace. This will enable the employer to assess which types of PPE are suitable to protect the workers from the hazard and allow the work to be completed in a safe manner.

Types of PPE include gloves, masks and goggles. These will be dealt with in more detail later.

The Fire Precautions (Workplace) Regulations 1997, amended 1999

The Fire Precautions (Workplace) Regulations 1997, as amended, cover places of work where one or more person is employed, e.g. commercial premises, universities, hospitals, shops, hotels and offices.

The regulations state that premises with five or more workers must have a written fire risk assessment detailing the appropriate fire safety work required, although some premises can be exempt.

Following the fire risk assessment the employer must, where necessary in order to safeguard the safety of employees in case of fire and to the extent that it is appropriate, provide:

- emergency exit routes and doors – the final emergency exit doors must open outwards and not be sliding or revolving
- emergency lighting to cover the exit routes, where necessary
- fire-fighting equipment, fire alarms and, where necessary, fire detectors.

Fire exit signs, fire alarms and fire-fighting equipment must be provided with pictograph signs, in accordance with the Health and Safety (Safety Signs and Signals) Regulations.

Employers must train employees in fire safety following the written risk assessment.

An emergency plan may have to be prepared and sufficient workers trained and equipped to carry out their functions within any such plan.

All equipment and facilities such as fire extinguishers, alarm systems and emergency doors should be regularly maintained and faults rectified as soon as possible. Defects and repairs must be recorded.

Employers must plan, organize, control, monitor and review the measures taken to protect employees from fire while at work. If there are five or more employees, then a record must be maintained.

Employers must appoint an adequate number of competent people to assist them to comply with their obligations.

Provision and Use of Work Equipment Regulations 1998

These regulations require risks to people's health and safety, from equipment that they use at work, to be prevented or controlled.

The regulations require that the equipment provided for use at work is:

- suitable for the job it has to do
- regularly maintained to ensure that it is safe to use.

They also require that:

- training is provided for those who use it.

In general, any equipment that is used by an employee at work is covered.

Potential hazards relating to mechanical equipment

All electrical and mechanical tools, equipment and machinery are potentially hazardous if misused, worn or damaged.

During your career in the construction industry you will come into contact with a wide variety of mechanical equipment designed to be used by a variety of trades.

Some machines are very specialized and only likely to be operated by trained and certificated workers.

Static plant should be sited on firm, level ground, with brakes on and chocks in place, as shown in Figure 2.4.

Main types of plant and equipment

Mobile plant

Mobile plant is defined as plant that moves under its own power. It is used to transport materials to or around building sites, and in construction operations.

On a large site, mobile plant could be controlled by restricting speeds, confining vehicles to site roads and authorized (one-way) routes. Places may also be set aside for unloading, turning and manoeuvring.

FIGURE 2.4
Small mixer ready for use

FIGURE 2.5
Mobile plant

One of the most common items of plant for moving materials around the site is the fork-lift truck (Figure 2.5).

Mobile plant should as far as possible be kept clear of structures, building activities, cranes, excavation and other hazards.

Static plant

Compressors, generators, pumps and concrete mixers are examples of static plant (Figure 2.6).

This type of equipment works in a fixed location, although the location may change as work proceeds and from site to site.

If static plant is powered by electricity, care must be taken to site the cables

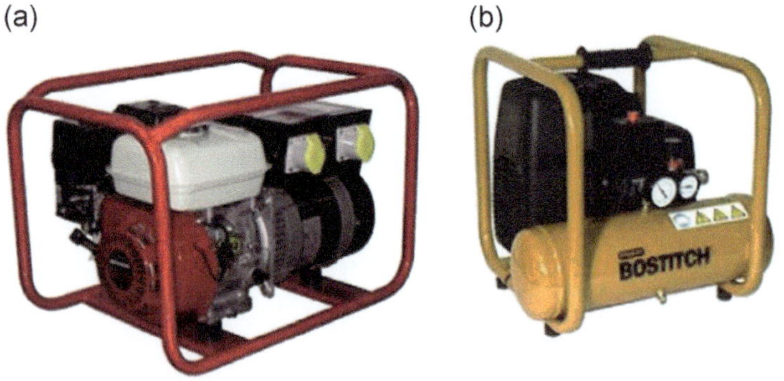

FIGURE 2.6
Static plant: (a) generator; (b) compressor

where they cannot be damaged by vehicles, and to avoid laying them on rough or sharp projections.

Mechanical tools and equipment

These have exposed moving parts such as grinding wheels, sanders, drills, chipping hammers, portable saws, rotary wire brushes and air compressors. A typical tool and sign are shown in Figure 2.7.

There are special regulations concerning grinding wheels and portable saws, and people must be authorized to use them.

- Never use any mechanical equipment that is unfamiliar.
- First read the manufacturer's instructions or seek advice.
- Never lay a tool down while it is still rotating.
- Never wear loose fitting clothes when using tools with fast moving parts.

Most accidents are caused by lack of knowledge, misuse, makeshift repairs or using faulty tools and equipment.

FIGURE 2.7
Abrasive wheels sign

FIGURE 2.8
Caution sign for electricity

All brickwork apprentices have to achieve the Abrasive Wheels Certificate as part of their programme.

The Electricity at Work Regulations 1989

Any equipment that uses electricity is covered by the Electricity at Work Regulations. Appropriate signs should be positioned as shown in Figure 2.8.

Your employer has a duty to make sure that the whole construction site is a safe area and that there is no chance of coming into contact with a live electrical current.

Electricity

Electricity is something that you cannot see or hear. Strict rules are laid down for its use and must be rigidly obeyed by everyone.

All tools should comply with British Standard 2769 and, except for 'double-insulated' tools, must be effectively earthed. Double-insulated tools, which have their own builtin safety system and bear the 'kite mark' and 'squares symbol' (Figure 2.9), do not require an earth lead.

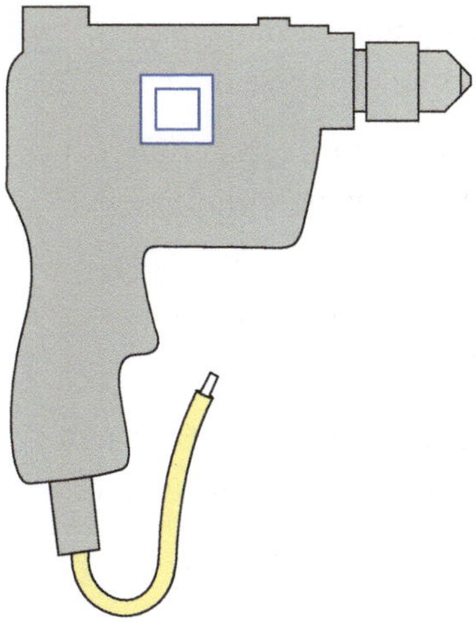

FIGURE 2.9
Double-insulated drill

Any electrical tool or equipment must be operated at a reduced voltage of 110 V (volts), or lower if possible. Below 60 V the risk of death is greatly reduced.

Before using electrical plant you should always:

- inspect for damage or defects
- check cables are not worn or frayed or have exposed wires
- check cables are connected to equipment correctly
- ensure cables are not in contact with water
- make sure your equipment is connected to a 110 V supply
- disconnect from the supply if you have to make adjustments
- remember all repairs and maintenance must be carried out by a qualified person.

It is not only tools and equipment that create dangers; on many building sites overhead high-voltage power lines criss-cross the working area. These can be fatal to people carrying or moving metal ladders and scaffolding. If you ever carry out this task, check first that your route is clear of overhead power lines.

For certain workers on site similar dangers come from underground power cables and exposed wiring.

To protect you from electrocution 'goal posts' must be erected either side of access and egress points (Figure 2.10). All other areas where power lines run should be sectioned off and large warning notices posted.

Remember

As a non-skilled or certificatec person, you may only carry ou safety checks and MUST NO attempt to repair any piece c electrical equipment.

If faulty electrical equipment is found you should return it to you supervisor who will see that the necessary repair work is carriec out by a competent person.

Most electrical equipment has tc be checked and tested annually.

Each tool or piece of equipment is made to be used at a certair voltage.

Check that the tool operating vol tage and the supply voltage are the same before connecting the tool to the supply.

Transformers

A transformer (Figure 2.11) may be used if the supply voltage and the operating voltage of the tool or equipment are different.

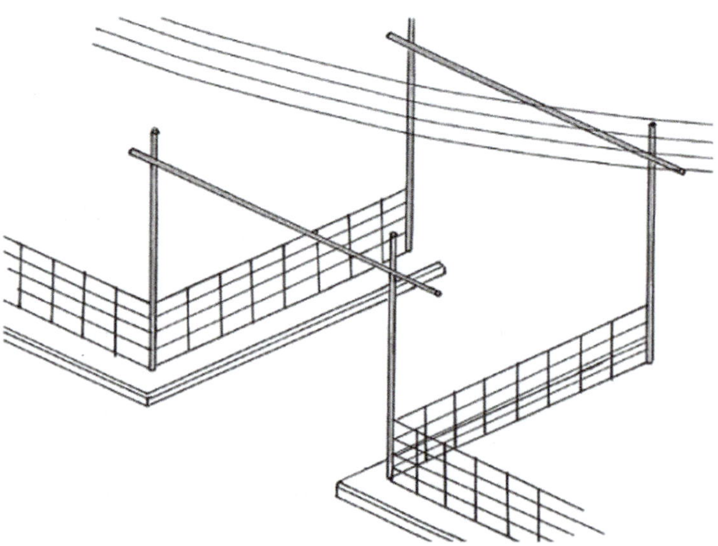

FIGURE 2.10
Goal posts

FIGURE 2.11
Transformer

Perhaps the most common transformer is the 110/240 V unit. This can be used to raise a 110 V supply to 240 V, or reduce a 240 V supply to 110 V.

Always check that the correct transformer is being used, and that it is of sufficient capacity to take the current flowing.

If you are using equipment operating at 230 V or higher a residual current device (RCD) can provide additional safety and rapidly switches off the current. An RCD plug is shown in Figure 2.12. If the RCD trips, this is a sign of a fault.

Connectors

The most common 110 V plugs and couplers are normally of the splash-proof type and are designed to make the connection to an incorrect voltage supply impossible.

Plug

The plug (Figure 2.13), connected to the end of the flex, fits into the electricity supply socket.

Electric fuses

A fuse is a safety device. It is a deliberate weak link in a convenient part of the circuit, usually in a form of a cartridge containing a wire which melts and therefore breaks the circuit when an excessive current flows.

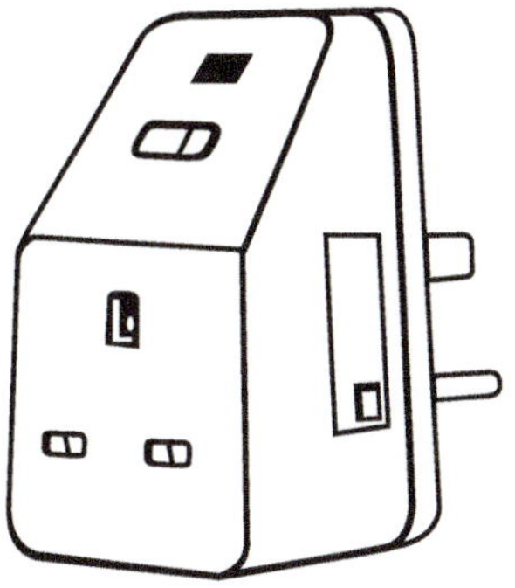

FIGURE 2.12
RCD

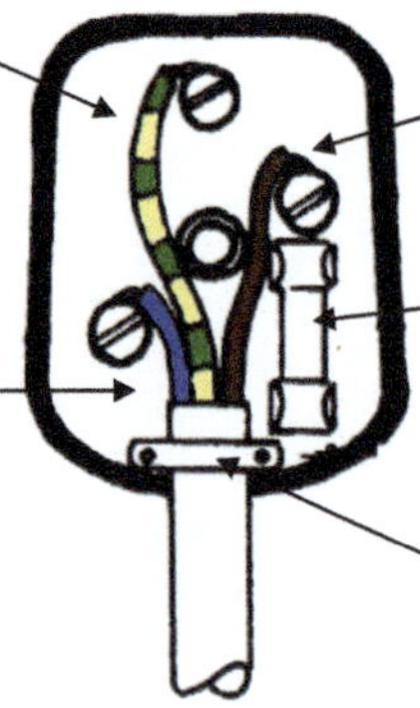

FIGURE 2.13
Wiring to a 240 V three-pin plug

PAT Testing Regulations

PAT testing was put into place to ensure that all companies conform to the Health and Safety at Work Act, Electricity at Work Regulations, Provision and Use of Work Equipment regulations and the Management of Health and Safety at Work regulations of 1999.

Conforming to PAT testing regulations will help you to ensure that the portable appliances in your business are safe and help you to avoid violations. As an employer your insurance company will expect you to carry out PAT testing to ensure that you are compliant with all the above construction regulations.

There are recommendations that stationary equipment should be tested every 48 months and moveable equipment such as extension leads and portable equipment every 24 months. Its purpose is to prevent electrical accidents in the workplace.

A full PAT test should include both a visual inspection and a more in-depth check using specialist PAT testing equipment. This test includes earth continuity, lead polarity, and insulation resistance checks.

At the end of a PAT test, every appliance should be marked 'passed' or 'failed'. Ideally there should be a record of the results (Figure 2.14).

Electric shock

- Cause: contact with low-voltage (240 V) supply.
- Action to be taken:
  1. Do not touch the casualty with hands.
  2. Break the electrical circuit by switching off or removing the plug from the socket.
  3. If this is not possible, immediately break the casualty's contact with the supply by standing on dry insulating material and – using

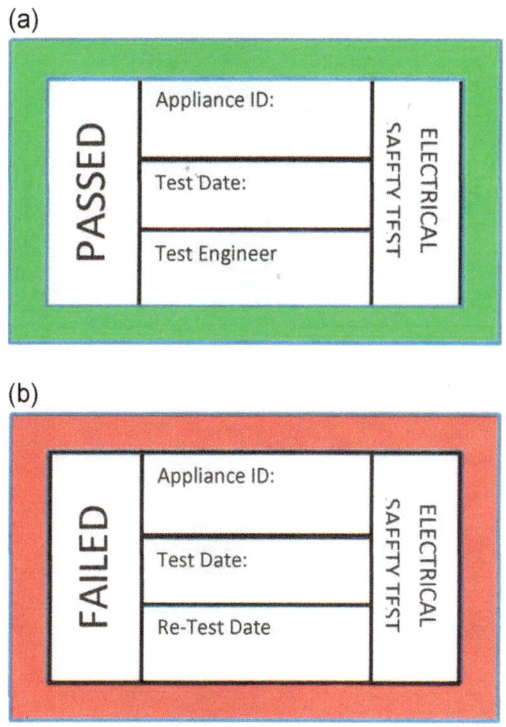

FIGURE 2.14
Pat test stickers to be placed on the appliance

dry wood, a folded newspaper or a rubber object – push the casualty out of contact with the supply.

4. Treat burns.
5. Apply respiratory resuscitation if breathing has stopped.
6. Apply cardiac massage if the heart has stopped. Ensure that there is no heartbeat before commencing this procedure.

Emergency resuscitation

In the event of someone collapsing through injury, it may be necessary to attempt emergency resuscitation while a medical team or ambulance is being called. This is known as CPR – Cardio Pulmonary Resuscitation.

Two methods can be used: the mouth-to-mouth with chest compressions method (Figures 2.15 and 2.16) and the Silvester method. The mouth-to-mouth method of resuscitation should be used if the casualty's mouth and face are not damaged. The Silvester method should be used if the casualty's mouth or face is damaged, but the chest is not damaged.

Lifting Operations and Lifting Equipment Regulations 1998 (LOLER)

These Regulations place duties on companies and people who own, operate or have control of lifting equipment.

FIGURE 2.15
Mouth-to-mouth resuscitation

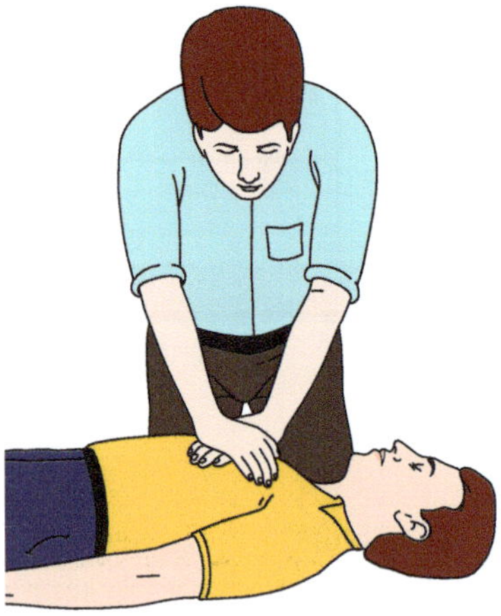

FIGURE 2.16
Chest compressions

In most cases lifting equipment is also work equipment so the Provision and Use of Work Equipment Regulations, (PUWER) will also apply.

All lifting operations involving lifting equipment must be properly planned by a competent person, correctly supervised and carried out in a safe manner.

LOLER also requires that all lifting equipment is fit for purpose, appropriate for the task in hand, suitably marked and subject to periodic examination.

Records must be kept of all examinations and any defects found must be reported to both the person responsible for the equipment and the relevant enforcing authority.

Control of Vibration at Work Regulations

The Control of Vibration at Work Regulations 2005 are based on a European Union Directive requiring similar basic laws throughout the Union on protecting workers from risks to their health and safety from vibration.

Vibration White Finger (VWF) or hand-arm vibration syndrome (HAVS) is caused by using vibrating tools.

Hand-arm vibration is vibration transmitted from work processes into workers' hands and arms. It can be caused by operating hand-held power tools, such as road breakers, and hand-guided equipment, such as powered lawnmowers, or by holding materials being processed by machines, such as pedestal grinders.

When is it hazardous?

Regular and frequent exposure to hand-arm vibration can lead to permanent health effects. This is most likely when contact with a vibrating tool or work process is a regular part of a person's job. Occasional exposure is unlikely to cause ill health.

Hand-arm vibration can cause a range of conditions collectively known as hand-arm vibration syndrome (HAVS), as well as specific diseases such as carpal tunnel syndrome.

The Manual Handling Operations Regulations 1992

Employers must comply with the Manual Handling Operations Regulations 1992, as amended by the Health and Safety (Miscellaneous Amendments) Regulations 2002.

The Regulations explain how to avoid, assess and reduce the risk of injury from manual handling.

The Regulations define manual handling as:

> *"...any transporting or supporting of a load (including the lifting, putting down, pushing, pulling, carrying or moving thereof) by hand or bodily force."*

The load can be an object, person or animal.

The MHOR 1992 set out a clear ranking of measures for dealing with risks from manual handling, these are:

- first: avoid hazardous manual handling operations so far as is reasonably practicable;
- second: assess any hazardous manual handling operations that cannot be avoided; and
- third: reduce the risk of injury so far as is reasonably practicable.

The Construction (Design and Management) Regulations 2015

The aim of the regulation is for Health and Safety consideration to be integrated into the development of a construction project from the beginning.

The CDM regulations reduce the risk of harm to all those personnel who work on a building site

Whatever your role in construction, CDM aims to improve health and safety in the industry by helping you to:

- Sensibly plan the work so the risks involved are managed from start to finish
- Have the right people for the right job at the right time
- Cooperate and coordinate your work with others
- Have the right information about the risks and how they are being managed
- Communicate this information effectively to those who need to know
- Consult and engage with workers about the risks and how they are being managed
- Under the CDM Regulations the HSE must be notified about certain projects before they can commence.

Site welfare facilities that must be provided by employers are as follows:

> *Now, all sites, regardless of size, must make sure there are not only enough facilities for the number of workers on site but that they are appropriate for the type of work taking place. Both the client and contractor share this legal responsibility.*

Toilets

Toilets should be suitable and sufficient, ventilated, lit and kept in a clean and orderly condition.

Washing facilities must be provided so that workers can use them immediately after using the toilet or urinal, even if they are provided elsewhere.

Washing facilities

General washing facilities must be suitable and sufficient, kept clean and orderly and with basins or sinks large enough for people to wash their face, hands and forearms.

The facilities should include: clean hot and cold, or warm running water; soap or other suitable means of cleaning; towels or other suitable means of drying; and showers where the nature of work is particularly dirty or there is a need to decontaminate.

Drinking water

Drinking water must be provided or made available at readily accessible and suitable places. Cups are required unless the supply is in a jet from which people can drink easily.

Changing rooms and lockers

Changing rooms are needed where workers have to wear special clothing for the purposes of their work and cannot be expected to change elsewhere.

The rooms must be provided with seating, means of drying and keeping clothing and personal effects secure.

Facilities for rest

Rest rooms or rest areas are required equipped with tables and seating (with backs) sufficient for the number of persons likely to use them at any one time.

There should be arrangements for preparing and eating meals, also for boiling water. In cold weather, heating should be provided.

Workplace safety

Accidents

When joining the construction industry it is important to remember that you will be joining an industry with one of the highest injury and accident ratings. It therefore it cannot be stressed enough that you could be at constant risk unless you start as you intend to continue, with a good safety attitude.

Any type of work carried out by the construction industry is often difficult and hazardous. Every site will be different, and therefore every site will bring possible new dangers.

It is of utmost importance that all trainees are capable of using hand tools and equipment efficiently and safely at an early stage in their development.

Furthermore, they should be aware of the causes of accidents and be able to take action and deal with any accident that may occur.

An accident is an unexpected or unplanned happening which results in personal injury or damage, sometimes death.

Reported accidents are those which result in death, major injury and more than three days' absence from work or are caused by dangerous occurrences reported to the HSE.

Every day a large number of the industrial accidents that are reported involve construction workers. Of course nobody wants to be one of those involved. Nobody wants to spend long, painful months in hospital, or to see a family deprived of its breadwinner.

Clearly, it is in everyone's interests to try to reduce the number of accidents. This is not some impossible task. Make no mistake about it; the vast majority of accidents could be prevented.

Accidents do not just happen, they are caused.

Finding out what causes them is the first step towards preventing them. Usually, an accident is the last link in a chain of events consisting of a series of dangerous conditions and dangerous actions.

Accident prevention is something that everyone can practise. It consists of being able to recognize when a condition has become dangerous and knowing what steps to take to remove the danger. This is everyone's business. It is not just the concern of management, or the supervisor. It is up to everyone, every trainee, every operative, anyone engaged in any way on the construction site.

Learning to spot a dangerous situation is not as difficult as it sounds, because accidents follow a regular pattern. The same kind of accident happens over and over again. Every day of the year, all over the country, the same set of dangerous conditions build up and the same unsafe acts take place.

Do any of the things you normally see and do at work add up to a source of danger? Next time you are tempted to take a risk - STOP and THINK again.

Causes of accidents

Accidents are caused in various ways and may well be attributed to:

- trying to get the job done to quickly
- too little preparation before commencing
- taking short cuts
- distraction by others causing a lapse in concentration
- lack of concentration due to lack of interest in the job
- failure to observe the rules of safety; not wearing the correct PPE
- horseplay, that is acting irresponsibly, creating a danger and a hazard to yourself and others.

Work areas

There are various types of work areas. The most common are listed below.

Construction site

The construction site is potentially the most dangerous area you could encounter.

It is essential to ensure that access and egress to the working areas are kept clear and materials are stored away from access routes.

Materials cause the most obstructions on construction sites.

Ensure that all routs are correctly marked and parking places displayed.

Workshop

The workshop floor should be marked out to show safe areas.

These are areas where materials and plant should not be stored, even for a short time.

The safe areas should lead to the fire exit, which should always be kept clear and well marked, with the regulation lights and signs.

Client's property

When you are working in the property of the client it is still referred to as a working area and must be safe.

Ensure that it is kept as clean as possible and access is kept clear and unobstructed.

Remember

It is your responsibility to act in a safe manner.

Types of hazard

Everyone involved in the construction industry should be aware of the possible dangers and hazards in the construction site.

Site safety will be improved if everyone is safety conscious. Types of hazard include the following:

- falling objects
- falls of operatives
- transportation of plant and materials
- electricity
- machinery and equipment
- fire and explosions.

Hazard spotting

Look at accidents that have occurred. The law requires that your company keeps records of all accidents to employees.

- When did you last look at your accident book?
- What are the most common accidents?
- What action has been taken to prevent similar accidents happening in the future?

Even in the best companies accidents still happen.

The next time there is an accident try looking at it more closely. An analysis procedure can help you to prevent accidents in the future. Start off by finding the facts. After analysing several accidents a pattern may emerge.

The causes of accidents may be similar, although these happen in different workshops or on different sites. Looking at the whole company may highlight accident prevention training needs.

How to spot hazards

Looking at any situation may be fun, but the action in spotting hazards in your own area is deadly serious. So too is training others to spot them.

As you walk around your work area use your senses: look, touch, hear and smell.

Practice in looking at hazards in more depth will help you to become more proficient at hazard spotting.

Personal Protective Equipment at Work Regulations 2002 (PPER)

Depending on the type of workshop or site situation, the wearing of correct safety clothing and safe working practices are the best methods of avoiding accidents or injury. On some sites certain PPE is compulsory.

All construction operatives have a responsibility to safeguard themselves and others. Making provision to protect oneself often means wearing the correct protective clothing and safety equipment.

Your employer is obliged by law to provide the following:

- suitable protective clothing for working in the rain, snow, sleet, etc.
- eye protection or eye shields for dust, sparks or chipping
- respirators to avoid breathing dangerous dust and fumes
- shelter accommodation for use when sheltering from bad weather
- storage accommodation for protective clothing and equipment when not in use
- ear protectors where noise levels cannot be reduced below 90 dB(A)
- adequate protective clothing when exposed to high levels of lead, lead dust fumes or paint
- safety helmets for protection against falls of materials or protruding objects
- industrial gloves for handling rough abrasives, sharp and coarse materials

Although our skin is not proof against knocks, bumps, cuts, acid, alkalis or boiling liquids, it is waterproof. Even so, we do have to cover up at times to protect ourselves.

Workers in the construction industry are liable to injury or even death if they are not protected. Because of this, protective clothing has been developed to help prevent injury.

Protective equipment

The following is a list of various PPE, some items of which are shown in Figure 2.17.

- *Safety helmets* – When you are on site there is always a danger of materials or objects falling into excavations or from scaffolds and there is also a danger that you will hit your head on protruding objects. Always wear your personal safety helmet, which you will have to adjust to fit your head correctly. Remember, the life of a helmet is two years.
- *Safety footwear* – You need to protect your feet against various hazards, including damp, cold, sharp objects, uneven ground and crushing. A good pair of boots with steel toecaps will be required for most conditions.
- *Safety goggles* – The law requires you to wear safety goggles, safety glasses or eye shields if there is a risk of injuring your eyes during your

FIGURE 2.17
A selection of personal protective equipment

work. You must wear safety goggles when cutting, grinding, drilling or chipping any hard materials where particles are likely to be thrown off at high speed.

- *Ear protectors* - Ear muffs should be used when the noise levels are unacceptable and cannot be reduced.
- *Overalls* - There are numerous types of clothing produced to wear over your normal clothes for protection from dust, dirt and grime. Certain tradespeople, such as decorators, have to wear them all the time.
- *Respirators* - These should be used when dangerous fumes and dust are present and adequate ventilation is not practicable.
- *Gloves/gauntlets* - There are many types of industrial gloves that should be used when handling certain materials, especially in the cold.
- *Jackets* - These should be worn for protection during inclement weather. High-visibility jackets and waistcoats are compulsory on certain sites.
- *Knee pads* - Essential equipment for the protection of your knees when spending a great deal of time kneeling on hard floors.
- *Sunscreen* - During the summer sunburn is a risk. Ensure that you apply sunscreen to avoid your skin burning.

CE marking

Ensure any PPE you buy is 'CE' marked and complies with the requirements of the Personal Protective Equipment Regulations 2002.

The CE marking signifies that the PPE satisfies certain basic safety requirements and in some cases will have been tested and certified by an independent body.

Safety signs

As you go about your work on the building or construction site you will see various signs and notices. Your employer will give you instruction on what they mean and what you should do when you see one.

Safety signs fall into four separate categories, which can be recognized by their shape and colour. Sometimes they may be just a symbol; others may include letters or figures and provide extra information such as the clearance height of an obstacle or the safe working load of a crane.

The four categories are:

Prohibition signs

- Shape – circular
- colour – red border and cross bar; black symbol on white background
- meaning – shows what must not be done
- example – no smoking.

Mandatory signs

- Shape – circular
- colour – white symbol on blue background
- meaning – shows what must be done
- example – wear hand protection.

Warning signs

- Shape – triangular
- colour – yellow background with black border and symbol
- meaning – warns of hazard or danger
- example – caution, fork-lift truck working.

Information signs

- Shape – square or oblong
- colour – white symbols on green background
- meaning – indicates or gives information of safety provision
- example – first aid point.

Signs with supplementary text

Any of the symbols shown on the last page may also contain text. A few examples of these are shown in Figure 2.18.

Signs may be produced and erected according to the situation, as shown in Figure 2.19. For example, where there is a dangerous hazard a sign can be produced and placed in a prominent position to warn anyone approaching to take care.

Security arrangements

It is the responsibility of everyone on the work site to ensure that the security of that site is maintained. Security can take many forms and they are all equally important.

Visual security

- Alarms – positioned in an accessible place within view of the general public
- bars, mesh and locks – fitted to glass panelled doors and windows

FIGURE 2.18
A selection of signs: (a) prohibition sign; (b) mandatory sign; (c) warning sign; (d) information sign

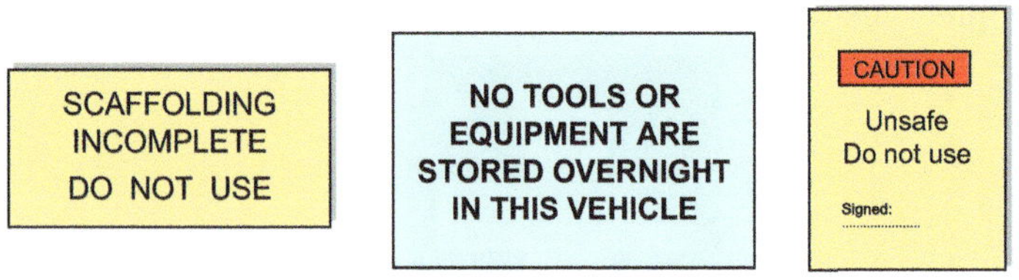

FIGURE 2.19
Types of notice

- padlocks, padlock and chains – fitted to compound gates, pieces of plant and machinery
- lighting – flood lights and movement-activated lights
- security firms.

Individual security

It is the responsibility of all employees to contribute to the overall security of the firm, for example:

- Tidiness – do not invite crime by leaving tools and equipment where they may be easily seen. If there is a secure store, lock them away.
- Plant – if possible, return all plant to a secure compound, or if necessary immobilize.
- Unauthorized access – it is the responsibility of all employees on site to challenge anyone who they feel has no authorization to be within a particular area. (Politeness is the best approach.)

If, despite the security measures taken, your site is breached, there are certain procedures you should follow. These should be given to you by your line manager, and may include:

- reporting the incident to the site supervisor
- reporting the incident to the police
- checking the inventory to find out what has been taken
- recording damage done to the premises and/or equipment.

Remember

Vandalism and theft do not always happen at night.

Always take care of tools and equipment you are using and report any suspicious people or events to your line manager.

Site access

It is important on larger sites to have a signing in and signing out procedure to increase the security of the site and its workers.

Site security

Site security is an important factor in any construction operation. There are three main aspects to be considered:

Loss of goods and materials

It is difficult to quantify the cost of these losses as distinct from materials misuse and wastage, but estimates as high as 3 per cent of total material cost have been made.

It should be noted that attention is required to prevent theft from within a site complex coupled with potential outsider risks.

Small, valuable items should be stored in locked cabins with either no windows or a facility for boarding-up at the end of each working day.

Large, bulkier items should be placed in a lockable compound if no protection from climatic damage is required.

Other perishable items should be placed inside a waterproof, lockable storage area.

The whole site should have a secure fence (Figure 2.20) and deliveries should be correctly phased.

Night watchmen may be employed on larger sites.

FIGURE 2.20
Site security

Vandalism

This is usually carried out by other people, rather than employees.

This makes it all the more important to provide the area with a secure perimeter fence.

Injury to children

The builder must provide some barrier that can reasonably be expected to prevent the entry of children. This may include hoardings, locked huts and secure storage areas.

Emergency procedures

Responding to emergencies

To save life and reduce the risk of injury occurring, all new people at work must be told of the current safety and emergency procedures.

From day one you should be aware of what to do in the event of a fire or an accident. Should an emergency happen you should be able to:

- Know what to do – acting quickly and calmly, carry out the correct procedure
- Follow the fire procedure – take the correct action in the event of discovering a fire:
    1. Select and use the correct type of fire extinguisher. (Only if the fire is small enough for you to put it out.)
    2. Call for help, sound the alarm.
    3. Telephone the Fire Service: 999. Give the correct address of the building.
    4. Leave the building by the nearest exit.
    5. Go directly to the assembly area. Await the roll call.

Accidents

An accident involving injury to a person can happen at any time. It may be a workmate who has fallen off a ladder or someone with a burn or a cut, or who has fainted. To help them when they most need it you should know what to do!

Immediate action

1. Unless you are a fully trained first aider – do not attempt to treat the injured person. (Only move an injured person if their life is in danger, e.g. danger from fire.)
2. GET HELP. Report the accident to a person in charge or:
3. Telephone the Emergency Services: 999.

When your call is answered you should have the following information at hand:

- type or types of services required – fire, ambulance, police

- type of accident
- location/address at which it has happened
- telephone number you are calling from
- your name

Remember

An accident is an event causing injury or damage that could have been avoided by following correct methods and procedures.

An accident is an unplanned event or occurrence which is outside the planned or expected procedure of work.

Although every worker should do their best to prevent accidents happening, there is still that unfortunate time when the worst happens.

First aid

If you are a qualified first aider you will know what to do. If you are not qualified in first aid, what can you do?

Do you know where the first aid box is kept?

According to the Health and Safety (First Aid) Regulations 1981 your employer must provide first aid equipment and facilities and appoint a qualified first aider or, on a smaller site where only a small number of people are working, should arrange for someone to take charge in the event of an injury.

On very large sites, there may be a first aid room with a qualified first aider in charge.

First aid boxes should be situated at various points around the site and clearly marked

with a white cross on a green background.

A small portable first aid kit (Figure 2.21) should be carried by an employee when working alone or in a small group well away from the main site.

If you have to use the first aid kit, report it and state what you have used so that it can be replaced.

Accident reporting

Every accident should be reported – an accident report book should be on every site or workshop, usually with the site supervisor, or whoever is in charge

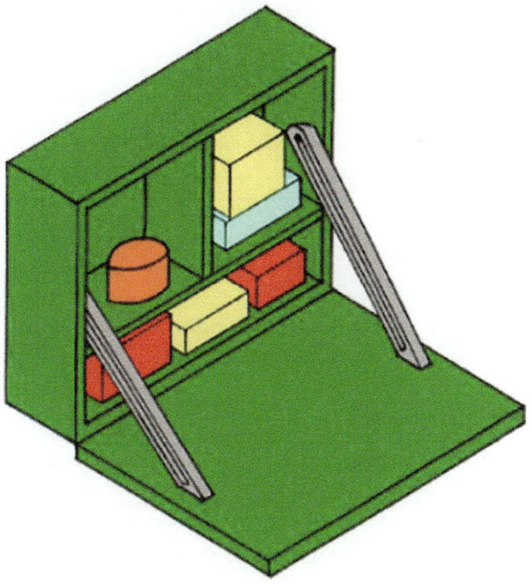

FIGURE 2.21
First aid box

of the site or workshop. A typical page from an accident book is shown in Figure 2.22.

Make sure that you report any accident in which you are involved as soon as possible.

Obviously some accidents are more serious than others. Accidents that result in death, major injury or more than three days' absence from work are called 'reported accidents'. Any such accident should be reported to the HSE.

Accidents where people require hospital treatment must be recorded at the place of work, even if no treatment was given there.

Fire and emergency procedures

All fires must be taken seriously and action taken immediately to prevent harm to people and damage to property.

1 **About the person who had the accident**	2 **About the person filling in this page** If you did not have the accident, complete the following
Full Name	Full Name
Address	Address
Postcode	Postcode
Occupation	Occupation

3 Please sign and date
Signature: Date:
The person who has had the accident should sign and date if they have not filled in the book (as confirmation that they agree the accident recorded is a true and accurate record)
Signature: Date:

4 About the accident– When and where it happened
Date: Time :
In what room or place did the accident happen?

5 About the accident– What happened?
How did the accident happen?
Materials used in the treatment

6 Reporting of injuries, diseases and dangerous occurrences
For the employer only – complete the box provided if the accident is reportable under RIDDOR
How reported
Date reported:
Employer's name and initials:

FIGURE 2.22
Typical page from an accident book

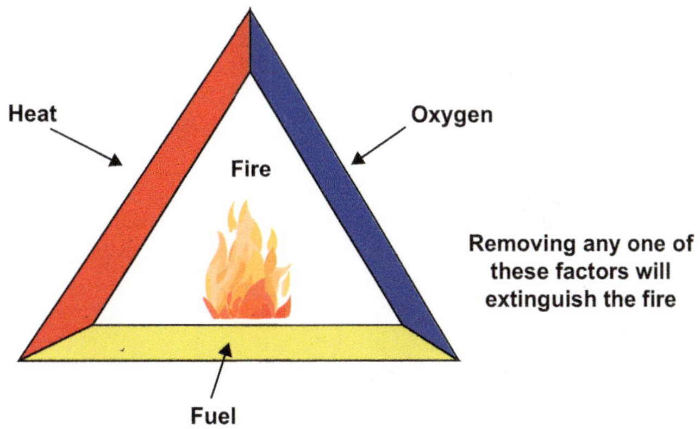

FIGURE 2.23
Factors in a fire

If a small fire cannot be controlled quickly, the building must be evacuated.

Fires require three factors to start. The removal of any one of these factors will extinguish the fire. Check out these three factors in Figure 2.23.

Fires, if they occur, need to be extinguished. Only tackle a fire if it is safe to do so or it is small enough to extinguish with a fire extinguisher.

The fighting of a fire should only be carried out by people who are fully trained. In some situations inexperienced site personnel are involved with the early stages of fighting a fire.

The fire-fighting should not continue if:

- the fire becomes too dangerous
- there is the possibility that the escape route might be cut off
- the fire continues to spread
- there are gas cylinders or explosives in the immediate area that cannot be removed or soaked continually with water.

Exits and oxygen

Never work in a position where a fire may block the exit.

Since all fires need an ample supply of oxygen, if a fire does break out try to close all windows and doors.

Fire-fighting equipment

To prevent fires the workplace should be kept clean and tidy. Rubbish should not be allowed to accumulate in rooms and storage areas. Fire prevention is therefore mainly good housekeeping.

When using flammable materials a suitable fire extinguisher must always be kept handy and ready for use. Before starting work on any job, make sure that the extinguisher operating instructions are fully understood.

Make sure that the extinguisher is the right type for the fire that may occur. Use of the wrong type can be disastrous. Fires can be classified according to the type of material involved:

- class A - wood, cloth, paper, etc.
- class B - flammable liquids, petrol, oil, etc.
- class C - flammable gases, liquefied petroleum gas, propane, etc.
- class D - metal, metal powder, etc.
- class E - electrical equipment.

The following types of fire-fighting equipment are available:

- Fire blankets (Figure 2.24) are useful for wrapping around a person whose clothing is on fire. They may also be used to smother a small, isolated fire.
- Sand (Figure 2.24) is also useful for dealing with small isolated fires, such as burning paint droppings.
- Water hoses (Figure 2.24) are fitted in some buildings.
- Fire extinguishers - pressurized extinguishers can be filled with various substances to put out a fire. Fire extinguishers are now all the same colour (red), but they have a colour band which identifies the substance inside:
    - Water: the colour band is red and the extinguisher can be used on class A fires.
    - Foam: the colour band is cream and the extinguisher can be used on class A fires. A foam extinguisher can also be used on class B fires if the liquid is not flowing and on class C fires if the gas is in liquid form.
    - Carbon dioxide: the colour band is black and the extinguisher can be used on class A, B, C and E fires.
    - Dry powder: the colour band is blue and the extinguisher can be used on all classes of fire.

A range of extinguishers is shown in Figure 2.25.

FIGURE 2.24
(a) Sand bucket; (b) fire blanket; (c) hose.

FIGURE 2.25
A range of pressurized extinguishers.

Risk assessments

The employer has a duty to protect the workforce as far as is reasonably practicable. Risk assessment is a very important part of protecting all site operatives.

Risk assessment is simply a careful examination of what could cause harm to the workforce. The workforce has a right to be protected from harm caused by a failure to take reasonable control measures.

In the construction industry risk assessments are carried out by experienced people who have been taught to identify what risks are possible when carrying out tasks.

There are five main steps to risk assessments:

- Step 1 – Identify the hazard.
- Step 2 – Decide who might be at risk and how.
- Step 3 – Evaluate the risks and decide on the best precautions.
- Step 4 – Record your findings and implement them.
- Step 5 – Review your assessment and update if necessary.

Management of Health and Safety at Work Regulations requires that risks are controlled in the workplace, including construction sites.

Where an employer employs five or more people, the significant findings of risk assessments must be recorded.

This involves thinking about what might cause harm to people and deciding what steps to take to prevent that harm. This is known as risk assessment and might include:

General assessments of health and safety risks to which employees and others are exposed on construction sites.

Specific assessments for specific hazards such as: work at height; hazardous substances (COSHH); manual handling; noise; vibration and lead.

Remember

Do not overcomplicate the process.

If you run a small business and you are confident you understand what is involved, you could do the assessment yourself. You do not have to be a health and safety expert.

Most of the risks come from tripping, slipping and moving heavy loads.

If you are an employer of a large company you should ask a health and safety advisor to help out.

On construction projects, this may relate not only to the construction of the works, but also to temporary works and to the operation, maintenance, cleaning, alteration or demolition of the completed development. It may also relate to inherent risks in the site or its surroundings.

It is not necessary to be a health and safety expert to write a risk assessment, simply a 'competent person', which according to the Health and Safety Executive (HSE) is 'someone who has sufficient training and experience or knowledge'.

Risk assessment should not be a bureaucratic exercise but should focus on the identification, elimination and management of risks (Figure 2.26).

<table>
<tr><td rowspan="4">RISK ASSESSMENT</td><td>Dept. Ref:</td></tr>
<tr><td>Date:</td></tr>
<tr><td>Prepared by:</td></tr>
<tr><td>Safety Unit Ref:</td></tr>
<tr><td colspan="2">Material or Process</td></tr>
<tr><td colspan="2">Trade Names</td></tr>
<tr><td colspan="2">Risk to Health

Hazardous Contents

Site of Exposure Skin ☐ Eyes ☐ Breathing in ☐ Cuts etc. ☐ Swallowing ☐

Degree of Risk

Maximum Exposure Limit / Occupational Exposure Standard</td></tr>
<tr><td colspan="2">Limitation of use

Persons authorised to use</td></tr>
<tr><td colspan="2">Handling Precautions

1. Work Methods

2. Controls</td></tr>
<tr><td>Protective Equipment</td><td>Type to be work</td></tr>
<tr><td>Eye Protection</td><td></td></tr>
<tr><td>Gloves</td><td></td></tr>
<tr><td>Overalls</td><td></td></tr>
<tr><td>Mask / Respirator</td><td></td></tr>
<tr><td>Other</td><td></td></tr>
</table>

FIGURE 2.26
Risk assessment forms

COSHH Assessment		Estimate of Risk to Health
Risk to Health adequately controlled if above Precautions are in force		
Risk to Health requires above precautions and further Action detailed overleaf		
COSHH Action required		
Handling precautions		Check to be made on Contractors
Maintenance of Exposure		
Monitoring Exposure		
Information to staff		Specify
Written Instructions issued		Specify
Training Required		Specify
Health Surveillance		
Other controls		
Further Information The following information is provided to assist in the safe handling, use and disposal of the materials described		
First Aid / Hygiene Arrangements		
Fire Precautions		
Storage		
Transport		
Spillage		
Waste Disposal		
For Further Information Contact:		

FIGURE 2.26
Risk assessment forms

Remember

A hazard is anything that can cause harm, such as chemicals, electricity or working from ladders.

The risk is the chance, high or low, that an employee could be harmed by these and other hazards.

Health and hygiene

Certain precautions must be taken to ensure that the health of employees in construction firms is protected against hazards, as mentioned in the previous section. As far as is practicable, their health must also be protected.

Vulnerable parts of the body

The health of the site operatives can be divided into the following areas of the human body:

- Skin – one of the most common problems with the skin is dermatitis. This is caused by contact between the skin and the many cements and plasters on site. To reduce the problem barrier creams could be used or appropriate gloves worn. Most construction sites now provide numerous types of gloves for every situation.
- Eyes – protection of the eyes has been mentioned previously, but as they are the only ones you have it is important to take extra care and use the appropriate glasses or goggles for the job.
- Ears – again, most sites provide a selection of ear protectors or plugs to be used when working with or close to noise.
- Lungs – many construction operations involve dust. It is therefore very important to protect yourself against inhaling any harmful dust. Protective breathing apparatus or simple disposable masks should be available.

Other areas to consider

Cold

This is most damaging to health when it is associated with damp weather.

In adverse conditions the body cannot maintain normal body temperature. Construction workers who get cold and wet frequently and for substantial periods, and who wear inadequate clothing, will have an increased risk of illnesses such as bronchitis and arthritis.

Workers are particularly at risk from cold when the temperature around them is below 10°C. It is important to realize that the wind-chill factor could lower this temperature considerably.

If you have to work outside in these conditions, take the following steps:

- Shelter from cold winds (using screens or sheeting).
- Get warm at intervals (accommodation should be provided).
- Dry out wet clothes.
- Consume warm food and drinks regularly.
- Wear good-quality, warm and waterproof clothes.
- Wear suitable footwear for the conditions.

Note

If you are very cold, your skin turns blue. If you suffer from shivering or dizziness or mental confusion then you may be in the early stages of hypothermia.

Remember

You are not being the tough guy in trainers, jeans and lightweight jacket. You are being stupid.

Heat

Most workers tend to discount any problems of this kind on construction sites in the UK. However, high temperatures (above 25°C) are common during summer.

Heat exhaustion could occur, which is dangerous if working on ladders, trestles, etc. Sunburn and sunstroke are also possible.

Sweating excessively causes a loss of body fluids and salt, creating severe muscular cramps.

Personal hygiene

Always keep yourself clean and tidy; just because you are in one of the dirtiest occupations, there is no need to look untidy.

If you are working in a client's home you need to present yourself correctly. Always wash regularly and have your work clothes regularly washed.

Wash your hands after going to the toilet and before eating and drinking.

General

Most construction sites undertake regular blood checks for drugs and alcohol.

It important that you arrive at your place of work in a fit state to carry out your duties in a manner safe to yourself and your workmates.

All the major sites have permanent health and safety officers, and one of their roles is to give 'toolbox talks' on various areas of health and safety.

All personnel new to the site should have a health and safety induction before they are allowed to work on the site. This will include the rules and regulations involved and the procedures to be followed if any are broken. This could result in someone being banned from entering the site again.

Construction Skills Certification Scheme

As part of the certification process each candidate has to pass the Construction Health and Safety test. This is a compulsory test consisting of a number of multiple-choice questions.

New entrants to construction will also have to apply for a Construction Skills Certification Scheme (CSCS) card. Most large sites now require proof of certification before a worker is allowed on site.

A trainee will be given a red card if he or she is registered for a certificate but has not achieved Level 2 or 3. The cards are valid for 3 years.

The aim of the scheme is to raise the standards of health and safety. It also provides a record of all workers in the construction industry who have achieved a recognized level of competence, and provides a means of identification.

Handling materials and components

The Manual Handling Operations Regulations 1992 outline how to deal with risks to the safety and health of construction site operatives.

The site operative should be able to select and use appropriate safety equipment and protective clothing when handling different materials.

If the item to be moved is an awkward size or shape the site operative should be able to select and use appropriate equipment or aids to carry materials.

They should also be able to demonstrate safe manual handling techniques.

Handling materials

Lifting or carrying heavy or awkward objects, such as bags of cement, can cause injury if performed incorrectly. The most common injury is to the back.

To avoid injury the following principles should be followed:

Lifting from the floor

Crouch down in front of the item to be lifted with the feet apart and one foot alongside the object in front of the other (Figure 2.27).

Keep the back straight at all times, letting the leg muscles do all the work.

Push off with the rear foot and move off in the direction you wish to go, in one smooth movement.

Carrying

Lifting an item from the back of a lorry or a bench and carrying

Keep your arms close to the body and take a good hold of the item.

Grip with the palms and the roots of the fingers. Make sure that the load does not impair your view.

Take care when positioning the fingers so they do not get trapped when landing the item.

Carrying on the back

Keep your back straight by raising the top of the head slightly and tucking in the chin. Figure 2.28 shows examples of correct lifting techniques.

Note

Always use lifting gear if it available.

FIGURE 2.27
Correct lifting

FIGURE 2.28
Correct lifting techniques

Awkwardly shaped and very heavy objects

Any awkwardly shaped objects should be analysed first to decide on a method of moving. Does the procedure require extra equipment and assistance from one or more workmates? If more workmates are required ensure that one is the team leader.

Special items of equipment are available to assist in lifting awkward objects (Figure 2.29).

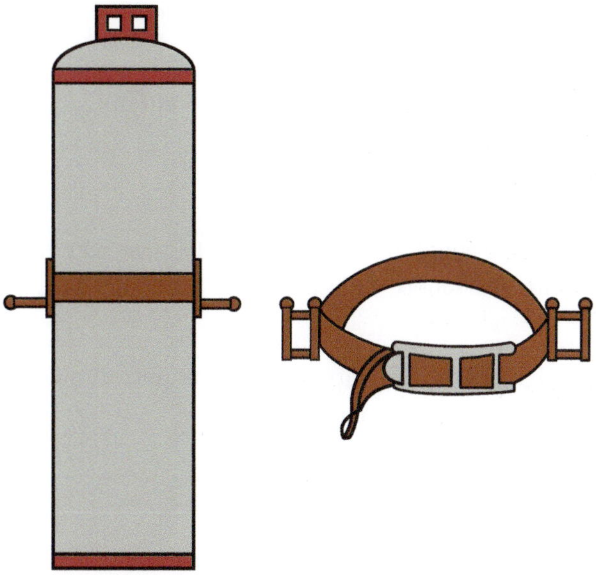

FIGURE 2.29
Equipment for lifting gas cylinders

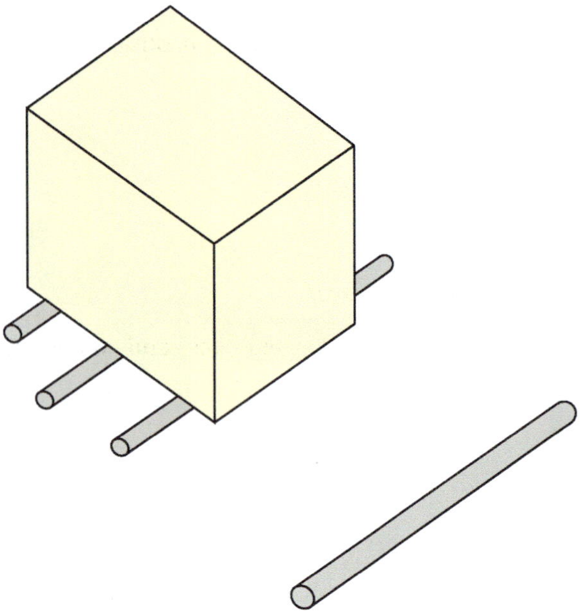

FIGURE 2.30
Correct rolling technique

Heavy loads that cannot be moved by lifting can sometimes be moved on rollers, and short scaffold poles are ideal for this purpose (Figure 2.30).

To move them at a later date the front end should be lifted up and a steel tube placed underneath. The rear end is then lifted and another steel tube placed under the load.

The load is then gently pushed forward by the helpers. As the object moves forward another roller should be ready to place under the front end of the load and the one freed at the rear should then be brought to the front.

The steel tubes can be slanted slightly to alter the direction of travel. This sequence should continue until the load reaches its destination.

Remember

The sequence for lifting heavy and awkward loads:

1. Plan the task.
2. Bend your knees.
3. Get a good grip.
4. Lift, with your legs taking the strain.
5. Place the load down.

Lifting gear

Numerous items of small lifting equipment are available to assist with handling materials on site and in the workshop. Only use these if you are qualified to do so.

They range from small brick lifts, slings, barrows and dumpers to mechanical fork-lift trucks. A selection is shown in Figure 2.31.

Mechanical equipment

On larger sites more use will be made of mechanical equipment.

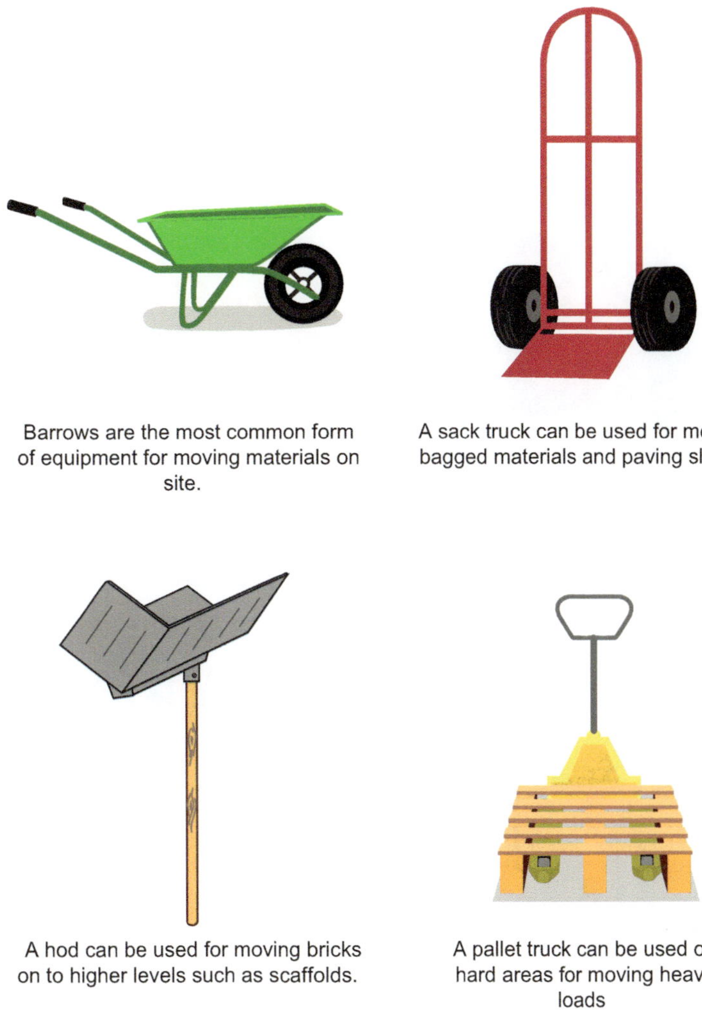

FIGURE 2.31
Types of moving equipment: (a) barrow; (b) sack truck; (c) hod; (d) pallet truck

If materials are stored in a compound they will be required to be transported around the site as required.

- *Scaffold crane* - these are useful for small jobs and are attached to the scaffold and can lift small amounts of materials to higher levels.
- *Dumper* - there are many sizes and designs of dumper. These are only available for movement at ground level.
- *Forklift truck* - these again are available in numerous designs and sizes and are excellent for moving materials both on the ground and at small heights.
- *Cranes* -these are available in all makes and sizes according to the job in hand. Cranage can be either mobile or static. Cranes that span the whole construction site are excellent for the placing of all materials wet or dry.
- Barrows are the most common form of equipment for moving materials on site.
- A sack truck can be used for moving bagged materials and paving slabs.

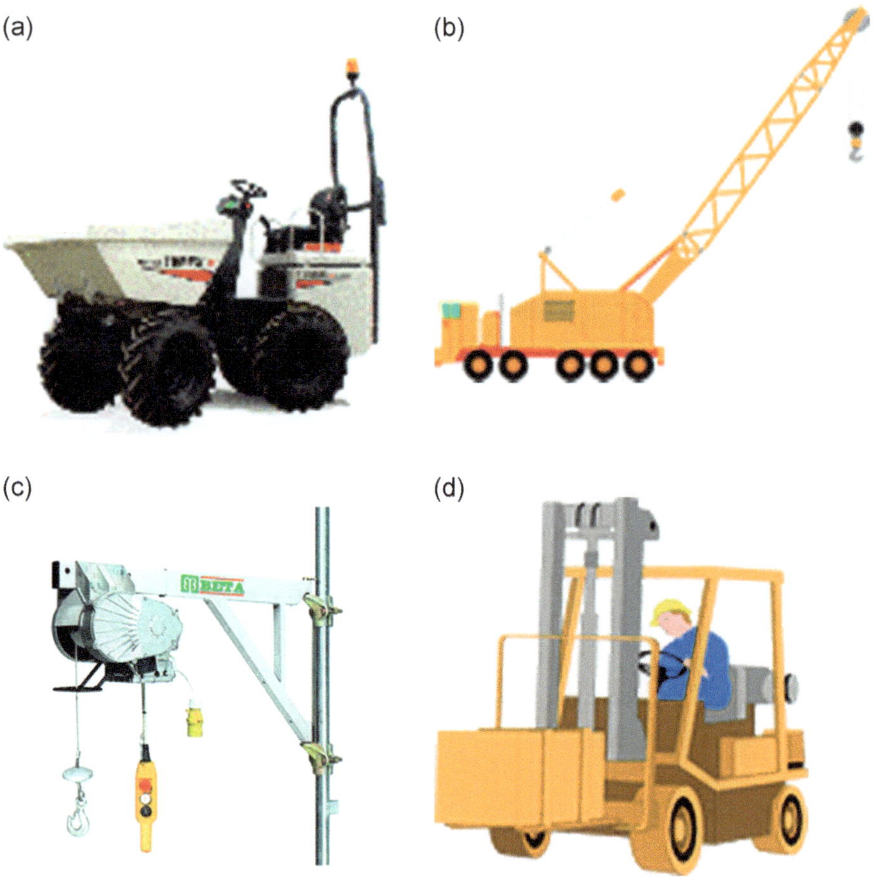

FIGURE 2.32
Types of mechanical lifting equipment

- A hod can be used for moving bricks on to higher levels such as scaffolds.
- A pallet truck can be used on hard areas for moving heavy loads.

Many materials are delivered to the site on lorries equipped with mechanical offloaders. Once the material has been off-loaded it is the builder's responsibility to move the materials to a secure place until required for use.

Working platforms

The Work at Height Regulations 2005 apply to all work at height where there is a risk of a fall liable to cause personal injury.

There are on average 65 fatal accidents and over 4000 major injuries in the construction industry each year. They remain the single largest cause of workplace deaths and one of the main causes of injury.

This unit will only deal with working platforms up to 2 m high.

These regulations place a duty on all employers to provide a risk-free environment in which to work.

- Duty holders must avoid working at height where possible.

- Where working at height cannot be avoided, equipment or measures to prevent falls must be in place.
- If there is a risk of a fall, equipment to minimize the distance of the fall must be used.

A scaffold is a temporary staging to assist bricklayers and other tradespeople to construct a building.

The scaffold must be spacious and strong enough to support people and materials during construction.

As explained before, many accidents are due to simple faults such as misuse of tools, untied ladders, a missing toeboard, etc.

The three basic requirements for scaffolds are:

- They should be suitable for the purpose.
- They should be safe.
- They should comply with the regulations.

Remember

Before any work is carried out at height a risk assessment must be produced.

Before work starts

Scaffolding should not be erected, substantially added to, altered or dismantled except under the immediate supervision of a competent person and, as far as possible, by competent workers possessing adequate experience of such work.

The competent person should also be given sufficient and sound materials for the job. It is false economy and highly dangerous to skimp on materials, and if faulty materials are provided the dangers may be hidden from those who use them.

Before they are used, tubes, couplers and boards should be inspected by someone who knows what defects to look for. Tubes that are bent, or weakened by rust, and damaged couplers and boards with bad splits or knots should be discarded.

Hop-ups

Usually purpose made, the hop-up is a low-level form of timber platform (Figure 2.33). Purpose-made hop-ups are also available and are usually called platforms (Figure 2.34).

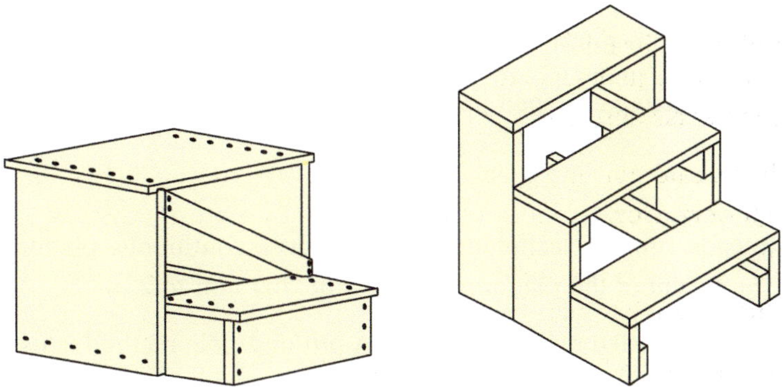

FIGURE 2.33
Types of wooden hop-ups

FIGURE 2.34
Purpose-made platform

They should stand on a firm and level base, up to 600 mm high, and have a standing area of 600 × 500 mm.

Used independently or in pairs, they allow the tradesperson to reach a wall working height of approximately 2.4 m.

A single hop-up provides an isolated working platform, whereas two hop-ups spanned by scaffold boards enable an entire wall to be covered.

Trestle platforms

There are several types of trestle scaffolds which are widely used by all trades to provide working platforms up to 2 m in height in confined spaces. All trestle platforms should stand on a firm level base.

Guard rails and toeboards are not normally required unless the platform exceeds 2 m in height.

They have the advantage over other forms of scaffold in that they are quickly and easily erected and dismantled.

Adjustable steel trestle

There are numerous designs of steel trestles; one is shown in Figure 2.35. Trestles are positioned to suit batten or staging thickness.

Boards normally used on sites are 225 × 38 × 3900 mm, which should be supported every 1.5 m for British Standard (BS) boards, or every 1.2 m for others.

Some patented staging can span up to 3 m.

Splitheads

Metal splitheads support scaffolding and provide a continuous platform for working. The height of the platform ranges from 700 mm to 2 m.

Splitheads are supported on a tripod base. A pin-and-hole method is used for main adjustment; fine adjustment can be achieved using a screwjack.

An example is shown in Figure 2.36.

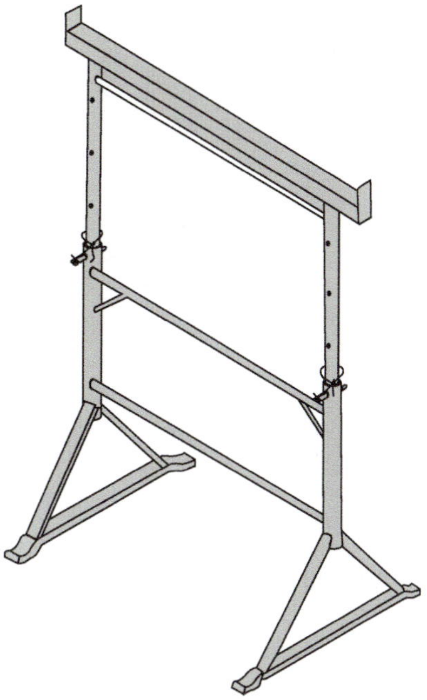

FIGURE 2.35
Typical steel trestle

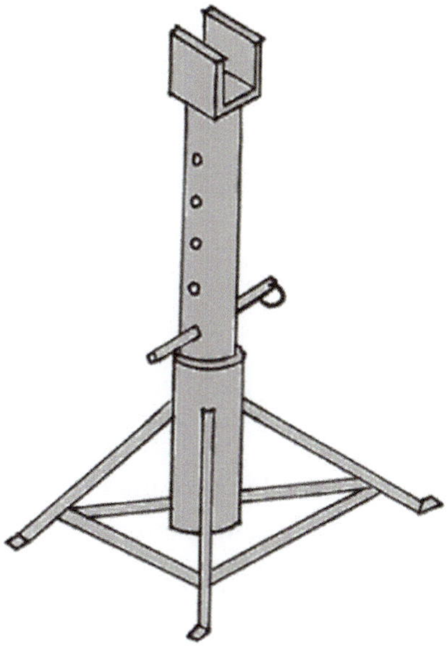

FIGURE 2.36
Splithead

Scaffold boards

All scaffold boards should be made to BS specifications or they should be 'specials'.

To prevent boards from splitting, the ends should be bound with a galvanized metal band (Figure 2.37). Sometimes the board ends are cut at an angle to reduce the risk of damage.

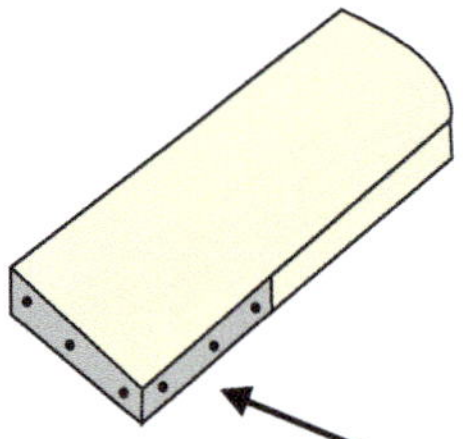

FIGURE 2.37
Boards prevented from splitting

Scaffold boards must be:

- made from straight-grained timber
- free from knots and shakes
- free from decay
- clean and free from grease and thick paint, etc.

Scaffold boards must not be twisted or warped or have split ends.

The distance between the supports governs the thickness of the board used:

- 1.2 m for graded boards
- 1.5 m for BS boards.

Overhang

No board should overhang its supports by more than four times the board thickness, or less than 50 mm.

Note

Two scaffold boards must not b
used one on top of the other.

Lightweight staging

These stages are 600 mm wide and designed to span a greater distance than normal scaffold boards.

Lightweight staging is designed to take the load of three people and is ideal to use with trestles to form a working platform (Figure 2.38).

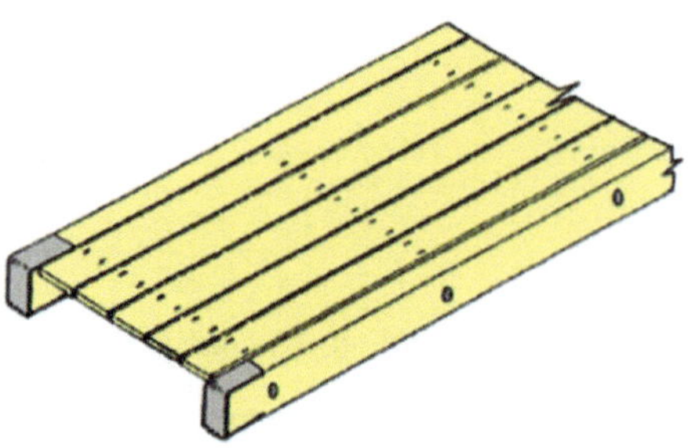

FIGURE 2.38
Typical staging

Stepladders

Stepladders are one of the most commonly used items of equipment in the construction industry.

Recently, the Work at Height Regulations have brought in stringent guidelines for their safe use. Ladders and stepladders have not been banned by these regulations, but require consideration to be given for their use.

They are only recommended for short duration work, i.e. a maximum of 30 minutes, and only for light work.

- Never work from the top two steps unless there is a safe handhold.
- Do not overreach.

The use of stepladders is also one of the most common subjects in a toolbox talk on site.

Every year an average of 14 people die and a further 1200 are injured by falling from a ladder or stepladder.

Stepladders are sold by the number of treads and the main sizes available have from five to 14 treads.

Wooden stepladders

These can be made from various redwoods (Figure 2.39).

You should always inspect a stepladder before using it.

Parts of wooden stepladders

Warning

Do not paint timber steps – paint can hide defects.

- Stiles – these usually taper towards the top and are wide enough to take one 250 mm wide scaffold board.

FIGURE 2.39
Wooden stepladder

- Treads – treads should be at least 90 mm deep and spaced at 250 mm intervals.
- Locking device – fitted to limit the degree of opening and prevent collapse.

Aluminium steps

Aluminium steps are made of aluminium alloy. They are lighter than timber steps, very strong and rot proof, and will not twist, warp, burn or rust. A typical aluminium stepladder is shown in Figure 2.40.

As with wooden stepladders, always check before using.

Parts of aluminium stepladders

- Treads – treads should be at least 90 mm deep and spaced at 250 mm intervals, and have a non-slip surface.
- Locking bar – this is fitted to limit the degree of opening and prevent collapse.

Standing ladders

Single section ladders are available up to 9 m long and made from timber or aluminium. They are the most common means of access to scaffolding.

Double extension ladders

These have two sections similar to standing ladders with a position for coupling them together:

FIGURE 2.40
Aluminium stepladder

- without ropes – up to 4.9 m long when closed, extending to approximately 9 m
- with ropes – up to 7.3 m long when closed, opening to approximately 12 m.

Triple extension ladders

These are similar to double extension ladders, but with three sections:

- without ropes – up to 7.3 m long when closed, extending to approximately 19 m
- with ropes – up to 3 m long when closed, extending to approximately 7.5 m.

A selection of ladders is shown in Figure 2.41.

Warning

Aluminium ladders should not be used near low-, medium- or high-voltage power cables.

Remember

Do not paint timber ladders as paint can hide defects.

Pole ladders

These are single-section ladders with the stiles made from one straight tree trunk cut down the middle. This ensures even strength and flexibility. They are available up to 12 m long and are used mainly for access to tubular scaffolding.

Parts of ladders

- Stiles – wooden ladders have stiles made from Douglas fir, whitewood, redwood or hemlock. The stiles of pole ladders are made from whitewood.
- Rungs – wooden ladders have round or rectangular rungs, made from oak, ash or hickory. Aluminium ladders have non-slip rungs.
- Ties – steel rods fitted below the second rung from each end on wooden ladders and at not less than nine-rung intervals. Ties can be fitted under every rung. They are not required on aluminium ladders.
- Reinforcing wires – these give wooden ladders extra strength. Galvanized wire or steel cable is fitted and secured, under tension, into grooves in the stiles. They are not required on aluminium ladders.

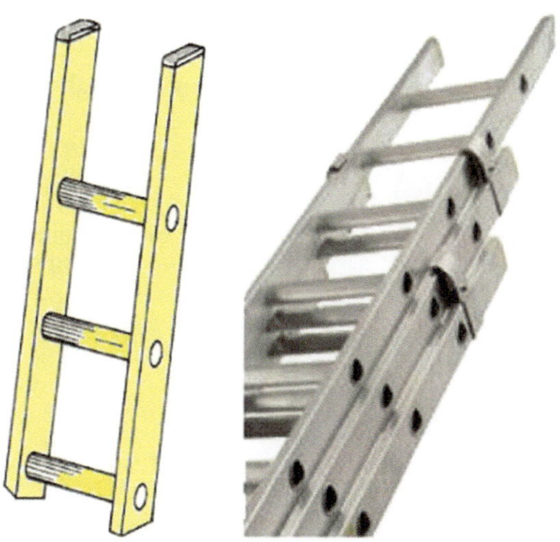

FIGURE 2.41
Types of ladders

- Ropes – both types of ladders use hemp sash cord or a material of equivalent strength. Ropes of artificial fibre must provide adequate hand grip.
- Guide brackets – fitted at the top of lower sections on both types to keep the sections together.
- Latching brackets or hooks – fitted to the bottom of extension sections on both types to hook over a rung of the section below.
- Pulley wheels – guide and facilitate the smooth running of the ropes of rope- operated wooden and aluminium ladders.

Safe use of ladders

Raising and lowering ladders

Ladders should be raised with the section closed.

Extension ladders with long sections are raised one section at a time and slotted into position before use.

Two site operatives are required to raise and lower heavier types of ladders.

When erected, the correct safety angle is 75° or a ratio of four up to one out (Figure 2.42).

Lighter ladders may be raised by one person, but the bottom must be placed against a firm stop before lifting begins.

Tying ladders

The appropriate regulations state that ladders must have a firm and level base on which to stand, and if over 3 m long, they must be fixed at the top, or if this is not possible at the bottom. If neither way is possible, a person must 'foot' the ladder, that is, stand with one foot on the bottom rung and the other firmly on the ground. The assistant must hold both stiles and pay attention all the time.

This prevents the base from slipping outwards and the ladder from falling sideways.

It is important to tie ladders when working at heights.

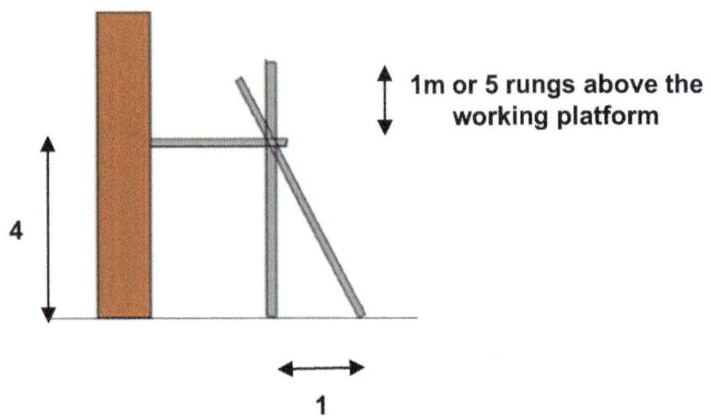

FIGURE 2.42
Correct angle of ladder

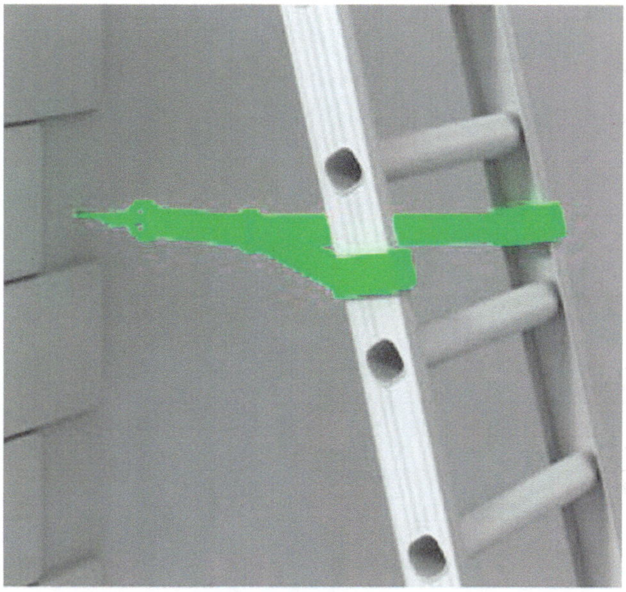

FIGURE 2.43
Method of tying a ladder

Bolts can be fixed to brickwork and a ladder tie secured.

Other methods could be used as shown in Figure 2.43.

Note

NEVER tie ladders to pipes or gutters.

Lifting and carrying ladders

To lift and carry ladders over short distances rest the ladders on the shoulder, and lift them on the shoulder.

Lift vertically by grasping the rung just below normal reach. The correct balance and angle must be found before moving.

Remember

Take care when raising to avoid over- head obstructions, such as electric cables.

Tower scaffolds

These may be either mobile or static and are suitable for both internal and external work up to approximately 6 m.

Towers may be constructed from individual scaffold components or proprietary units. This unit will only deal with proprietary mobile towers. A typical mobile tower is shown in Figure 2.44.

They usually support a single working platform, not projecting beyond the base area, and are provided with toeboards and handrails.

Access to the working platform is by a ladder, which may be fastened inside or outside the structure.

Light-duty access towers are very common in the building industry for lightweight work such as maintenance of gutters, painting, etc.

They will not support a load greater than 1.5 kN/m^2. This is approximately equivalent to two people standing per square metre.

FIGURE 2.44
Mobile tower scaffold

The safe working load should be clearly displayed on the working platform.

Tower scaffolds should always be vertical, even if erected on sloping ground. Mobile towers should only be moved on firm, level ground.

Towers should never be moved with people or materials on them. They should always be pushed at the lowest practical point.

If extra working height is required, then the base measurement can be increased by the use of outriggers. These are tubes or special units that connect to the bottom of the tower, at the corners, giving greater overall base measurements. Outriggers also help to stabilize a scaffold tower and are usually used for this purpose as well as giving extra working height.

Note

NEVER lean a ladder against th
tower scaffold.

Remember

Access must be provided throug
the working platform.

System scaffold

These systems are proprietary systems consisting of special units that fit into each other (Figure 2.45). They can be use to produce an access tower or a hop-up in a very short time.

This type of scaffold is increasingly being used for quick, straightforward platforms.

There are many patent types of frame available, but basically each consists of two short tubes which act as uprights, and these are joined near the top and bottom by tubes with a welded joint at each end.

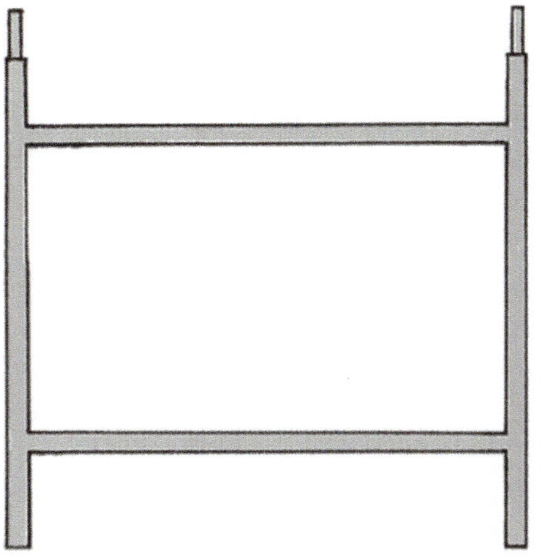

FIGURE 2.45
System tower scaffold

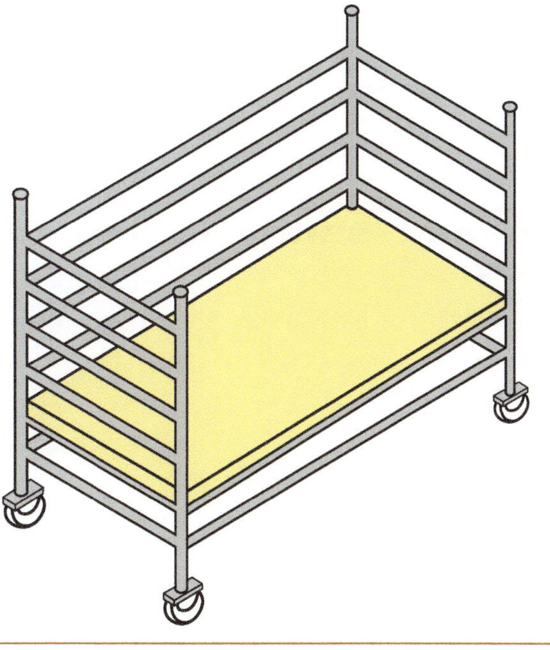

FIGURE 2.46
Portable low working platform

Low towers

Many manufacturers now produce various designs of working platforms to meet the current Work at Height Regulations.

They are restricted to a maximum of 2 m in height and usually have a safe working load of 150 kg. They arrive ready assembled, are very lightweight and are easy to move.

A low working platform is shown in Figure 2.46.

Multiple-choice questions

Self-assessment

This section of the book is designed to allow you to check your level of knowledge. The section consists of revision questions for this chapter. The questions are all multiple choice and have four possible answers. The answers are to be found at the end of the book.

The main type of multiple-choice question will be the four-option multiple-choice question. This will consist of a question or statement, known as the stem, followed by a choice of four different answers, called the responses. Only one of these responses is the correct answer; the others are incorrect and are known as distracters.

You should attempt to answer the questions by choosing either (a), (b), (c) or (d).

Example

The person employed by the local authority to ensure that the Building Regulations are observed is called the:

(a) clerk of works
(b) building control officer
(c) council inspector
(d) safety officer

Health and safety in the construction industry

Question 1 Which fire extinguisher should you use when dealing with a petrol fire?

(a) water
(b) foam
(c) vaporizing liquid
(d) carbon dioxide

Question 2 When should you wear safety boots on site?

(a) at all times
(b) when working in the rain
(c) when moving materials
(d) when told to

Question 3 The electric tool you have been given to use has a faulty switch. What action should you take?

(a) stop working and inform your supervisor
(b) stop working and find another similar machine
(c) try to fix the fault
(d) tape up the switch to keep it working

Question 4 What does the following double-square symbol represent on electrically operated hand tools?

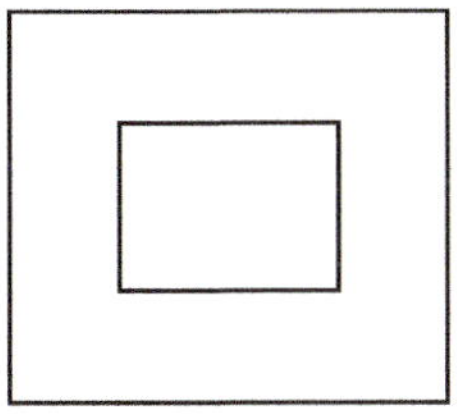

(a) it was manufactured in the UK
(b) it is safe to use in a vertical position
(c) it is double insulated
(d) it has been tested

Question 5 What shape should a warning sign be?

(a) circular
(b) rectangular
(c) square
(d) triangular

Question 6 Which of the following current regulations covers working platforms?

(a) Health and Safety at Work Act
(b) Control of Substances Hazardous to Health
(c) Work at Height Regulations
(d) Personal Protective Equipment Regulations

Question 7 What should you do before lifting a load?

(a) put on gloves
(b) keep your back straight
(c) bend your knees
(d) assess the weight

Question 8 When erecting a ladder to a working platform which is 4 m above the ground, how far out should the bottom of the ladder be?

(a) 1 m
(b) 2 m
(c) 3 m
(d) 4 m

CHAPTER 3

Communication

This chapter will cover the following City and Guilds units:

• Apprenticeship	Module 3	K1.1
		K2.1, K2.2, K2.3, K2.4, K2.5
		K3.1, K3.2, K3.3, K3.5, K3.6, K3.7
		K6.1
	Module 10	S1.1, S2.1
• Diploma	Unit 101/501	1.1, 1.2, 1.3, 1.4
	Unit 102	1.1,1.2, 1.3
		2.1
	Unit 103	1.3, 2.3
	Unit 104	1.2, 1.3, 2.3
	Unit 105	2.3

This chapter is about:

- Interpreting building information
- Determining quantities
- Relaying information

The following City and Guilds performance criteria will be covered:

- Understand how to communicate with colleagues and customers in the workplace
- Know company structures, professions and progression routes in the industry
- Know the information and documentation required for completing construction projects
- Known methods of calculating using formulae for construction

The following Diploma outcomes will be covered:

- Know how to interpret building information
- Know how to determine building quantities
- Know how to relay information to the workplace

Interpreting information

This chapter deals with extracting and interpreting information and then correctly relaying it to other people.

Throughout your working life you will have to consult various sources of information and during this chapter you will be required to make decisions and solve problems from the information given or extracted.

Information sources

The construction industry is almost unique in that the design process is separated from the production process.

In the construction industry:

The architect designs and the contractor builds

In most other industries the designer is employed by the producing firm, as in the motor industry.

It is essential that all parties have access to various sources of up-to-date information. No single person in isolation could satisfy the demands of all members of the design team and construction teams. Every designer and contractor must be able to call on a team of specialists and other back-up information.

It is also very important that the student understands where information can be found.

Storing information

When information has been found the last place to store it is in your head.

The relative information should be stored in a filing system which everyone can understand. The filing system could consist of something as simple as document files or a more elaborate computer program.

Any information found may be required over and over again for a particular contract, but can also be useful information for future programmes.

Sources of information

Before any contract is started it is important to ensure that all information is available and understood.

Information on materials can be obtained from numerous sources:

- manufacturer's literature
- specifications
- schedules
- simple drawings.

Manufacturer's technical information

Any efficient office should have up-to-date information regarding new and existing products from the various manufacturers who produce materials of interest to them. Many manufacturers produce technical information which is free for the asking.

Many manufacturers produce technical information which is free for the asking.

Manufacturer's technical information consists of drawings and text which are provided to give the user details of the product.

Specification

This is a document giving a written description of materials to be used, and construction methods to be employed in the construction of the building.

Schedule

This is a contract document that can be used to record repetitive design information about a range of similar components, such as:

- doors
- windows
- ironmongery
- decorative finishes
- inspection chambers (manholes).

Simple drawings

The design team will be required to produce working drawings for the builder to use on the site.

Drawings, schedules and specifications will have to be prepared, explaining how the design team requires the building to be constructed. To be able to read these drawings it is essential that the trainee is able to understand them.

Drawings should be produced according to:

British Standard Recommendations for Drawing Office Practice BS 1192.

BS 1192 is a British Standard that sets out a best practice method for development, organization and management of production information in architecture, engineering and construction.

In 2007 it was updated to BS 1192:2007 + A2:2016 to support PAS 1192.

These recommendations apply to the sizes of drawings, the thickness of lines, dimension of lettering, scales, various projections, graphical symbols, etc.

The person carrying out a task should be able to read drawings and extract the required information.

Information concerning a project is normally given on drawings and written on printed sheets.

Drawings should only contain information that is appropriate to the reader; other information should be produced on schedules, specifications or information sheets.

Understanding working drawings

Scale

Scales are used in drawings to enable large objects such as buildings to be drawn to a convenient size, which will fit onto the drawing paper, while still maintaining accurate proportions which can be drawn or measured as required.

Scales use ratios to relate measurements on a drawing to the real dimensions of the actual item being drawn.

Once these items have been drawn to a smaller size, which is known as a ratio to the real item, it is known as a scale drawing.

Scale rules

An ordinary ruler, 300 mm long (Figure 3.1), is useful for simple drawings, but scale drawings require a scale rule.

A scale rule (Figure 3.2) has a series of marks for measuring purposes. The scale rule is graduated in millimetres. Each scale represents a ratio of given units to one unit.

Table 3.1 lists of the preferred scales for building drawings.

Use of scales

The choice of scale will depend on two things.

- the size of the object to be drawn
- the amount of detail that needs to be shown.

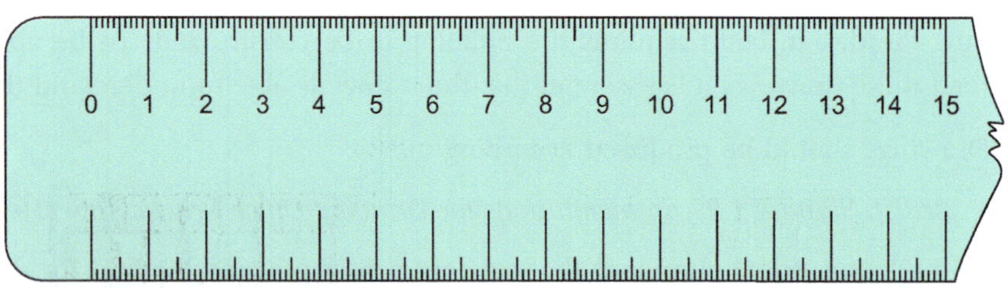

FIGURE 3.1
Ruler

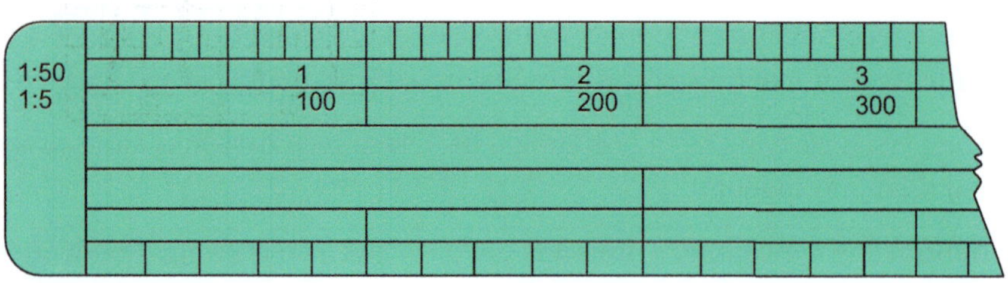

FIGURE 3.2
Scale rule

Table 3.1 Preferred scales for building drawings

Type of drawing	Scales
Block plans	1:2500, 1:1250
Site plans	1:500, 1:200
Location drawings	1:200, 1:100, 1:50
Range drawings	1:100, 1:50, 1:20
Detail drawings	1:10, 1:5, 1:1
Assembly drawings	1:20, 1:10, 1:5

A scale is used to measure distances on drawings and for taking measurements of a drawing. If scaled dimensions and written dimensions disagree, then the written dimension should always be used.

The ratio shows how many times bigger one quantity is than the other.

If the drawing is made to a scale of 1:100, 1 mm will represent 100 mm.

If the drawing is made to a scale of 1:20, 1 mm will represent 20 mm, and so on.

If the following drawing was drawn 150 mm long it would represent something much bigger in size:

To a scale of 1:10 it would represent 1500 mm.

If it is drawn to a scale of 1:20 it would represent 3000 mm.

All you have to do is multiply the scale measurement by the scale ratio.

Scale rules can be used for both reading and preparing a drawing.

Remember

It is recommended that written dimensions on a drawing should always be used, rather than taking dimensions off drawings with a scale rule.

Remember

Too much hatching could complicate a drawing.

Hatchings

It is often convenient to emphasize the differences between materials when drawn in section. This need is most common on large-scale details, such as sections through parts of the construction (Figure 3.3).

The British Standard shows a number of materials in common use and the manner in which these should be drawn. All construction materials can thus be identified by the type of hatching (Figure 3.4).

Abbreviations

Abbreviations are a simple way of conveying information on drawings, reducing words to initial letters, e.g. rain water pipe becomes RWP. This saves cluttering up the drawing.

Abbreviations have to be used in context, e.g. MS stands for mild steel in the context of construction but could also be an abbreviation for some other words in another situation.

A few of the most common abbreviations are shown in Table 3.2. Avoid making up your own abbreviations as these can lead to confusion.

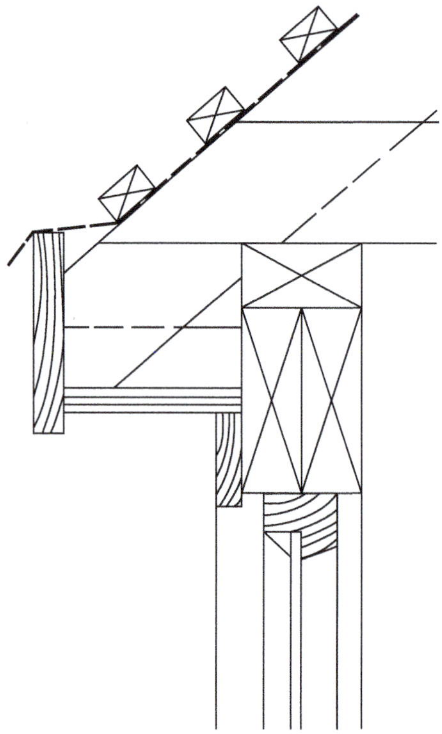

FIGURE 3.3
Section through the eaves showing the hatching of the materials required

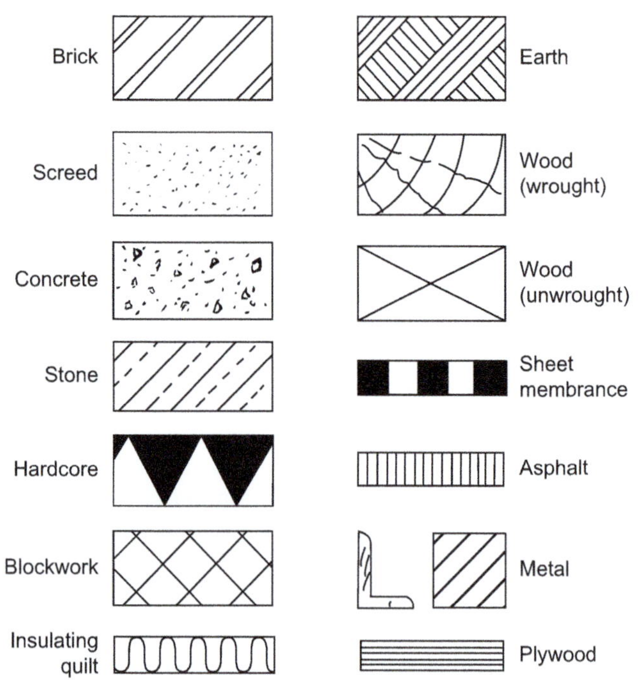

FIGURE 3.4
Hatching of common building materials

Table 3.2 Standard abbreviations for building materials

Item	Abbreviation	Item	Abbreviation
Airbrick	AB	Glazed pipe	GP
Aggregate	agg	Hardcore	hc
Brickwork	bwk	Hardwood	hwd
Building	bldg	Inspection chamber	IC
Cement	ct	Insulation	insul
Damp-proof course	DPC	Joist	jst
Damp-proof membrane	DPM	Mild steel	MS
Drawing	dwg	Plasterboard	pbd
Foundation	fnd	Reinforced concrete	RC
Fresh air inlet	FAI	Vent pipe	VP

Graphical symbols

These are small standard pictures used to reduce the amount of drawing detail required on individual drawings.

Abbreviations and graphical symbols are often used together to give complete information, as shown in Table 3.3.

Colour

A further classification for representation of materials is the use of colour (Table 3.4).

It used to be customary to colour drawings, particularly for submission to local authorities, but this practice has fallen into disuse, for it is a slow and costly process.

The exception is the application of a red line to mark the site or area of land on block plans.

Production drawings

These are known as working drawings and are used for the purpose of relaying information to the building contractor and other members of the building team.

The working drawings can be classified as:

- location drawings
- component drawings
- assembly drawings.

Location drawings

These are further classified into:

- block plans
- site plans
- location plans.

Table 3.3 Common graphical symbols

Representation of components: fittings	
Name	Symbol
Shower stall	
Rectangular bath	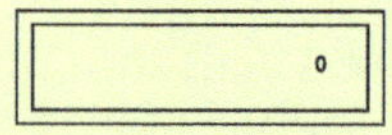
Hot and cold water draw-off	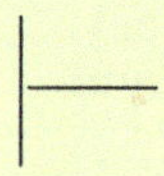
Safety valve	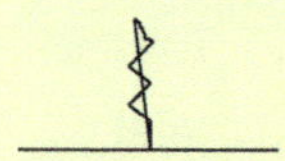
Stop valve	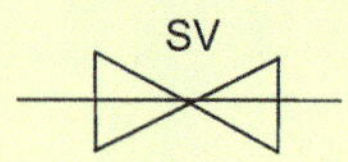
Hot water tank	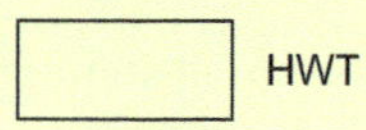
Rainwater head	
New tree	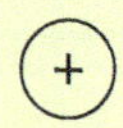
Tree to be removed	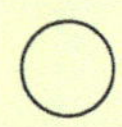
Existing tree	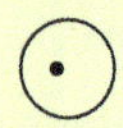
Rainwater pipe	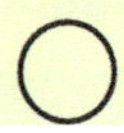

Table 3.4 Colour of materials

Materials	Colour
Brick	Vermilion
Earth	Sepia
Steel	Purple
Glass	Paleblue
Concrete	Hooker's green
Plaster	Terra-verde
Unwrought timber	Raw sienna
Wrought timber	Burnt sienna

Block plans

These are used to identify the site in relation to the surrounding area (Figure 3.5).

The scale is usually 1:2500 or 1:250 and is too small to allow much more than an outline of the site and boundaries, road layouts and other buildings in the near vicinity.

The orientation of the site is always shown with a suitable logo depicting north. The actual site should be outlined in red.

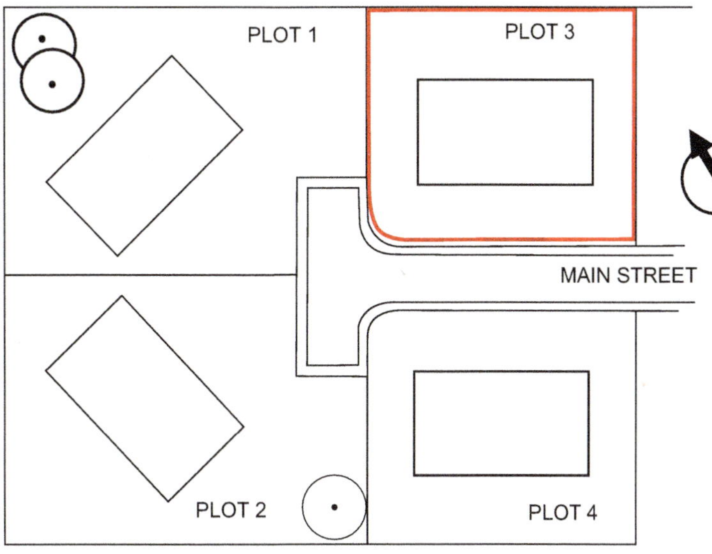

FIGURE 3.5
Block plan

It is unlikely that dimensions would be added to these drawings.

Plans of this sort are usually based on the Ordnance Survey sheet for the area; however, if such a source is used, permission should be obtained for its reproduction.

Site plans

These are used to show the position of the proposed building on the site, together with information on proposed road, drainage and service layouts, and other site information such as levels. A typical site plan is shown in Figure 3.6.

Again the orientation of the site should be shown.

Scales of 1:500 and 1:200 are often used.

The information on this drawing is used by both the design team and the contractors.

The drainage layout should be used in conjunction with the drainage schedule which will give details of inspection chamber construction, cover and invert levels, and other relevant information.

Location plans

These are used to show the size and position of the various rooms within the buildings and to position the principal elements and components. A typical location plan is shown in Figure 3.7.

Plans are usually drawn to a scale of 1:200, 1:100 or 1:50.

These scales are sufficiently large to allow for dimensions to be added.

Further information from the drawing includes the wall construction and position and type of doors and windows. This information is used in conjunction with door and window schedules. The way the doors are hung is also shown.

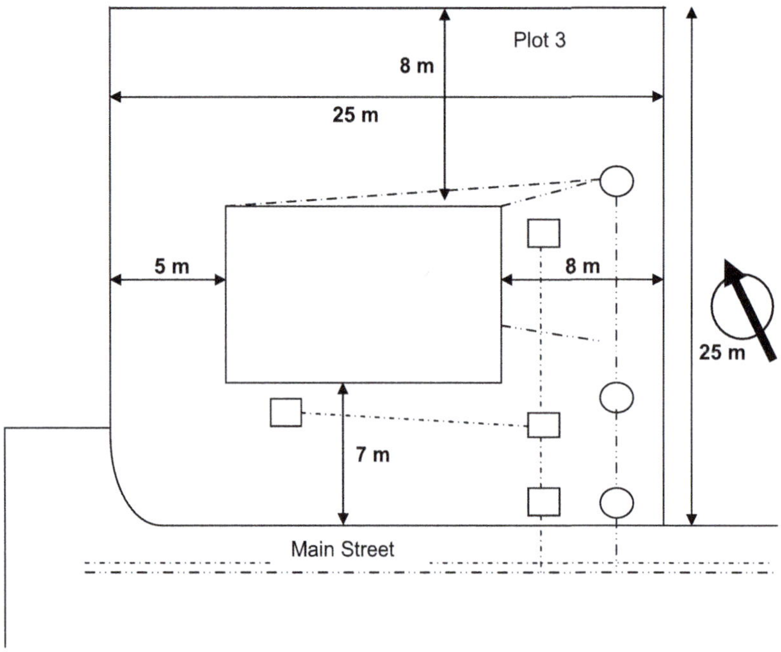

FIGURE 3.6
Site plan

Component drawings

These are further classified as:

- range
- detail.

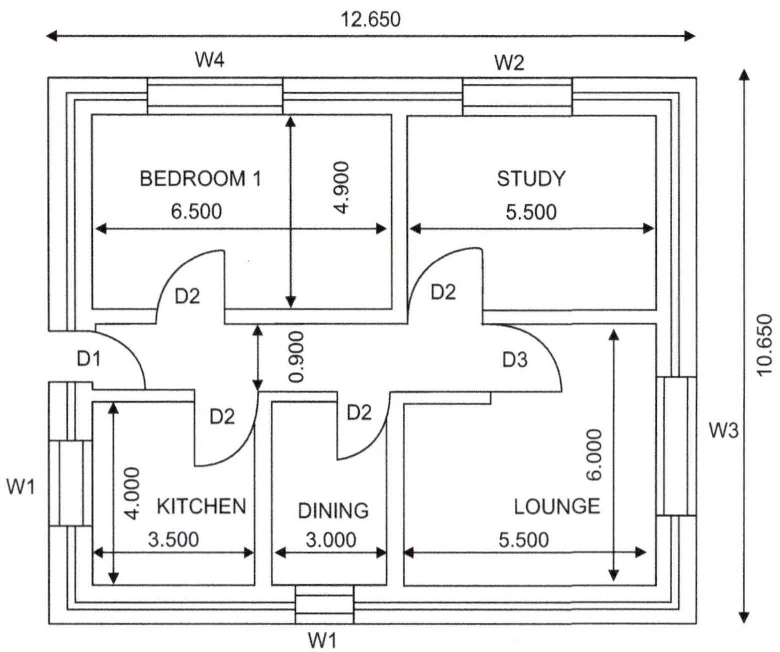

FIGURE 3.7
Location plan

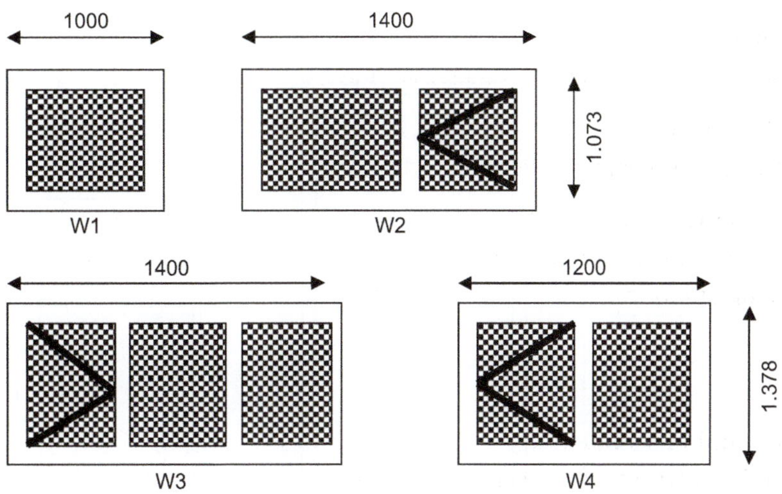

FIGURE 3.8
Windows range

Range drawings

These show the basic sizes and reference system of standard components. Typical range drawings are of doors and windows. A range of windows is shown in Figure 3.8.

They are usually drawn to a scale of 1:100, 1:50 or 1:20.

Drawings of windows can be obtained from any manufacturer's literature with their references, i.e. W1, W2, W3, etc.

This reference is all that is required by the architect when completing the working drawings as the builder is provided with a window schedule for cross-referencing.

Door range details are shown in Figure 3.9.

Detail drawings

These are used to show the information necessary for the manufacture of the various components, i.e. doors, windows, concrete units, cupboard units, etc. Figure 3.10 shows a typical detail.

The information from these details is only for the manufacturer – the architect will produce these details either in full size, i.e. 1:1, or at 1:5, 1:10, etc. They should include every tiny detail required for the manufacture of the component.

Assembly drawings

These are used by the architect to show in detail the junction between the various elements and components of the buildings. A typical detail is shown in Figure 3.11.

These details are necessary for the builder to know exactly how the architect requires the construction to be completed.

The scales are usually 1:20, 1:10 and 1:5 and should be fully dimensioned and annotated.

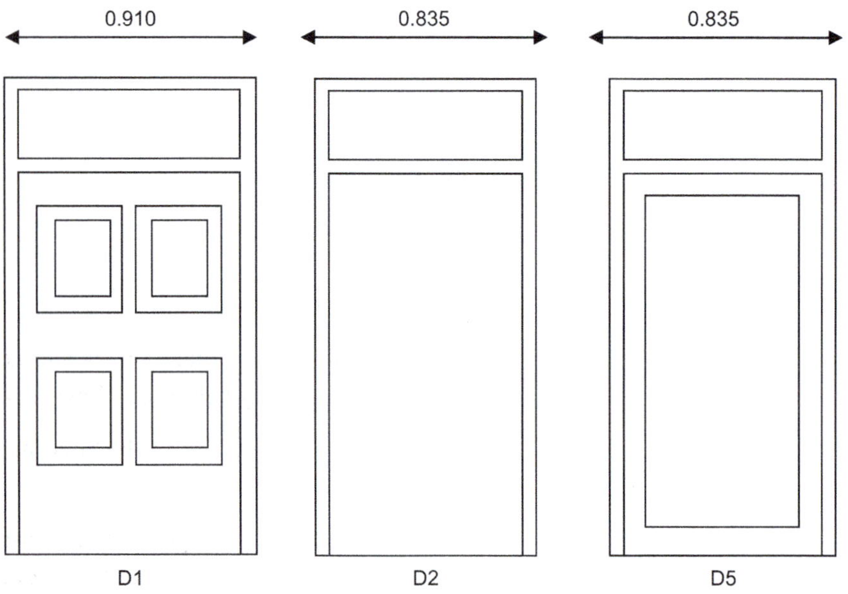

FIGURE 3.9
Door range

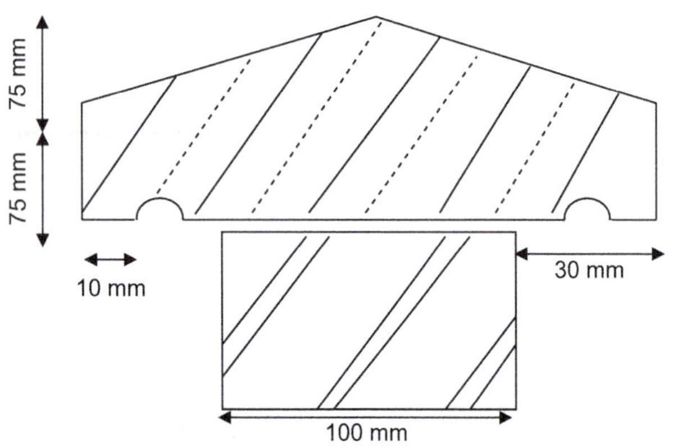

FIGURE 3.10
Detail of coping stone

Understanding drawings

Selecting information from simple drawings, specifications and schedules

It is important that you can understand drawings and extract the correct information from them. The following section gives you a chance to practise reading and understanding drawings, specifications and schedules.

Drawings

You have received the block plan shown in Figure 3.12. Explain the items marked 1–6 on the plan by completing Table 3.5.

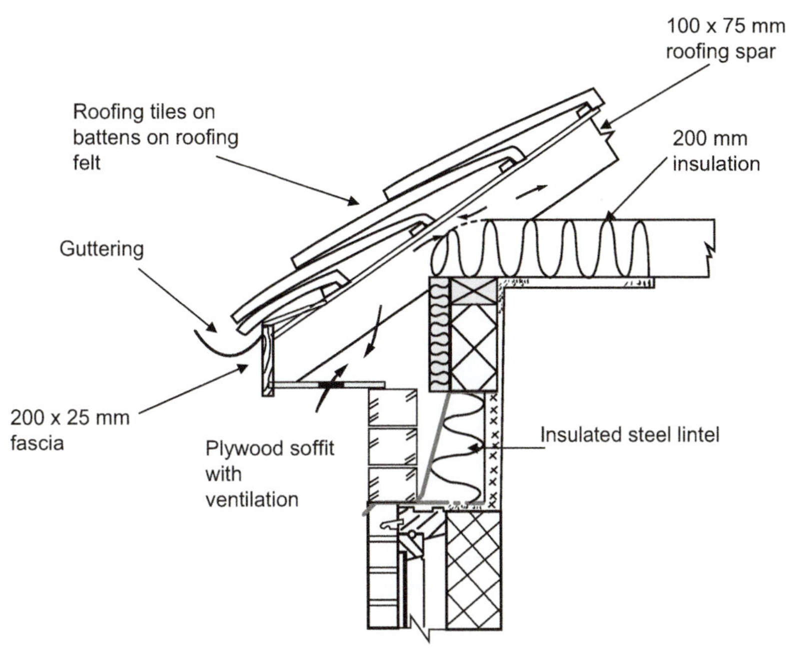

FIGURE 3.11
Eaves detail

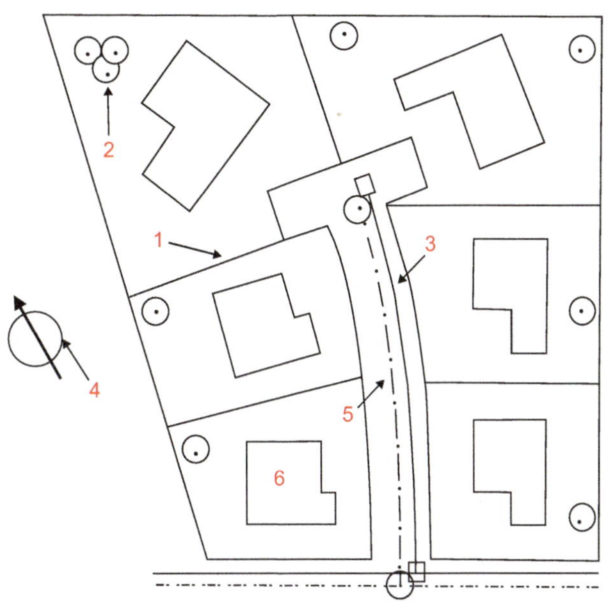

FIGURE 3.12
Block plan

Specifications

These documents are prepared to specify the exact quality of materials and skill required throughout the contract.

It is essential to read and understand exactly what is required on a contract as this will affect the price of the work.

Table 3.5 Definitions of items in block plan

No.	Definition
1.	
2.	
3.	
4.	
5.	
6.	

Table 3.6 Specification

Item no.	Description
1	Block partition walls, 102.5 in stretcher bond in cement mortar (1:4)
2	Brick facework, 102.5, in stretcher bond using multicoloured facings in cement mortar (1:4) Pointing with half round joint as the work proceeds
3	Blockwork to inner skin of cavity wall, 102.5, in cement mortar (1.5)stretcher bond in
4	Form cavity in hollow wall Insert 50 mm polystyrene batts
5	Damp-proof course, 102.5, of single-layer hessian base bitumen felt and bedded in cement mortar

Specifications are used only on large contracts as the drawings cannot contain all the information required by the contractor.

Drawings for small building works usually have the specification written on the drawings.

A typical specification is shown in Table 3.6. Read it carefully and complete the following questions.

Specification details

To show that you can understand specifications in Table 3.6, complete Table 3.7.

Schedules

As mentioned previously, schedules form an important method of communication for the design team.

Table 3.7 Specification details

Question no.	Question	Answer
1	What type of cavity insulation will you order?	
2	What type of damp-proof course is specified?	
3	How many m^2 of blockwork is required?	
4	If three wall ties are required per m^2 how many will you order for the cavity?	

Table 3.8 Door schedule

Description	D1	D2	D3	D4	D5
External panelled	•				
Internal flush		•4			
Internal panelled					
Internal half-glazed					
Internal glazed					•

Table 3.9 Window schedule

Description				

They are prepared for large contracts to simplify tabulated information about a range of components.

It is important to be able to understand schedules, to extract and order the correct item for the correct position in the building.

Using the location plan in Figure 3.7, showing the ground floor, you can see five windows and 6 doors, marked W1–W4 and D1–D3, respectively.

The design of a schedule can take various formats, but normally the component types are across the top and the various finishes are listed below. A typical door schedule is shown in Table 3.8.

Details relevant to a particular door opening are indicated in the schedule by a bullet (•).

Where more than one item is required a number is inserted.

To show that you can read and understand a schedule, complete the window schedule shown in Table 3.9, using information shown on the ground floor location drawing in Figure 3.7 and the windows range shown in Figure 3.8.

Quantities of materials

You will at some stage in your career have to determine quantities of materials required to carry out an activity. This will entail having to calculate areas and volumes.

The buyer will have the responsibility of taking off and scheduling materials from the bills of quantities, or on smaller jobs, the working drawings.

When the site supervisor is given this role he or she must be careful to ensure that correct quantities and details of quality of materials are extracted from the drawings and specifications.

Allowances should be made for materials wastage which, from experience, is usually 5–10 per cent.

Taking off materials measurements from drawings should be carried out with care to allow for extra lengths or quantities, especially on joists, to give a suitable margin of safety.

Too many examples are noticed on building sites of long offcuts from joists, rafters and other structural members owing to generous allowances having been made at the taking-off stage. Extra members can also be seen lying around the site unused.

Types of calculation

Calculations in construction can be:

- numbers – one-off items such as chimney pots
- linear – lengths of materials, etc.
- superficial – areas of walls and floors, etc.
- cubic – volume of excavations, etc.

Number

Many building materials are measured by number.

Chimney pots, air bricks, doors and windows are all examples of materials that are measured by number.

Linear

Many building materials are measured by the length, but stating the width and thickness.

Timber is measured by length, for Example 3 m length of 50 × 50 mm.

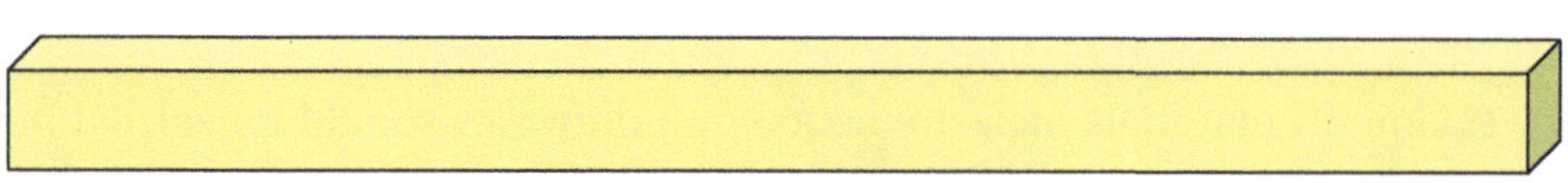

Superficial

This is also known as square measurement, as the length and depth are multiplied to achieve the superficial area of the material required.

Examples are brick and blockwork.

The length of the wall is multiplied by the height to give the square area of brickwork required.

Cubic

Cubic measurements are taken when there are three dimensions: length, width and depth.

Examples are found in excavating trenches and concrete for foundations.

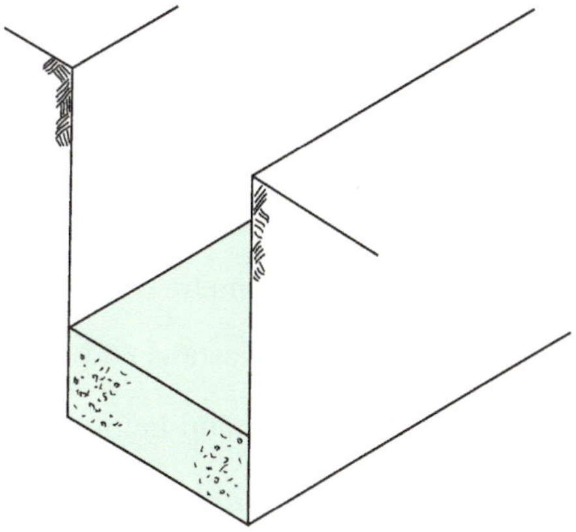

Examples

Bricks

The drawing below shows the nominal sizes of a brick as:

215 mm × 102:5 mm × 65 mm

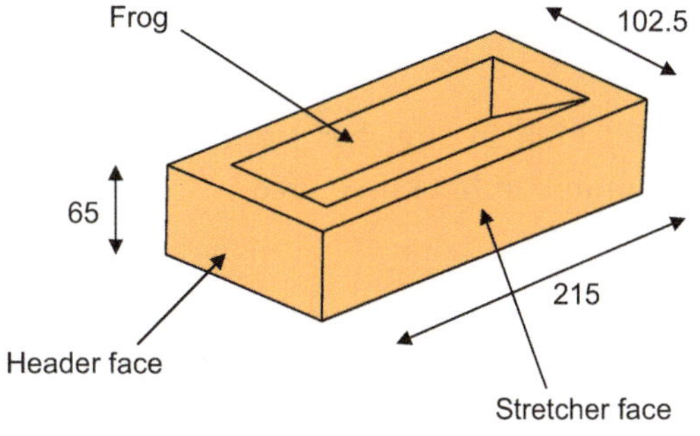

These are the net sizes of the average brick and it is necessary to add 10 mm to the length, width and depth to produce a size when installed in the wall.

Therefore: length = 225 mm; depth = 75 mm and width = 112.5 mm:

Bricks are laid to a gauge (number of courses to a set measurement) of four courses to 300 mm.

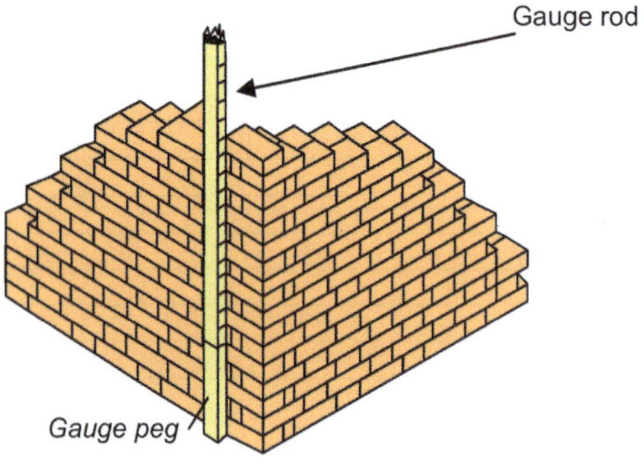

To calculate the number of bricks required per square metre, in stretcher bond, it is necessary to calculate the face area of one brick, and then divide that into 1 m^2.

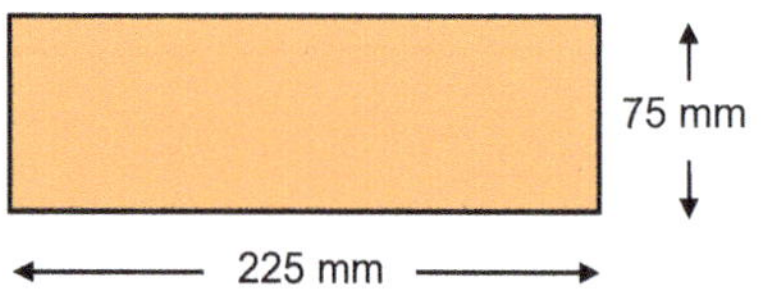

$$0.225 \times 0.075 = 0.016875$$

Number of bricks per square metre = 1.0000/0.0169 = 59 Bricks

A more realistic figure allowing for waste is 65 bricks.

These figures alter according to the bond being used:

- half-brick wall in stretcher bond = 65 bricks
- one-brick wall in stretcher bond = 125 bricks
- half-brick wall in English bond = 90 bricks
- one-brick wall in English bond = 180 bricks
- half-brick wall in Flemish bond = 80 bricks
- one-brick wall in Flemish bond = 160 bricks.

Example 1

A one-brick wall built in stretcher bond is 6.00 m long and 1.50 m high. Calculate the number of bricks required.

Answer 1

$$6.00 \times 1.50 = 9.00 \text{ m}^2$$

$$9.00 \times 125 = 1125 \text{ bricks required}$$

Blocks

The dimensions for blocks are shown as:

440 mm long × 215 mm high × 100 mm wide

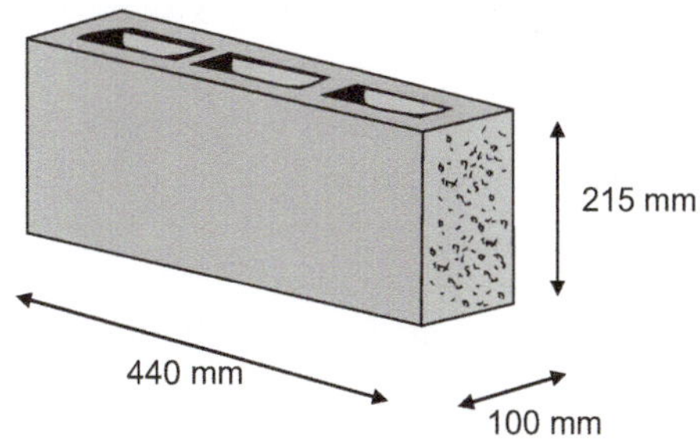

These are net dimensions and the thickness of the joint is added to obtain the following dimensions:

450 mm long × 225 mm high × 100 mm wide

Blocks when laid are equal to three courses of bricks.

The calculations for blocks are very similar to brick calculations.

First, calculate the area of one block, then divide it into 1 m^2.

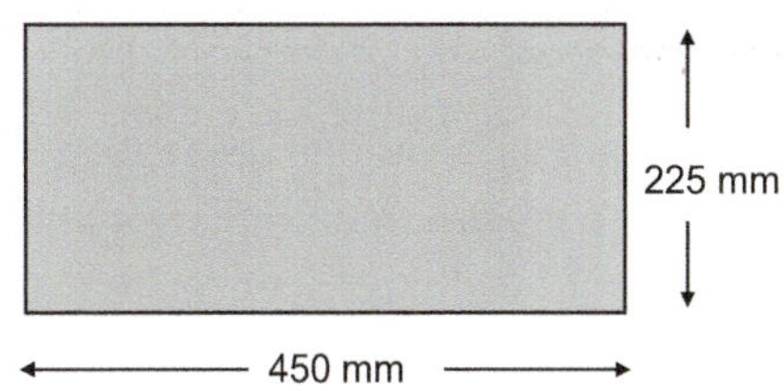

$$450 \text{ mm} \times 225 \text{ mm} = 0.10125 \text{ m}^2$$

Number of blocks per square metre = 1.000/0.10125 = 9.87 blocks

It is usual to include waste and allow 11 blocks per square metre.

Example 2

Calculate the number of blocks required to construct a block wall 2.690 m long and 0.675 m high.

Answer 2

$$2.690 \times 0.675 = 1.815 \text{ m}^2$$

$$1.815 \times 11 = 19.9; \text{ say 20 blocks required}$$

Concrete

To calculate the volume of concrete foundations you need three dimensions: the length of the trench, the width of the trench and the thickness of the concrete foundations.

Example 3

Calculate the concrete required for a foundation if the trench is 39.00 m long and 0.75 m wide if the concrete is 0.15 m thick.

Answer 3

$$39.00 \times 0.75 \times 0.15 = 4.387 \text{ m}^3$$

Drainage

The number of pipes can be calculated by dividing the length of the trench by the length of each individual pipe.

Example 4

Calculate the number of drain pipes required for a trench 36.00 m long if 0.900 mm pipes are being used.

Answer 4

$$36.00/0.900 = 40 \text{ pipes}$$

Calculations for areas

It will be necessary to calculate various areas to find the required materials. The shapes may include triangles, quadrilaterals and circles.

Triangles

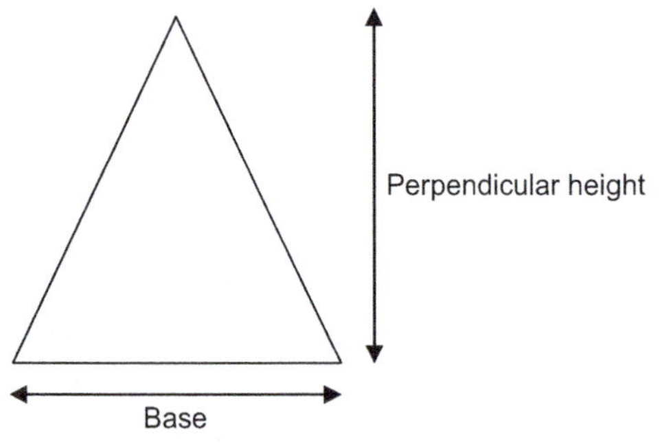

$$\text{Areaa of a triangle} = \frac{\text{Base} \times \text{Perpendicular height}}{2}$$

Example 5

Calculate the area of a triangular piece of land if the base is 25 m and the perpendicular height is 35 m.

Answer 5

$$\text{Area} = \frac{\text{Base} \times \text{Perpendicular height}}{2}$$
$$= \frac{25 \times 35}{2}$$
$$= \frac{875}{2}$$
$$\text{Area} = 437.5\ \text{m}^2$$

Quadrilaterals

There are five types of quadrilateral: rectangle, square, parallelogram, rhombus and trapezium.

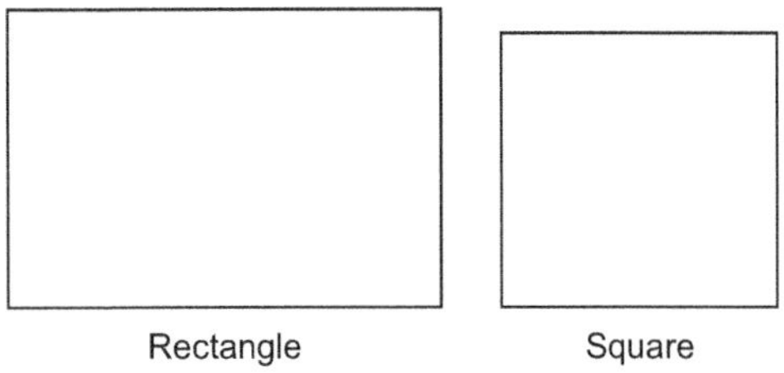

Area of a rectangle and square = Length × Width

Example 6

A rectangular plot of land measures 18 m long and 15 m wide. Calculate the area of the plot.

Answer 6

$$\text{Area} = \text{Length} \times \text{Width}$$
$$= 18 \times 15$$
$$\text{Area} = 270\ \text{m}^2$$

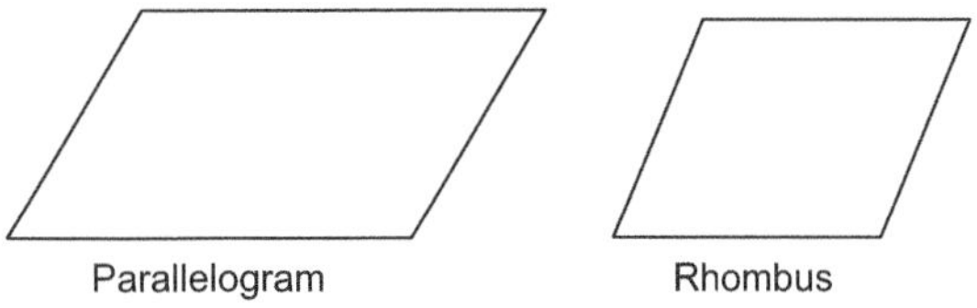

Area of a parallelogram and rhombus = Length × Perpendicular height.

Example 7

Calculate the area of a parallelogram with a 35 m length and 12 m perpendicular height.

Answer 7

$$\text{Area} = \text{Length} \times \text{Perpendicular height}$$
$$= 35 \times 12$$
$$\text{Area} = 420\ \text{m}^2$$

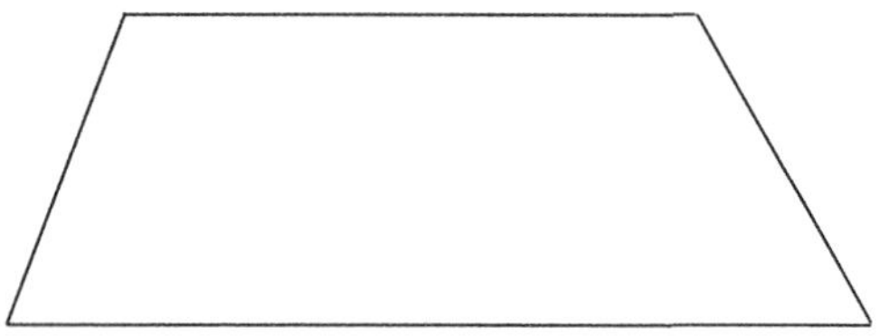

Trapezium

$$\text{Area of a trapezium} = \frac{1}{2}(a + b) \times \text{Height}$$

Example 8

Calculate the area of a trapezium with small length of 15 m and a long length of 35 m. The height of the plot of land is 11 m.

Answer 8

$$\text{Area of a trapezium} = \frac{1}{2}(a + b) \times \text{Height}$$
$$\frac{(15 + 35) \times 11}{2}$$
$$= \frac{550}{2}$$
$$\text{Area} = 275\ \text{m}^2$$

Circles

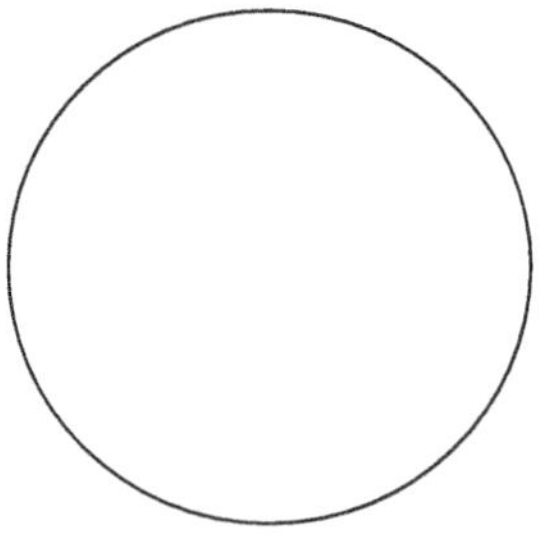

$$\text{Area of a circle} = \pi r^2$$

Example 9

Calculate the area of a circle with a diameter of 18 m.

Answer 9

$$\text{Area of a circle} = \pi r^2;\ \pi = 3.142$$

$$\pi \times 9 \times 9$$

$$= 3.142 \times 9 \times 9$$

$$\text{Area} = 254.5\ \text{m}^2$$

Calculation of perimeters

Rectangles

$$\text{Perimeter of a rectangle} = 2 \times \text{Length} + 2 \times \text{Width}$$

Example 10

Find the perimeter of the building shown below.

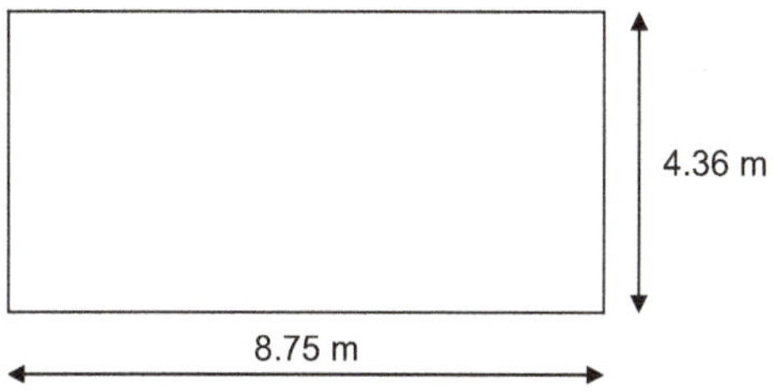

Answer 10

$$\text{Perimeter} = 2 \times L + 2 \times W$$

$$= (2 \times 8.75) + (2 \times 4.36)$$

$$= 17.50 + 8.72$$

$$= 26.22\ \text{m}$$

$$\text{Perimeter} = 26.22\ \text{m}$$

Example 11

Find the perimeter of the building shown below.

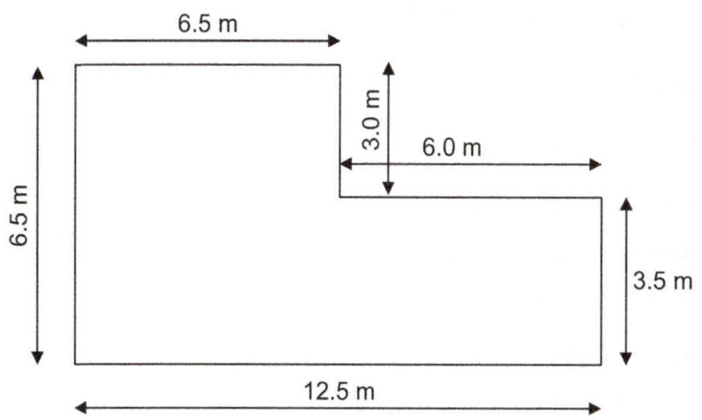

Answer 11

$$\text{Perimeter} = 2\text{ L} \times 2\text{ W}$$
$$= (2 \times 12.5) + (2 \times 6.5)$$
$$= 25 + 13$$
$$= 38\text{ m}$$
$$\text{Perimeter} = 38\text{ m}$$

There is no need to add all the individual sides when the overall length and width are known.

If the dimension is missing from the long side then the two smaller sides will have to be added together to give the overall length:

$$= 6.5 + 6$$
$$= 12.5\text{ m}$$

Mortar

The amount of mortar required will depend on the type of brick being used, e.g. wire cuts or bricks with deep frogs, the gauge and the thickness of the wall.

Approximately 1-1.5 m^3 will be required for 2500 bricks.

Therefore, for a wall requiring 6000 bricks, assuming it requires 1 m^3 of mortar per 2500 bricks, the volume of mortar required will be:

6000/2500 = 2.40 m^3

Percentages

When we say "percent" we are really saying "per 100".

So 1% means 1 per 100.

Therefore 50% means 50 per 100 and 25% means 25 per 100.

So to find 100% of 80 $= \frac{100}{100} \times 80 = 80$

Also to find 50% of 80 $= \frac{50}{100} \times 80 = 40$

Also to find 5% of 80 $= \frac{5}{100} \times 80 = 4$

Percent can be expressed as a decimal or a fraction.

A half can be written as 50%, as a decimal = 0.5 or as a fraction = ½

Example 12

15% of a load of 200 facing bricks were found to be damaged. How many facing bricks were usable?

Answer 12

$$15\% = \frac{15}{100}$$

$$\frac{15}{100} \times 200 = 30 \text{ damaged}$$

Therefore 170 facing bricks were usable.

Ratio

Ratios are used when mixing cement mortar or concrete.

So a mix of cement mortar may be 4 parts of sand to 1 part of cement.

So the ratio of sand to cement is 4:1.

To make a lot of cement mortar we might need 4 times the quantity, so we multiply the numbers by 4:

$4 \times 4 : 1 \times 4 = 16{:}4$

In other words, 16 parts of sand and 4 parts of cement.

On site these could be shovels or buckets full.

The ratio is still the same, so the cement mortar should be of the same consistency.

Relaying information

To be able to relay information gathered to the appropriate people, good communication skills are required. Good communication skills are important in every aspect of our lives; in the workplace they are essential.

You may at times be the first point of contact with the client or customer, whether you are self-employed or working for a company.

Meeting and talking to people outside your organization can be one of the more interesting aspects of any job, but it is also a challenge.

You may be talking to people on the telephone or face to face, but in each case you are acting as the representative of your company and therefore you should present a positive and friendly image.

Methods of communication

When entering the construction industry you will have to communicate well, irrespective of your position in the company.

It may be necessary to communicate with customers, visitors and potential employers in a number of different ways, for example:

- orally – face to face
- in writing
- by telephone
- through drawings.

Orally – face to face

When dealing with a customer or visitor face to face, always appear helpful and polite. Smile but do not appear to be over-friendly as this can make some people feel uncomfortable.

Never act in an aggressive or immature manner.

Remember, first impressions count. If your attitude comes across as 'couldn't care less', potential customers may think that this is your attitude towards your standard of work. Obviously this can frighten off potential customers.

Always offer as much advice and information as possible. If you cannot answer their questions or deal with their requests, find someone who can.

Verbal communication can include talking on the telephone, mobile or walkie-talkie. Verbal communication is instant and feedback can be obtained immediately.

Written

Written communication can take many forms and one or more of the following will commonly be used:

- letters
- memos
- reports
- records
- site diary
- handbooks and manufacturer's details
- e-mails and text.

The main advantage of written communication is the evidence it provides. Written communication may take longer than verbal communication, but with e-mails and mobile texting written communication is becoming much more rapid than in the past.

> **Remember**
>
> Ask rather than tell. Always be polit
> Keep instructions short and to th point.
> Be friendly.

Letter writing

Writing a letter is an effective form of communication. We need to write letters on a number of occasions to:

- give information
- ask for details
- confirm things
- make arrangements.

Letters provide a permanent record of communication between organizations and individuals.

They can be handwritten, but formal business letters give a better impression of the organization if they are typed.

They should be written using simple, concise language. The tone should be polite and business like, even if it is a letter of complaint.

The letter must be clearly constructed, with each fresh point contained in a separate paragraph for easy understanding. When you write a letter for any reason, remember the following basic structure:

- your own address

- the recipient's address
- date
- greetings
- endings
- signature.

Report writing

A report is written to pass on information quickly and accurately to another party.

When writing a report, sufficient information should be included to allow the reader to understand fully what the report is intended to say.

A report should be divided into five parts:

- headings
- introduction to the report
- body of the report
- conclusion, with any recommendations
- any data, drawings, contracts, etc.

Site diary

This is one of the most important reports that a site manager has to produce during a contract.

It will include the following:

- the weather, morning and afternoon
- any visitors on site
- delays by subcontractors
- instructions from the clerk of works
- materials delivered
- number of staff on the site.

Telephone

Telephones have a very important role to play in communications. Landline and mobile phones are shown in Figure 3.13.

Often, a telephone conversation may have taken place before a letter is written to confirm something. Its clear advantage over the written message is obviously the speed with which people are put in touch with one another.

A good telephone manner is essential. Remember, the person you are talking to cannot see you so you will not be able to use facial expressions and body language to help make yourself understood.

The speed, tone and volume of your voice are very important. Speak very

FIGURE 3.13
Types of telephone

clearly at a speed and volume that people can understand. Always be cheerful and remember that someone may be trying to write down your message.

Making notes

It is important when receiving information that some kind of record is kept. Very often information is not written down and then mistakes are made through not remembering correctly.

Instructions are very often given by word of mouth. They should be kept brief and to the point, without complicated and confusing detail. This should make it easier for the recipient to remember them.

It is always good practice to aid your memory by making notes of what is being said. In order to record and possibly pass information received to others correctly, keep a small notebook in your pocket and always use it if the oral information is complicated and has to be passed on to others.

You may not be able to write down the instructions word for word.

- Make lists.
- Use notes and sketches.
- Abbreviate notes using key words.

FIGURE 3.14
Typical computer set up

E-platforms

Most up to date construction firms are now making use of the Internet.

Web site design has been a benefit through the following:

- Improving advertising effectiveness
- Saving money on printing and distribution costs
- Easy access to new customers
- Easy to use and update
- Improving productivity
- Educating your customers
- Expanding your market
- Extending your local reach

The Internet has also made a huge impact on the construction industry through online research and buying.

The Design members of the construction industry are also advantaged with CAD (Computer Aided Design).

Positive and negative communications

For a company to function properly and make a profit, it is essential to maintain good working relationships within the workforce.

To achieve this, various sections and members of the workforce must communicate with one another as fully as possible.

When you first join a company you are often the odd one out and therefore you should try hard to co-operate with your new work colleagues. This will mean being polite, acting on request as quickly as possible and building a good working relationship with workmates.

You may be talking to people on the telephone or face to face, but in each case you must try to be both positive and friendly, both to give a good account of yourself and to promote good, effective working relationships.

Poor communications cause negative feedback and can lead to delays and mistakes which will cause a loss of profit for the company.

The source of construction work

The client

The client is the employer or building owner who employs an architect to design a building or a contractor to construct a building, and has overall responsibility for the financing of the project.

- A client can be an individual or an organization. For example:
- private individual - Mr Smith
- partnership - Mr and Mrs Smith
- limited company - Smith Ltd
- local authority - Oxford District Council
- government department - Department of the Environment
- statutory authority - police
- public undertaking - private hospital.

The above clients can also be divided into two main categories:

- the private sector
- the public sector.

The private sector

The private sector consists of work financed by an individual client such as:

- private individual
- partnership
- limited company.

Examples of the above are:

- Private individual - a small garage owner requires a boundary wall to be constructed around his premises.
- Partnership - two taxi cab proprietors need a new garage to be designed for their fleet of cars.

- Limited company – a large organization may contact a contractor to provide a large office accommodation block.

The public sector

The public sector consists of work engaged by publicly financed sources such as:

- local authority
- government department
- statutory authority
- public undertaking.

The following are examples of clients within each group:

- Local authority – Mansfield District Council may require a new library to be constructed.
- Government department – the Department of Health may require a new office accommodation block for their new department.
- Statutory undertaking – a police authority may require a police station to be built in a new area.
- Public undertaking – an open-cast coal mine may require a landscaping project.

Range of activities

The construction industry undertakes an enormous amount of work including a variety of activities. It would be impossible to find a firm in the industry that could carry out all of the activities required.

The following activities are undertaken in the construction industry (Figure 3.15):

- erecting buildings
- repairing and maintaining buildings
- constructing roads and bridges
- erecting steel and concrete structures
- civil engineering work, including sewers, gas and water mains, electricity cables, etc.
- open-cast developments l demolition work
- plant hire
- work carried out by specialist firms.

The range is very wide, which reinforces the fact that the industry is very complex.

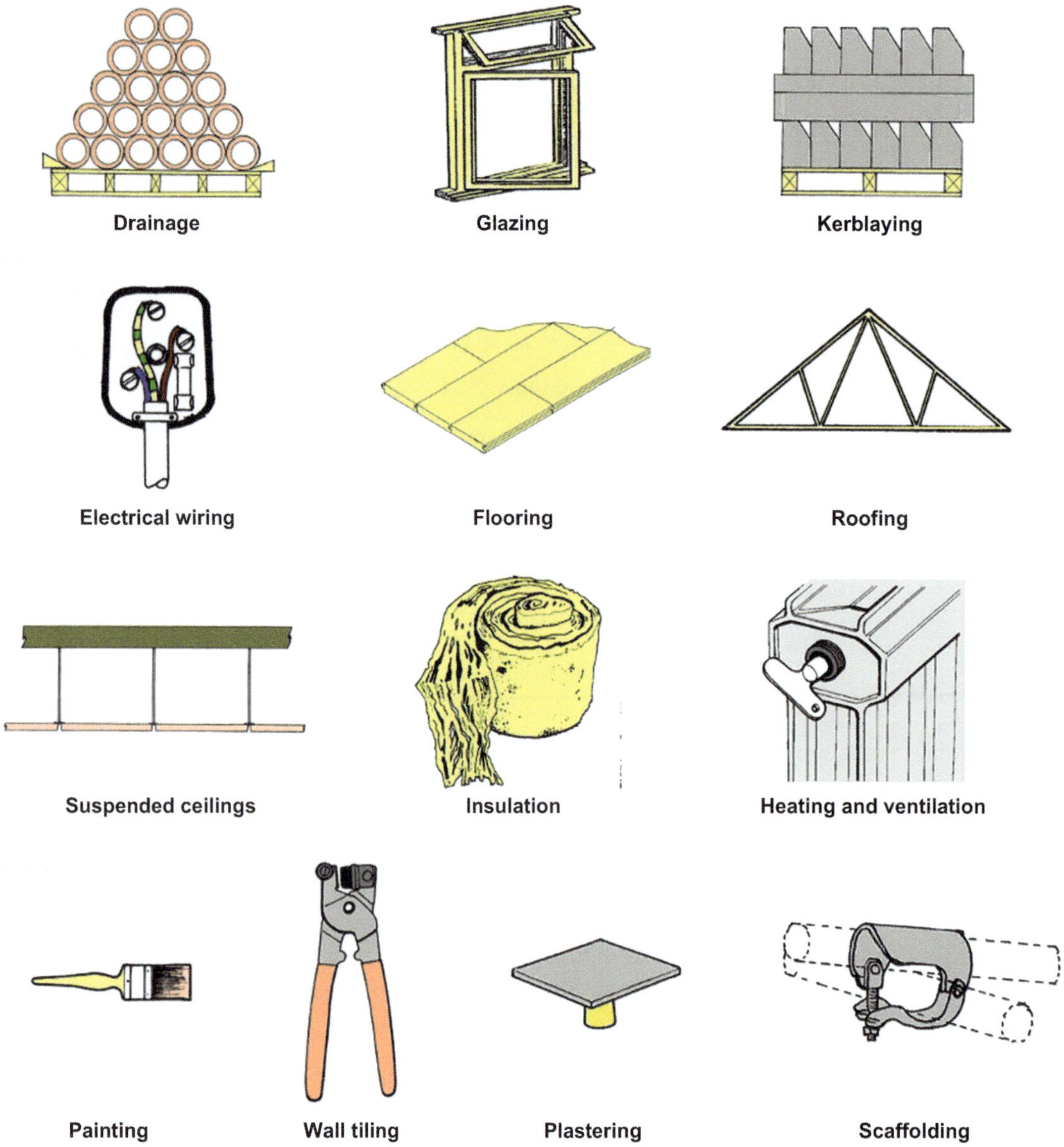

FIGURE 3.15
Range of activities

The building team

When you start work in the construction industry you will meet a lot of other people.

Constructing buildings is not a task that can be completed by one person alone; it takes a whole group of people, who are known as the building team. The members of the building team will change according to the type of building.

With residential buildings the main members are as follows:

Client

This person is the most important member of the building team. Without the client no buildings would be required. The client provides the money in exchange for a completed building.

Architect

The architect is the client's agent and is considered the leader of the design team. The main role of the architect is to put into reality the client's ideas for a building.

Private quantity surveyor

The private quantity surveyor is employed by the client, but generally on the recommendations of the architect.

The quantity surveyor is employed as early on in the contract as possible – normally at design stage – so that advice can be given to the client, with the approval of the architect, on the approximate cost of the contract.

During the work the private quantity surveyor will meet on site with the contractors' surveyor to agree interim payments for the work completed, and at the end of the contract agree a final account for presentation to the client.

Estimator

The role of the estimator is to arrive at an overall cost for the complete contract.

He or she will normally break down each item in the bill of quantities into the three main parts (labour, materials and plant) and apply the firm's rate to each item to produce the amount it will cost the contractor to complete each item. The estimator will also add to these prices the cost of overheads and profit.

Clerk of works

The clerk of works is nominated or approved by the architect, but is employed by the client.

The clerk of works acts as the client's representative on site under the direction of the architect. He or she may be resident on larger sites or be a regular visitor on smaller ones.

Their primary duty is to ensure that the constructed works conform to the specifications laid down in the contract documents.

Main contractor

In the construction industry a firm operates either as the main contractor, directing all the work on site, or as a nominated or domestic subcontractor.

The main contractor agrees to complete the building for an agreed sum of money.

The organization of a firm can grow as the workload is increased, i.e. extra posts can be created or departments can be introduced, such as joinery works.

The structure can therefore differ considerably between firms. The roles and responsibilities will therefore vary to a large degree from firm to firm and even from site to site.

The structure of a firm could be divided into office staff and site staff.

Office staff	**Site staff**
Office manager	Contracts manager
Estimator	Site agent
Quantity surveyor	General supervisor
Plant manager	Trades supervisor
Personnel officer	Tradespeople
Buyer	Safety officer

Site agent

Sometimes called a site manager, the site agent is the contractor's senior representative who is resident on site and has complete responsibility for the contract from start to finish.

The site agent carries out similar duties to those of the contracts manager, with the addition of:

- providing the head office with daily and weekly reports recording progress on site
- providing a safe site
- liaising with all visitors
- recording architects' variations and additional work.

Contracts manager

The contracts manager is classified as the leader of the site management team on a number of projects.

He or she normally travels around the various sites and is not normally resident on a particular site.

The contracts manager is responsible for overall planning, management and building operations, and should keep in constant contact with head office and site staff.

General supervisor

On large contracts a general supervisor may be employed who works under the contracts manager and is responsible for organizing and controlling the work of all the trades supervisors on site. This role is more supervisory when working under the contracts manager.

On small sites the role of the general supervisor becomes more important. He or she has overall control of the site and carries out both the duties of the contracts manager and his or her own duties.

Trades supervisors

It is normal for each one of the trades on site to have its own trades supervisor to assist in the day-to-day control of the various trade teams.

Each particular trade may have a supervisor to control them.

Building control officer

The building regulations set out legal requirements for specific aspects of building work and notification procedures that must be followed when starting, carrying out, and completing building work.

The building regulations require that a project's compliance is independently verified.

Historically, this verification could only be carried out by local authorities, however, it can now also be provided by privately appointed approved inspectors.

The individuals that check building works compliance are sometimes referred to as 'building inspectors', or 'building control surveyors'.

The term 'building control officer' generally (although not always) refers to an employee of the local authority, whereas 'approved inspector' is a private sector individual or organisation.

Planning officer

Planning officers are public sector planning practitioners.

Planning officers generally work for local planning authorities, involved in development management associated with the local town planning system.

However, they may also work in other public sector organisations.

Planning officers can be involved in a wide range of developments, from small changes to private dwellings through to large infrastructure projects. They must have a good knowledge of the local community, legislation, environmental issues and social responsibilities.

Civil engineer

Civil engineering is a professional engineering discipline that deals with the design, construction, and maintenance of the physical and naturally built environment, including public works such as roads, bridges, canals, dams, airports, sewerage systems, pipelines, structural components of buildings, and railways.

Civil engineering takes place in the public sector from municipal through to national governments, and in the private sector from individual homeowners through to international companies.

Ganger

On large construction sites with many general operatives it is advisable to employ a person to organize and control them.

The ganger has responsibility over all the semi-skilled labourers, plant operators, drain layers, concreters, etc., and performs a very similar job role to that of the trades supervisor.

Tradespeople

The operatives on site are those who actually carry out the physical work – bricklayers, joiners, plumbers, painters, etc.

It would be impossible to carry out any contract without some contribution from these people.

General operatives

These are employees who perform tasks that do not require the levels of skill that a trades person should possess.

Some of them are semi-skilled and perform such tasks as concreting and drain laying.

Safety officer

It is the responsibility of each employer who employs more than 20 personnel to appoint, in writing, a safety officer.

The safety officer ensures that the firm complies with various pieces of legislation.

This person must be experienced and knowledgeable on safety, and if possible should be allowed to devote all of his or her time to this area of work.

They should be able to advise on all matters of safety to the contracts manager, carry out safety inspections, keep records, investigate accidents and arrange for staff to have adequate safety training.

If it is uneconomical for firms to employ full-time safety officers, an alternative is to join a safety group where the safety officer is shared.

Subcontractor

The construction industry undertakes a wide variety of tasks.

The vast majority of main contractors only employ tradespeople in the main trades, such as bricklayers, labourers, painters and joiners.

If a contract has been won by a firm that requires other trades, then the main contractor will contact one of the many subcontractors who specialize in various trades. The subcontract work may account for a substantial amount of the contract. Subcontractors are a very important part of the industry and cover a wide range of activities not covered by the main contracting firms.

This system of subcontracting work out generally operates very well, because the firms become experts in their own field, resulting in a high standard of work at very competitive prices.

The main difficulty in subcontracting work is the problem of controlling and co-ordinating the numerous contractors into the main contractor's programme.

Note

This set of notes on personnel in the construction industry, either in the head office or on site, is by no means conclusive.

Please add to these notes any other personnel about which you have found information.

Jobs and careers

Because the production of the built environment is such a large-scale and complex operation, the construction industry demands a wide range of knowledge and skills.

Traditionally, employers recruit people for work in the industry by carefully defining the tasks to be performed and then matching them to the apparent abilities of candidates who present themselves for the appointment.

These abilities are reflected in the candidate's:

- qualifications - obtained through specific training programmes or educational courses
- experience - with evidence in the form of references provided by previous employers, describing their impressions of the candidate's performance at work.

According to the kind of qualification achieved, employees have traditionally played fixed and distant roles in the construction industry:

- as an operative, working on site constructing the built environment
- as a technician, communicating information about the built environment
- as a professional, making decisions about the built environment.

Through increased experience within each category it has traditionally been possible for an employee to improve the quality of their output so that they become more valuable to a firm. In return, they may be given an increase in wages and receive more responsibility.

Such development is described as a career in the construction industry.

Types of work

Apprentice

Apprentices are aged 16 or over and combine working with studying to gain skills and knowledge in the construction industry.

There is government funding to cover some of the cost of training and assessing an apprentice if you're in England.

As an apprentice you should be paid at least the minimum wage.

They combine work, training, and study, letting you 'earn while you learn'.

To become an apprentice you have to have the required qualifications and join a Training Programme.

Employee

At the completion of your apprentice and you have become qualified you employer may employ you as a tradesperson.

An employee is an individual who works part-time or full-time under a contract of employment, whether oral or written, express or implied, and has recognized rights and duties. Also called a worker.

Labourer

A construction labourer or construction worker does physical labour on construction sites. They may prepare sites by cleaning them, loading or unloading materials and removing hazards. A general labourer may also run some types of equipment, i.e. cement mixer, dumper.

A labourer is an essential part of a brickwork gang.

They are essential on building sites.

Self-employment

Some of the tradespeople who start off their employment working for others become self-employed and begin working for themselves.

Subcontractor

A subcontractor is a company or person who is hired by a general contractor (or prime contractor, or main contractor) to perform a specific task as part of the overall project and is normally paid for services provided to the project by the originating general contractor.

Remember all operatives and technicians have to start somewhere, such as a college course, and gradually increase their knowledge and certification until they become qualified.

Other roles in the construction industry

Among the most common construction trades are those of carpenter, electrician, heavy equipment operator, ironworker, labourer, mason, plasterer, plumber, pipefitter, sheet metal worker, steel fixer and welder.

Qualifications

In this day and age it is becoming very important to achieve qualifications and recognition for work done.

The construction industry employs millions of people and is one of the most exciting and diverse sectors to work in. It offers a wide variety of career options to suit both graduates and those who leave full-time education earlier and opt for vocational training.

There are ample opportunities for those with an interest in the design and management of construction works, and also for those who wish to develop and use more practical skills getting projects built on site.

Either way, you'll stand a better chance of career progression with the right qualifications.

When you have successfully passed certain examinations you are entitled to use various letters after your name. For example, after completing the Higher National Certificate in Building the letters HNC can be used.

Multiple-choice questions

Self-assessment

This section of the book is designed to allow you to check your level of knowledge. The section consists of revision questions for this chapter. The questions are all multiple choice and have four possible answers. The answers are to be found at the end of the book.

The main type of multiple-choice question will be the four-option multiple-choice question. This will consist of a question or statement, known as the stem, followed by a choice of four different answers, called the responses. Only one of these responses is the correct answer; the others are incorrect and are known as distracters.

You should attempt to answer the questions by choosing either (a), (b), (c) or (d).

Example

The person employed by the local authority to ensure that the Building Regulations are observed is called the:

(a) clerk of works
(b) building control officer
(c) council inspector
(d) safety officer

The correct answer is the building control officer, and therefore (b) would be the correct response.

Communication

Question 1 A site plan would be most likely to be drawn to a scale of:

(a) 1:100
(b) 1:250
(c) 1:500
(d) 1:2500

Question 2 Which construction material is shown below?

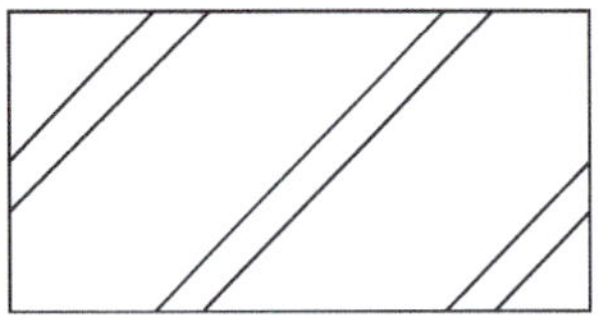

(a) brick
(b) blockwork
(c) concrete
(d) stone

Question 3 Which of the following is used for planning and ordering repetitive items?

(a) specifications
(b) schedules
(c) schemes
(d) contract plans

Question 4 Which of the following formulae can be used to calculate the area of the circle shown?

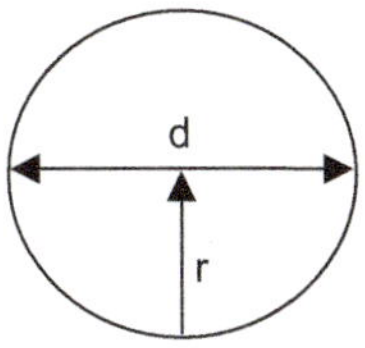

(a) πd^2
(b) d^2
(c) r^2
(d) πr^2

Question 5 Which of the following is the correct area for the rectangle shown?

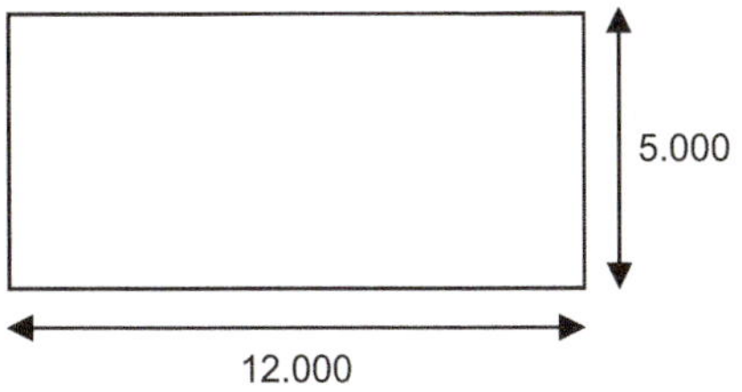

(a) 34 m^2
(b) 17 m^2
(c) 60 m^2
(d) 50 m^2

Question 6 Which of the following methods of communication would be verbal?

(a) e-mail
(b) letters
(c) site diary
(d) telephone

Question 7 The most important factor to remember when answering a telephone is to:

(a) speak slowly
(b) answer promptly
(c) take notes
(d) always be cheerful

Question 8 What do the initials RC on a drawing mean?

(a) reinforced concrete
(b) ready-mixed concrete
(c) rapid hardening cement
(d) ready-mixed cement

CHAPTER 4

Construction Technology

This chapter will cover the following City and Guilds units:

• Apprenticeship	Module 4	K1.1
• Diploma	Unit 101/501	3.1, 3.2, 3.3, 3.4
		4.1, 4.2, 4.3, 4.4, 4.6, 4.7
		5.1, 5.2, 5.3
		6.1, 6.2, 6.3, 6.4, 6.5

This chapter is about:

- Knowledge of building methods and construction technology

The following City and Guilds performance criteria will be covered:

- This chapter deals with methods used to excavate foundation trenches

The following Diploma outcomes will be covered:

- Know about foundations, walls and floor construction
- Know about construction of internal and external walls
- Know about roof construction

Site preparation

It is important the student understands the work involved before entering the building site.

The drawings will already have been produced by the architect to the client's requirements, or produced for a construction firm who will build them on a site as speculative housing to be sold to prospective clients.

Whichever method is used, the drawings have to be approved by the local authority. Once approval has been received building can start.

During the construction period a representative from the local authority will visit the site to check the various stages of building. This person is known as the building control officer.

Once approval has been given the contractor or supervisor visits the site to determine where best to place the cabins, materials, site roads, etc.

One of the first jobs to be done on the site before building operations can begin is clearing the site. There may be existing trees, hedges, old buildings and turf.

The top layer of soil has to be removed before setting out the buildings. This type of soil is unsuitable for building because of its instability and unreliable bearing capacity. Vegetable soil contains organisms and chemicals necessary for the growth of plants and if left under the proposed building would encourage plants to grow and could cause dry rot. For this reason the Building Regulations state that the part of the site to be covered by a building shall be effectively cleared of turf and other vegetable matter.

The top soil should be stored on site to be used for landscaping when the building is complete. It is important that it is stored in a position where it will not be in the way of any constructional activity because double handling is very costly.

When the site clearance has been completed the setting out of the work can begin.

Information from drawings

Students should be able to extract information required for setting out from drawings. For more information see Chapter 3, which covers drawings in more detail.

When drawings are received on site they should be carefully studied so that the work to be done is fully understood.

Groups of individual measurements should be added up and checked against overall dimensions.

It is also very important to find out how the measurements have been taken, e.g.

- over all the walls
- centre to centre
- between the walls.

Drawings that show the work to be carried out are drawn to scale, in one of the general scales used in the building industry. The larger the scale, the more detail can be shown on a drawing.

Site location

When planning to erect a new building, one of the first considerations is: 'where are we going to build on the plot of land'?

To determine this, a site location plan is drawn. This drawing, along with other plans and documentation, is submitted to the local authority for approval. You should be able to extract sufficient information from this drawing to be able to set out the building.

There may also be information regarding the drainage runs.

Setting out

The first task in setting out a building is to establish a base line to which all the setting out can be related.

The base line is very often the building line; this is an imaginary line that is established by the local authority. You may build behind the building line and even up to the building line.

Note

NEVER build in front of the building line.

It is usual for the building line to be given as a distance from one of the following:

- the centre line of the road
- the kerb line
- existing buildings.

The frontage line of the building must be on or behind the building line, never in front of it.

Ground works

In order that the foundations can be constructed on site, certain site processes have to be completed. These can be grouped under the heading of ground works.

The preparation of the site to receive the building is of vital importance, and even though the foundations have been architecturally designed, they may still fail without the correct ground works.

The ideal site very rarely exists. Building sites are all different and buildings are constructed on plots of land that require the contractor to carry out the work that precedes the construction of the building. This preparatory work is almost identical on each site.

Typical ground-work activities are:

- site clearance
- site preparation
- excavation for trenches
- foundations.

Site clearance

The work involved with site clearance will vary from site to site. It will depend on whether the site is an open site in the country or a confined site in the centre of town.

Before any work commences the contractor will erect hoardings or fences around the site to restrict the entry of all unauthorized personnel to the site.

As mentioned before, there may be some demolition of existing buildings, and trees and hedges to remove, unless there is a preservation order on the trees.

Site preparation

Top soil will need to be removed from the site, usually 150 mm deep. This is a primary function, undertaken before construction commences.

There are several items of plant that can carry out site clearance, depending on the type of site. One of the most common items of plant for removing top soil is the excavator (Figure 4.1).

Removal of the top soil sterilizes the ground, since the top 150 mm or so will contain plant life and decaying vegetation. This means that the top soil is easily compressed and would be unsuitable for foundations. Clearance of vegetable soil or topsoil from the construction area is mandatory and specified in the current Building Regulations.

The topsoil is valuable and could be either stored on site for later use or removed from site and possibly sold for landscaping purposes.

Any formation of temporary roads can be done at the same time as the removal

FIGURE 4.1
Excavator

of the top soil. Not all sites are level and after the removal of the topsoil further excavation may be required.

Levelling

Drawings for small buildings, including one-off houses, do not always show the level of the ground floor.

Where it is not stated the level is decided on site to suit the surrounding ground level and road kerb level.

Where the level is shown on a drawing it will be related either to ordnance datum or to a local datum.

Temporary bench marks

Temporary bench marks (TBMs) are set out on the site in a prominent but secure position where readings can easily be taken and where they are not likely to be disturbed by site operations.

To give a TBM peg some degree of protection it should be concreted in and fenced around (Figure 4.2). The peg should be painted blue and the level painted on it.

On large sites TBMs should not be more than 100 m apart. They should be checked from time to time by levelling back to the original ordnance bench mark (OBM) reference point or another TBM.

Ordnance bench marks

These are levels above sea level and are taken from a location at Newlyn, Cornwall (which is exactly at sea level).

These types of level are often shown on block plans, cut into stonework or brick-work and usually on the corner of old buildings and stone gate posts, so that it is convenient for levels to be transferred from them. A typical OBM is shown in Figure 4.3.

Datum

A datum is a fixed point, positioned on the building site, to which all other levels are related, and should be established at an early stage. If possible, the datum should be positioned where it will not be disturbed but can be easily reached.

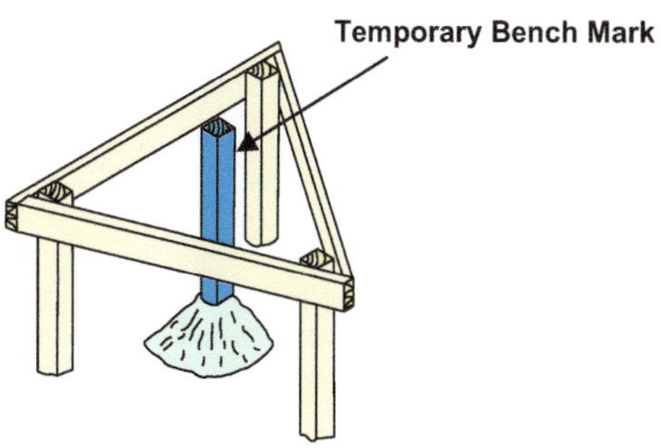

FIGURE 4.2
Temporary bench mark

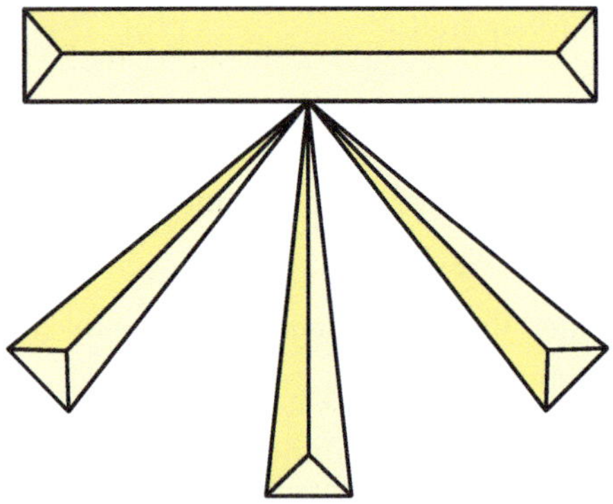

FIGURE 4.3
Ordnance bench mark

A peg is usually driven into the ground near the site cabin and protected with concrete as shown in Figure 4.2.

Other fixed points could be used, such as an inspection chamber (manhole) cover or a kerb stone. A typical small site is shown in Figure 4.4 with a TBM and possible fixed points on the existing main kerb and inspection chamber cover.

Excavations

On small sites such as extensions to existing properties; the excavations may have to be carried out by hand, using pick and shovel. On larger sites mechanical equipment will prove to be much quicker and cheaper.

The main item of machinery for trench excavations is the backactor (Figure 4.5).

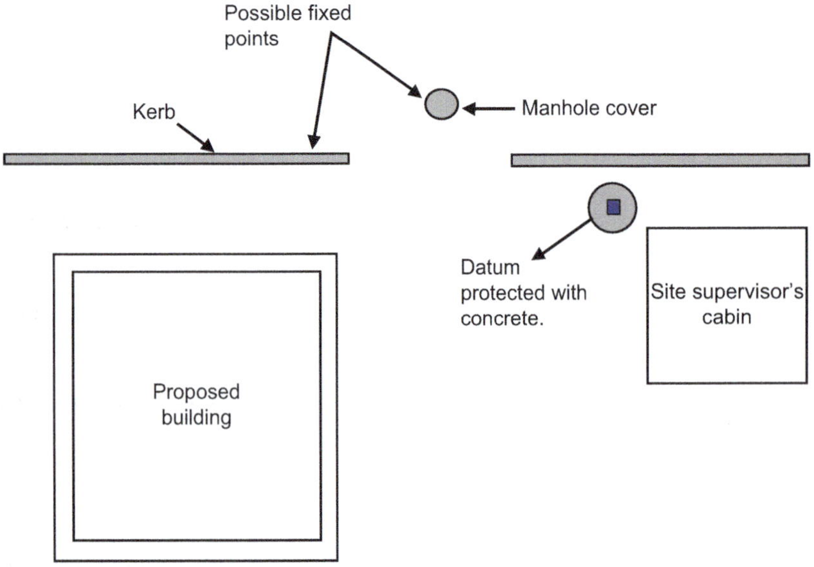

FIGURE 4.4
Small site with datum

FIGURE 4.5
Backactor

Foundations

Once the trench excavations have been completed and made safe with timbering, steel pegs can be positioned along the bottom of the trench to give the depth of the foundation.

The most common type of foundation is the strip foundation (Figure 4.6).

Concrete

Once the trench excavation has been completed and made safe the foundation concrete can be laid.

Mixing concrete will be dealt with in more detail in Chapter 6. Materials required are cements, aggregates, water and possibly reinforcement.

Concrete is, nowadays, the most important material used in the construction industry. This is because of its:

- strength
- durability
- low cost

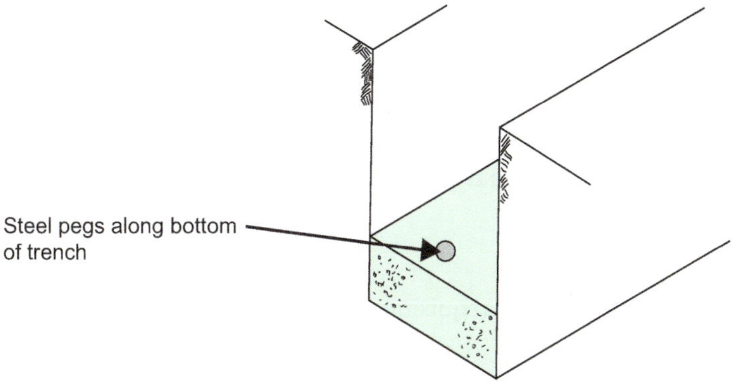

FIGURE 4.6
Strip foundation

and the ease with which it is moulded into any shape.

Concrete may be required to withstand:

- high loads
- all types of weather
- abrasion
- high temperatures
- water and chemical attack.

It can be used as either:

- plain concrete – for paths, drives, etc.
- reinforced concrete – if steel is included, for structural elements of the building such as lintels, columns, floors, etc.

Storing, stacking and protecting materials and components

Objectives of storage

The objectives of storage are to enable the quick and easy location of materials, plant and equipment.

Purpose of storage

Secure accommodation together with an effective storage system is essential on the construction site to prevent unnecessary waste and loss of materials due to vandalism and theft, and damage caused by constant handling.

Storage control

Irrespective of the amount of materials being stored, the following factors need to be considered.

Deliveries

Materials received need to be checked against suppliers' delivery advice notes for correct quantities and damage.

Stock rotation

Stocks of some materials (plaster, cement, paint, etc.) need to be used in the order of delivery as they only have a limited shelf-life.

Withdrawals

Materials drawn from stock by requisition need to be replaced.

Security

Storage facilities must be capable of being made secure against vandalism and theft.

Cleanliness

The store and unloading areas must be kept clean and free of debris at all times.

Safety

Materials must be handled and stacked in such a manner that injuries and damage are avoided. Hazardous materials need to be stored in separate accommodation constructed of non-flammable materials.

Site storage

Provisions for site storage will include the following.

Site compound

A fenced area is required for storing materials and plant, providing a means of control and protection against vandalism and theft. A typical site compound is shown in Figure 4.7.

Site stores

Site stores are needed for storing materials that must be protected against the weather and which are easily lost or stolen. Two examples are shown in Figure 4.8.

Storage areas

Predetermined areas are required on the site where materials are to be placed before use.

Materials such as bricks and blocks could be stored in the compound or placed close to where they will be used on the site, as shown in Figure 4.7.

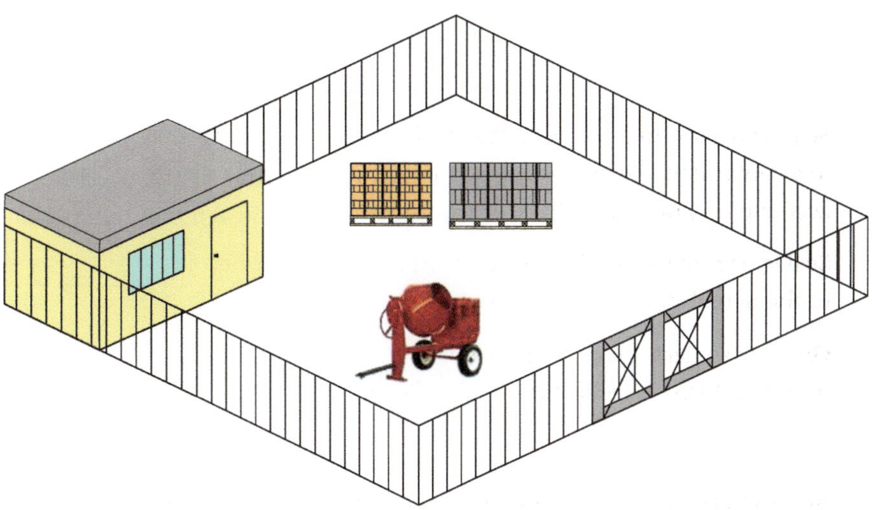

FIGURE 4.7
Typical site compound

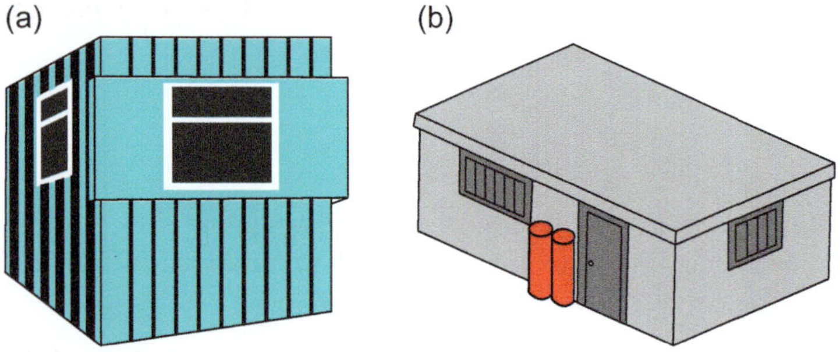

FIGURE 4.8
Types of material store: (a) metal; (b) wooden

Workshop/yard storage

Convenient workshop storage should be provided for tools and equipment.

Storage under cover should be provided for timber, sheet and insulation materials.

Separate workshop accommodation may be required for the storage of paint and inflammable fuel.

A maintenance area should also be provided so that plant and equipment can be maintained and cleaned before being stored or sent out to other sites.

Material storage

The amount of storage on site depends on the size of the contract and the amount of space available for storage.

On a large site storage facilities can be very comprehensive, but on small sites the materials are often delivered as required, materials being retained at the builder's merchant until called for.

Some firms will store small items at their main depot and the site supervisor will call them forward with requisition sheets.

Even on large sites some materials are delivered as required, such as ready-mixed concrete, mortar and hardcore.

Figure 4.9 shows a typical site for a factory unit. Areas have been set out for material storage and security. The whole site has a fence around it and an internal compound.

Aggregates

Aggregates should always be kept in a clean condition and should not become contaminated with undesirable substances.

Not only can soil, clay, organic or deleterious materials impair the strength of the concrete, but they can also stain the mix, resulting in patchiness or an unsightly finish.

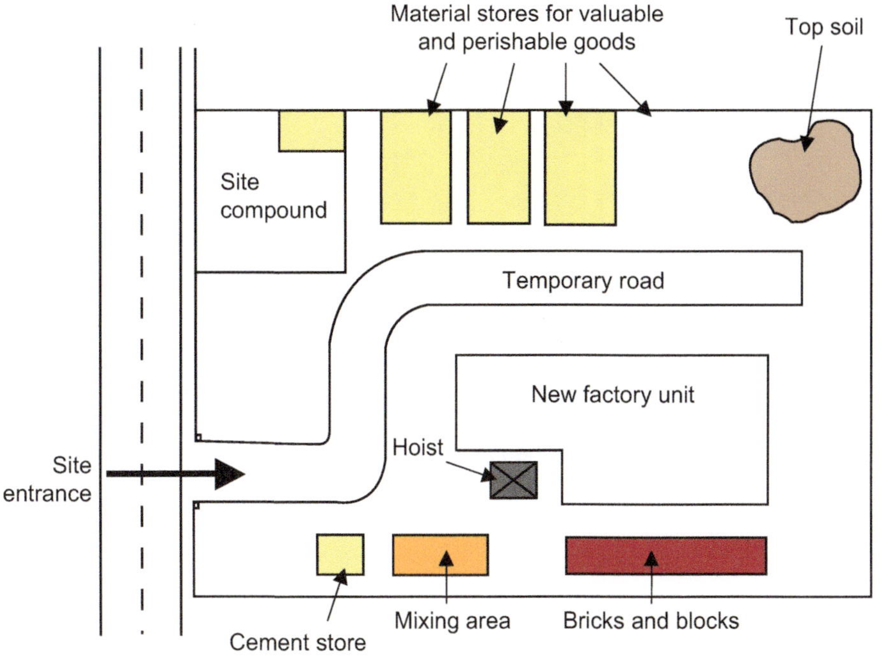

FIGURE 4.9
Typical site layout for materials

Aggregates stored in builder's yards are usually stored in specially prepared areas.

Figure 4.10 shows a typical area for storing aggregates on site. These areas are usually built with a concrete slab, sloping to the outside, and brick walls separating the various types of aggregates and sands.

On the site aggregates are very often stored directly on the ground. In this case the top soil should be removed and the pile should be as large as possible, with the bottom 500 mm never being used.

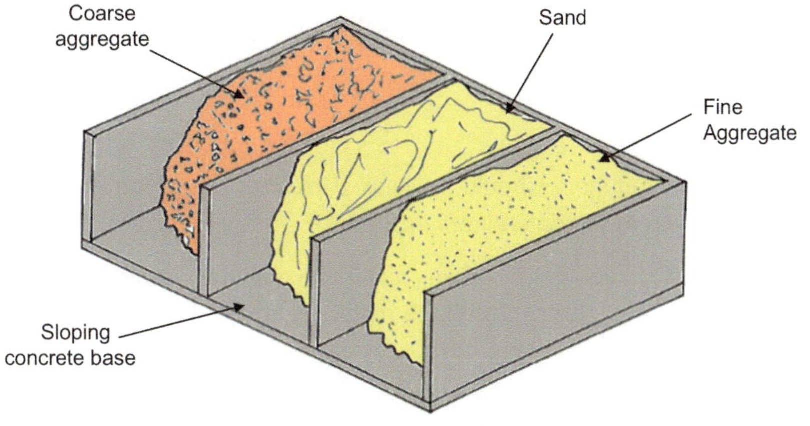

FIGURE 4.10
Site storage for aggregates

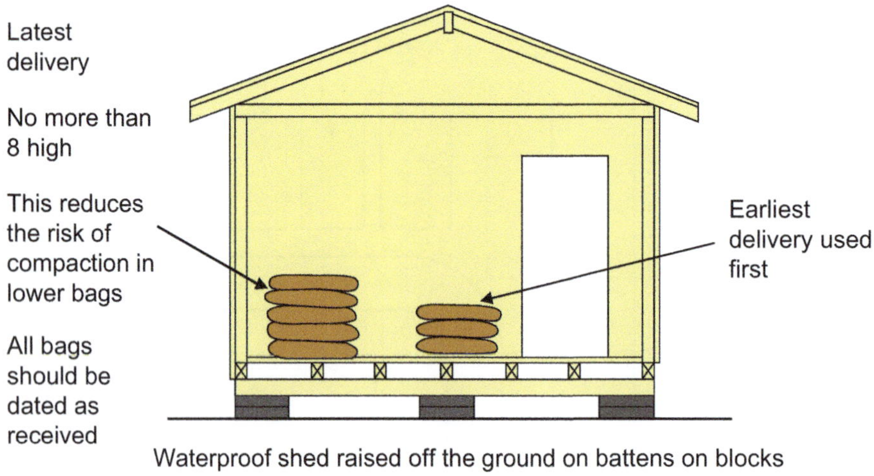

FIGURE 4.11
Cement storage

Aggregates and sands that have become contaminated should be removed from site.

In severe weather the stockpiles should be protected by sheeting materials.

Both very hot and cold materials will reduce the quality of the finished mix.

Bagged materials

These should be stored in a weatherproof shed, preferably without windows, and raised off the ground. An example is shown in Figure 4.11.

Cements, plasters and lime are the main bagged materials, all of which will deteriorate rapidly when damp.

The materials should be stored separately and clearly dated.

They must be stored in such a way that they can be used on a first-in, first-out rotation. This is to minimize the storage time and prevent the bag contents becoming stale or 'air setting'.

Care should be taken when using bagged materials as the paper can cause a problem on the site.

Bricks, blocks and stone

Bricks, blocks and stone should be stacked clear of the ground and covered completely with tarpaulins (Figure 4.12).

Some manufacturers deliver their materials already shrink-wrapped. The wrapping should be kept on until the materials are ready for use.

Drainage materials

Drain pipes should be stored in the open where they will not be accidentally damaged, on firm, level ground that has been covered with hardcore (Figure 4.13).

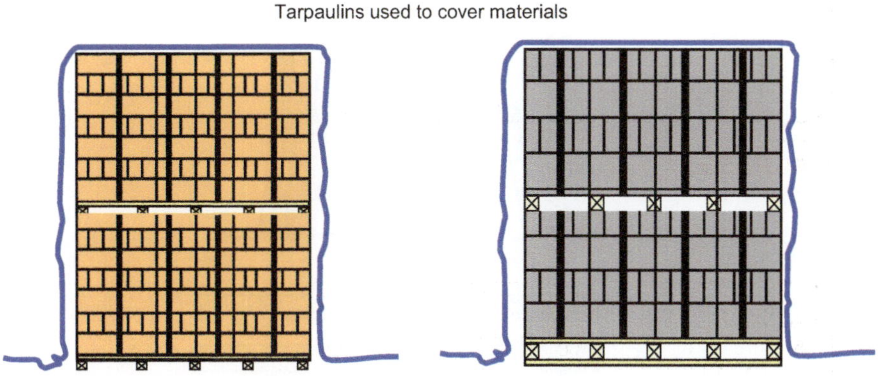

FIGURE 4.12
Protection of materials

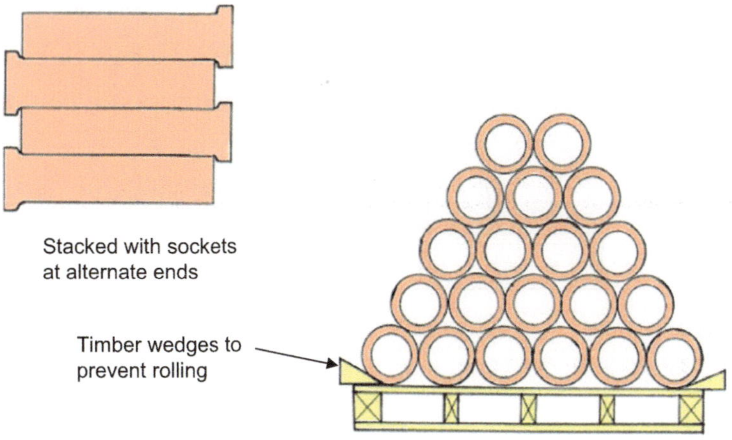

FIGURE 4.13
Stacking of drain pipes

Because of their shape they must be restrained from rolling about and occupying a large ground area.

If the drain pipes have socketed ends they should be stacked with sockets at alternative ends for protection.

Concrete goods

Most concrete goods can be stored outside on a level, well-drained site. Examples are shown in Figure 4.14.

Paving slabs should be stored on their edge, standing on timber bearers. Different sizes, shapes and colours should be stored separately.

Concrete lintels should be stored on timber bearers, but it is essential to ensure that they are stored in the way they are to be fixed, i.e. top side uppermost.

Kerbs should be stored on edge on timber bearers. They may be stored in bonded layers, but never more than three high.

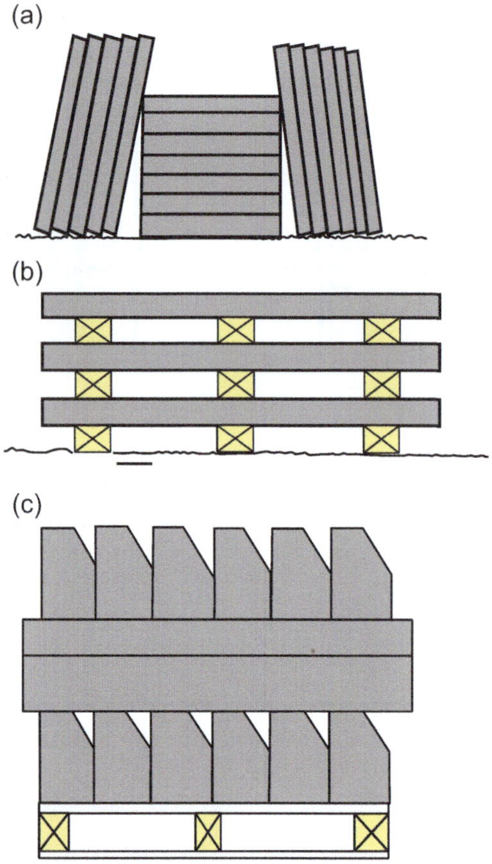

FIGURE 4.14
Storage of concrete goods: (a) paving slabs; (b) lintels; (c) kerbs

Roofing tiles

Roofing tiles and slates should be stored on level, well-drained ground, possibly stoned to keep the area clean. See the examples in Figure 4.15.

Care should be taken when storing roofing tiles and slates to avoid damage.

Roofing tiles are impervious to moisture and can be stored outside. However, on a building site they should be wrapped around with a tarpaulin to protect them from splashing.

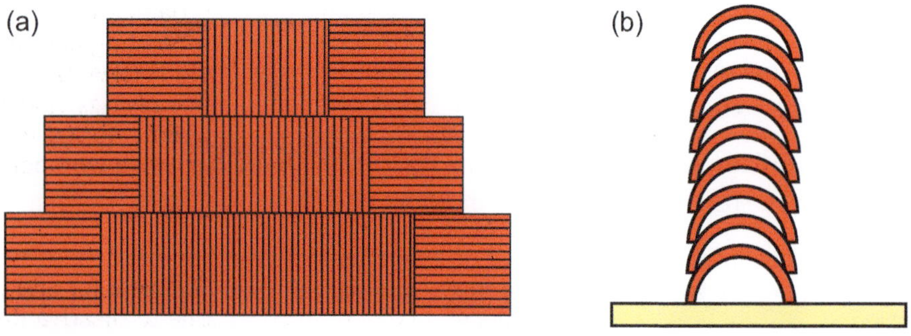

FIGURE 4.15
Stacking of roof coverings: (a) plain tiles; (b) ridge tiles

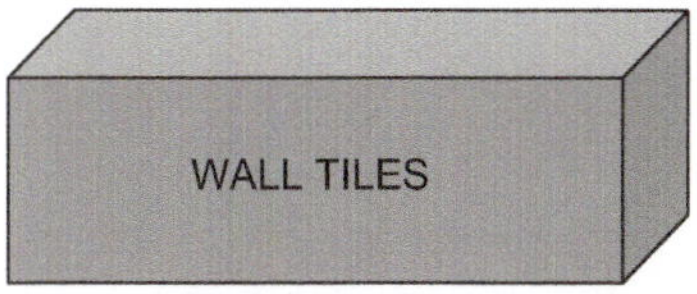

FIGURE 4.16
Storage of ceramics

Plain tiles are stacked on end, end tiles flat and taper stacked. The tiles are placed alternately to prevent damage to the nibs. Ridge tiles are stacked as shown - never more than 10 high.

Ceramic tiles

These are delivered to the site in boxes and after checking for damage should be stored in their boxes (Figure 4.16). They should be stored inside a store or the building where they are required.

Sheet materials

Plywood

Small quantities of plywood are best stored in racks, while large stocks can be stacked on bearers spaced at 400 mm centres to prevent distortion and covered with a tarpaulin sheet (Figure 4.17).

When handling care should be taken to avoid damage to corners and edges.

Plasterboard

Plasterboard should be stored in a similar way to other sheeting materials.

Plasterboard can be unloaded directly and stored in the building where it is required, laid flat on timber bearers (Figure 4.18).

Timber lengths

Poor storage of timber, wood products and units is a major cause of material wastage on building sites. If carcassing timber has to be stored outside then it should be stored off the ground on timber battens and protected against the weather with a tarpaulin cover (Figure 4.19).

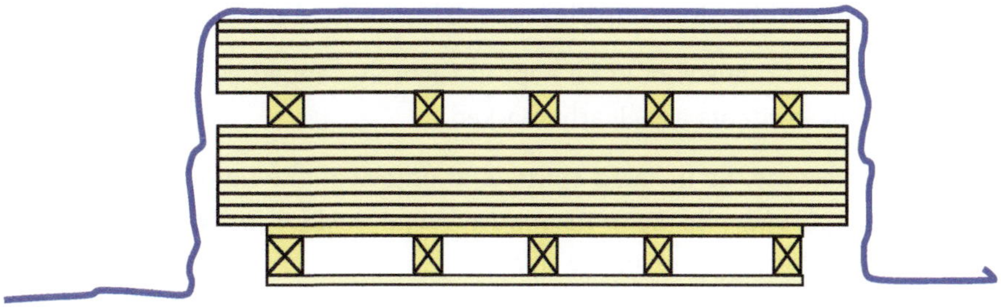

FIGURE 4.17
Storage and protection of sheet materials

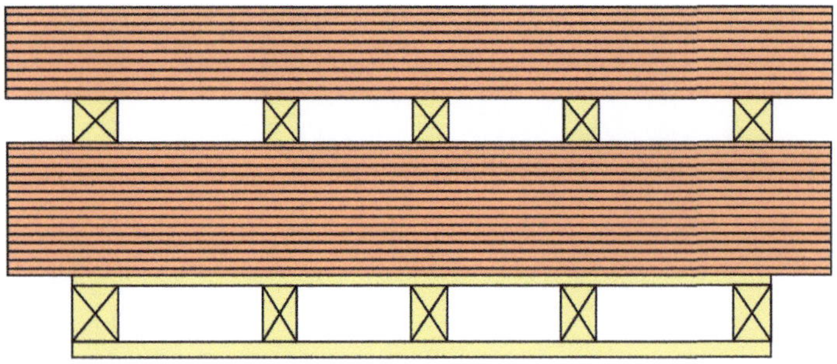

FIGURE 4.18
Storage of plasterboards

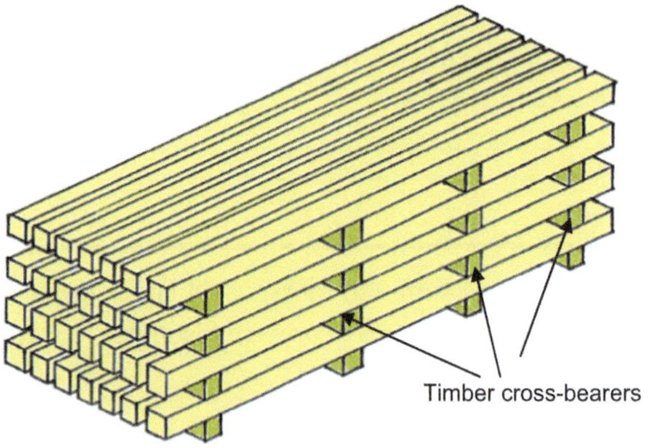

FIGURE 4.19
Storage of timber lengths

It is extremely important to ensure that the moisture content of the timber is kept at the right level for its use.

Timber components are usually delivered to site when required and should be stored inside a large, dry store or using part of the completed building as a store.

Always lay them flat and never lean them against a wall, etc., or they will bow, making them unusable (Figure 4.20).

Paints and flammable liquids

Paints should be stored inside a special store as shown in Figure 4.21.

Flammable liquids should be kept in securely capped cans or steel drums on which the contents are clearly marked.

Petrol, acetone, methylated spirits and other volatile liquids with flash-points below 32C may be kept in robust metal lockers or well-ventilated non-combustible cabins, or in the open.

Storage should be securely locked or fenced, and be situated at least 6 m away from any other huts, stores, site boundaries or buildings.

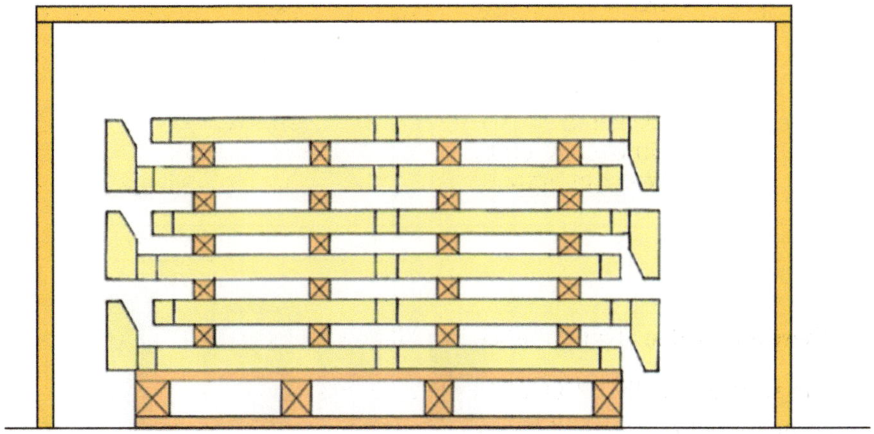

FIGURE 4.20
Storage of timber components laid flat under cover

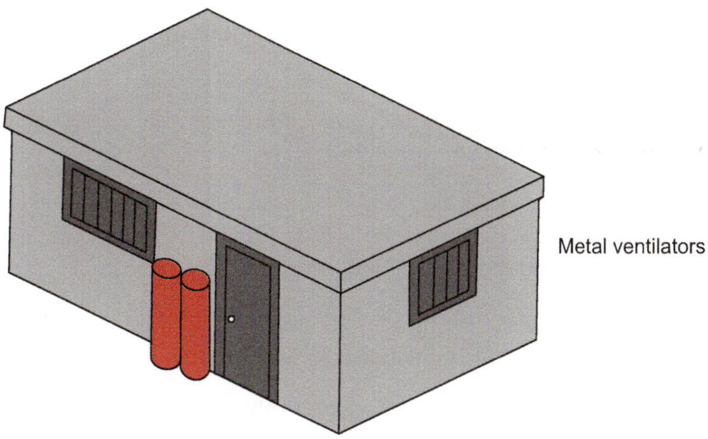

FIGURE 4.21
Storage of flammable liquids

Smoking should be prohibited within 6 m of the store.

No other materials should be stored with flammable liquids.

Containerized items

Any items delivered to the construction site in cardboard boxes should be ordered when required and stored in waterproof, dry conditions (Figure 4.22).

A secure store should be provided unless there are sections of the building that are complete and lockable which could be used.

Paint is delivered in tins, which also needs special arrangements for storage. Paint has a limited shelf-life so should only be ordered as and when required.

Shelving should be provided in the store so that items can be easily stored and retrieved.

FIGURE 4.22
Storage of containerized materials

Rolled materials

All rolled materials should be stored in an appropriate store, on end to retain their shape (Figure 4.23). They should be stored in their appropriate sizes for ease of identification.

They should be stored away from any source of heat as they could melt and become useless.

Note

NEVER stack DPCs flat as thi causes the rolls to flatten an crack.

Roofing felt should be stacked in similar fashion, on end and i special types for ease of identif cation.

Rolls of lead should also be store on end in their separate codes. is important to protect lead fror damage, as damaged lead is use less. Lead should be kept in lockable shed or building.

Minimizing damage and waste

With the present trend of material costs rising at a faster rate than labour costs, the control of waste is very important both on and off site.

The design of the building can add to the waste problem by involving cutting of materials.

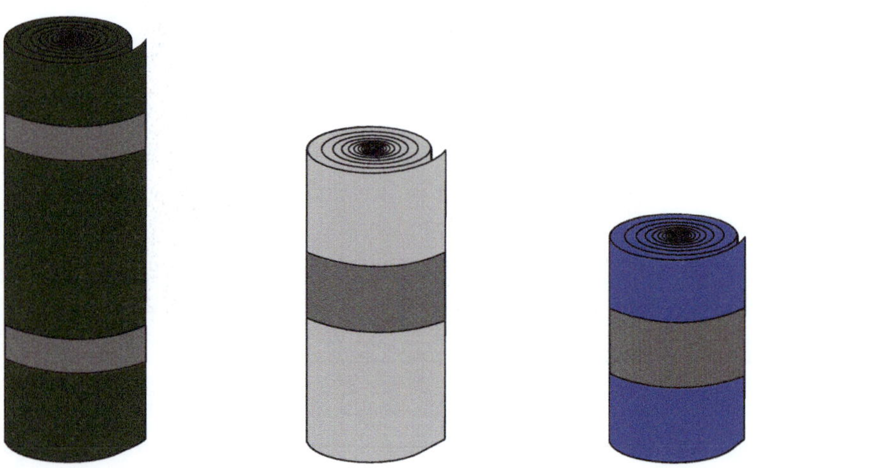

FIGURE 4.23
Storage of rolled materials

Waste can be divided into:

- waste from mixing and cutting
- waste from poorly stored and protected materials
- waste from overordering material supplies
- waste from theft
- waste from the nature of the application of the material, e.g. mortar dropped on the floor by the bricklayer.

This chapter will only deal with the management of the resources on site.

As described previously, materials can be delivered to the site in various ways and the site manager is responsible for the best handling and storage to protect the materials from damage.

Materials should be ordered in exact quantities. In some cases this may not be economical owing to transport costs imposed by the supplier.

Loose materials such as building sand and ballast are ordered by the load but invoiced by the tonne.

Only order the required amount, allowing only 5 per cent for wastage, as the remainder is always left behind. A central mixing area, or using ready-mixed materials, can help to minimize wastage in this area.

Bagged materials are also a source of wastage, especially if part-open bags are left out at night and get damp.

The storage of materials on site is the direct responsibility of the site management and has a very strong influence in the control of waste. A careless regard for the value and utility of the materials on the part of management could lead to a progressive deterioration in the operatives' regard for the material.

Areas allocated on site for the storage of materials should be determined after considering the following questions:

- Will construction take place in that area? If so, when?
- Is the storage for a long or a short time?
- Can delivery vehicles safely and easily reach the area?
- Can on-site movement from storage areas to point of use be safely and economically carried out?
- Are the materials as near to their point of use as practically possible?
- If the materials have considerable value are they in the secure area?
- Will the storage area create problems in routing site transport and personnel?

The answers to the above questions will, of course, vary from site to site.

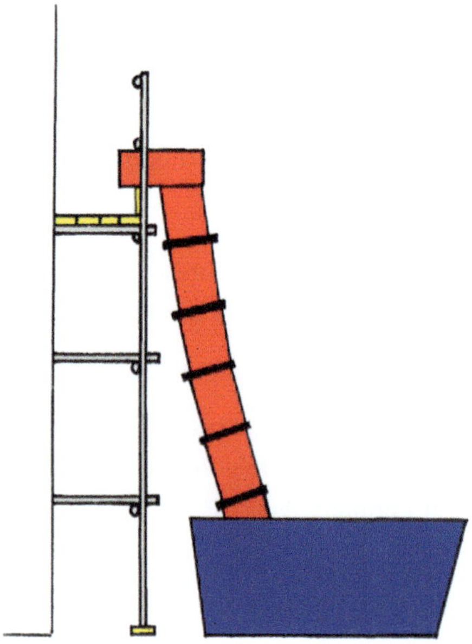

FIGURE 4.24
Removing waste from a building

Please refer back to Figure 4.9, which shows a typical site layout with areas set out for material storage.

Disposal of materials from heights

Careful disposal of materials from heights is essential.

They should always be lowered safely and never thrown or dropped from scaffolds and window openings, etc.

Chutes can be employed as shown in Figure 4.24. These can be connected together to form any reasonable length. They should terminate in a tarpaulin-covered skip.

Waste can be kept in a skip until collected and taken to a suitable tipping area.

Remember

The protection of materials w depend on:

- susceptibility to climatic condition
- value of materials
- size of the units
- scarcity of materials.

Hazardous waste, non-hazardous waste and debris

Waste resulting from work materials can include sand, gravel, cement, concrete, bituminous waste, sealants, bricks, stone, steel mesh, nails, timbers, fuels, oils, etc.

Waste resulting from work activities can include soils, vegetation, cuttings, inorganic materials, etc.

Depending on the sector in which the operative works, the materials will differ. Some operatives will work in a specific trade occupation, relating to bricks or timber, while others will be involved in general construction activities, relating to excavation materials.

Most substances and materials are safe provided they are handled or worked on sensibly and with proper precautions. Almost anything can be dangerous if handled or used irresponsibly. Some materials require extra care; a few need extreme caution.

It is important to know what the hazards are, when they occur and how they can be prevented. Usually prevention will involve the use of protective clothing and equipment.

Removal of waste

Since 1 April 1992, the law on waste has included a duty of care that means you must take all reasonable steps to look after any waste you have and prevent its illegal disposal by others.

If you break the law, you could be fined an unlimited amount.

'Controlled waste' is any household, commercial or industrial waste, such as waste from a house, shop, office, factory, building site or any other business premises.

How do I know what to do?

The law requires you to complete certain paperwork and to take all reasonable steps to meet the duty of care. What is reasonable depends on who you are and on the circumstances. *The Duty of Care, A Code of Practice*, is published by HMSO (ISBN 011 75 25 57 X).

Steps to be taken if the duty of care affects you

- When you have waste you have a duty to stop it escaping. Store it safely and securely.
- If you hand the waste to someone else:

1. Secure it.

- Most waste should be in suitable containers.
- Loose material loaded into a vehicle or skip should be covered.

2. Check that the person taking your waste away is legally authorized to do so.

Selection of resources

When building solid one-brick walls the resources will simply be tools, bricks and mortar.

This chapter will assume that the work has been completed below ground level, so the first procedure will be setting out the materials at ground level.

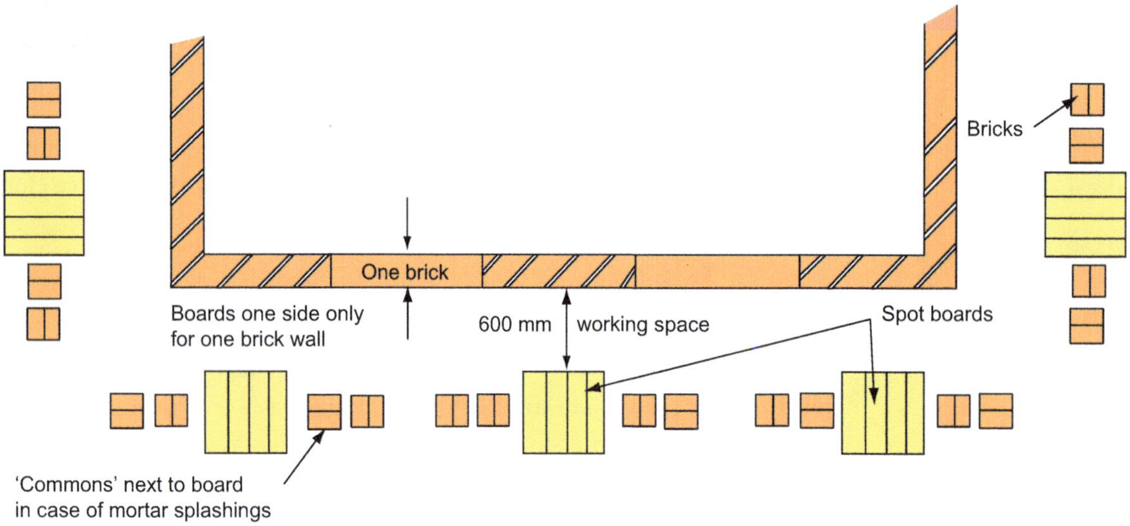

FIGURE 4.25
Setting out materials before building

Setting out materials

Before brickwork commences the working area must be stacked out with bricks and blocks as in Figure 4.25.

Keep stacks away from edges trenches. Keep walkways and runways clear. Do not stack materials too high. Do not stack wider at the top than at the bottom.

Stack materials near the job to avoid double handling. Keep materials at working height.

Protect materials from breakages, dampness and frost. Cover bricks with tarpaulin, hessian or polythene.

Take care when opening packs of bricks. The polythene wrapping or banding should be left in place until the bricks or blocks are required. The correct way to open packs is by cutting the straps with snips. Do not chop the straps with a brick hammer or trowel or any tool that will damage the bricks. For safety reasons, take the banding to the skip immediately, or make into safe bundles and remove from the work area later.

Polythene wrapping, if removed carefully and placed to one side, could be used elsewhere for protection. If it is not required remove to the skip as soon as possible.

When loading out for the bricklayer place the bricks with the frog or perforations uppermost, for easy handling by the bricklayer.

Source the bricks from as many packs as possible, but at least three. This will prevent banding along the wall of different coloured or textured bricks. Banding of bricks can occur in the face of walls if bricks are taken from only one pack at a time (Figure 4.26).

Remember

Stack bricks and blocks safely within easy reach of the bricklayer.

Space mortar boards at the corners of the building and not more than 3 m apart along the wall.

Level the ground off for easy walking and block up the mortar boards about 600 mm from the face of the wall.

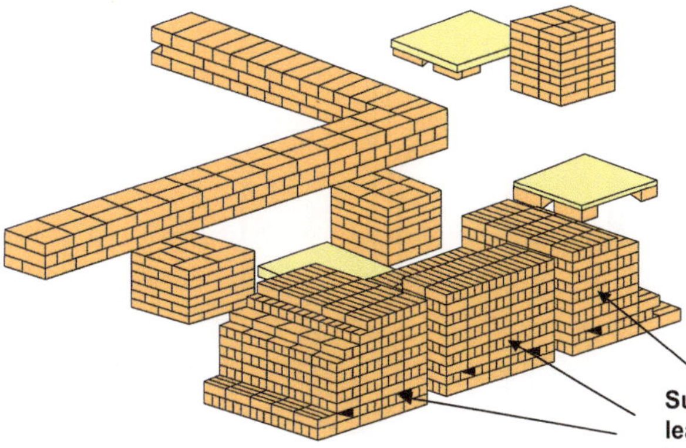

FIGURE 4.26
Selecting bricks when setting out

Minimizing risks

It is important that the trainee is aware of the dangers on site and takes the necessary precautions such as wearing the appropriate PPE.

One of the most dangerous aspects of solid brick walling, power tools, has already been covered. Another important task to watch is the movement of materials around the site.

Transporting bricks

Bricks can be carried under the arm from the stack to where they are required, but this is laborious and slow.

The use of brick clamps which can pick up six bricks on edge saves the hands and is quicker.

Barrows are much quicker, although barrow runs will have to be laid over rough ground.

Brick trolleys which carry 48 bricks held by a clamp are best used on the scaffold or on smooth ground. Bricks must be stacked on edge ready for the brick trolley.

The most common method used for moving bricks on larger sites is by forklift truck (Figure 4.27).

Protection of the surrounding area is important so as to minimize risks to the public.

The building site should be kept clean and free from as much dust and noise as possible. All waste materials should be collected in a skip, which can be moved off site when full.

The new walling should be protected with sheeting as the work proceeds, until fully set.

FIGURE 4.27
Moving bricks with the fork-lift truck

Damp-proof course

The current Building Regulations state that no wall or pier shall permit the passage of moisture from the ground to the inner surface of a building

Damp-proof courses (DPCs) are classified under the following headings:

- rigid
- semi-rigid
- flexible.

Rigid

Materials selected for DPCs must be capable of resisting the passage of moisture, and when applied to the wall must be continuous throughout. This continuity may be broken either by bad workmanship or, where a rigid type of DPC is used, e.g. slate, by settlement of the building. Failures will therefore be prevented by the selection of a suitable material to supplement good craftsmanship and a reliable foundation.

Slates must be hard and rough in texture and be laid in 3:1 cement mortar, in two bonded layers (Figure 4.28).

Bricks

Two courses of dense engineering bricks, laid bonded in cement mortar, act as a very effective DPC, especially for boundary walls and piers (Figure 4.29).

Semi-rigid

Mastic asphalt

This is used nowadays mainly as a damp-proof membrane. When used as a DPC it was spread hot in one or two coats to a thickness of approximately 13 mm and was impervious to moisture. It could be affected by moderate settlement in a wall, which could cause cracking. It is an expensive form of DPC (Figure 4.30).

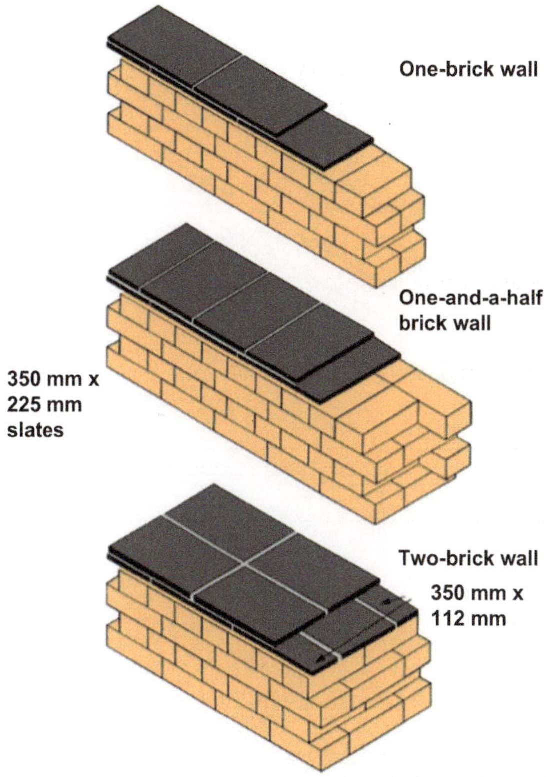

FIGURE 4.28
Slate damp-proof courses

Flexible

There are many flexible materials that can be used as a DPC.

Lead and copper

Lead and copper are very good but expensive. Lead is used mainly by plumbers for flashings.

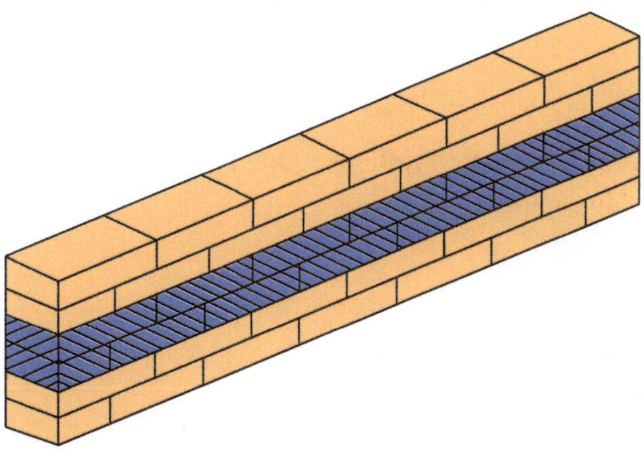

FIGURE 4.29
Engineering brick damp-proof course

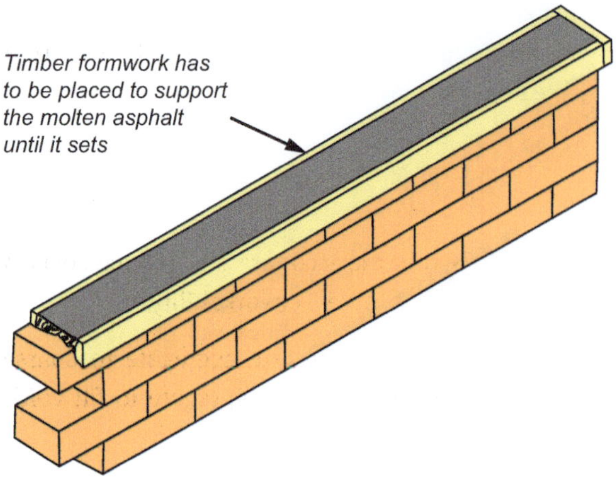

FIGURE 4.30
Mastic asphalt damp-proof course

Bitumen-based products

There are many types of bitumen DPC, all of which are flexible but may squash under pressure.

Bitumen DPCs should be used in rolls to suit the wall widths. They are bedded on mortar and lapped at joints and intersections by a minimum of 100 mm.

All rolls should be stored on end to prevent distortion.

Polythene

Black low-density polythene, like bituminized felt, should be laid in mortar with laps at least equal to the width of the DPC (Figure 4.31).

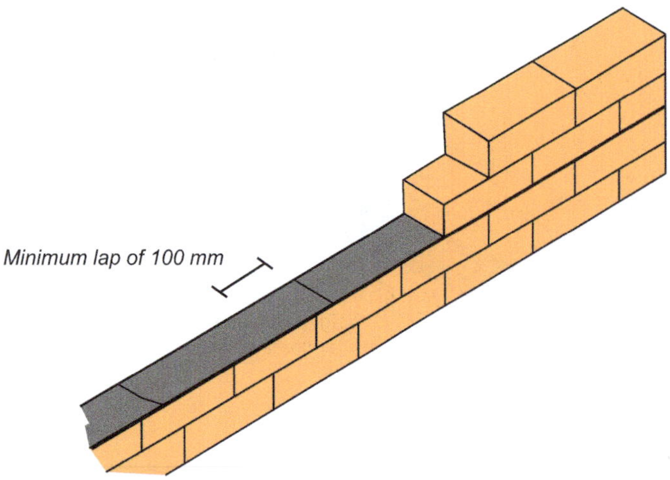

FIGURE 4.31
Flexible damp-proof courses

Roofs

The roof is that part of the external envelope that spans the external walls at their highest level and, being part of the envelope, it must fulfil the same functions.

Basic roof forms

These may be either flat or pitched (Figure 4.32).

- Pitched roof – any roof having a sloping surface in excess of 10 degrees pitch. Those with a single sloping surface are known as mono pitch and those with two opposing sloping surfaces are known as double pitched.
- Gable roof – a double pitched roof with one or more gable end.
- Hipped roof – a double pitched roof where the slope is returned around the shorter sides of the building to form a sloping triangular end.
- Flat roof – any roof having a pitch or slope of up to 10 degrees to the horizontal.
- Lean-to roof – a single pitched roof, which butts up to an existing building.

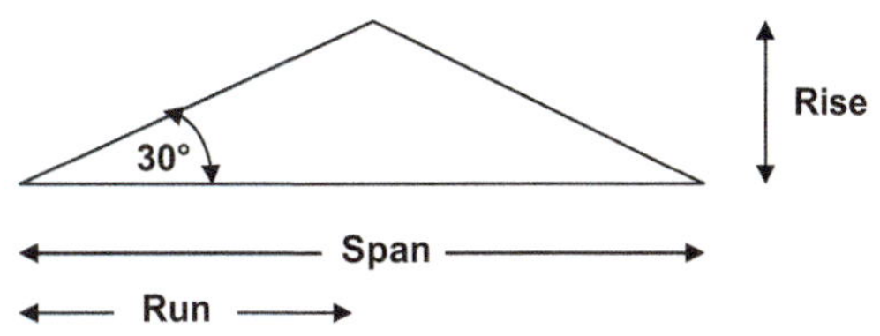

Pitch = 300 (degrees) or rise/span

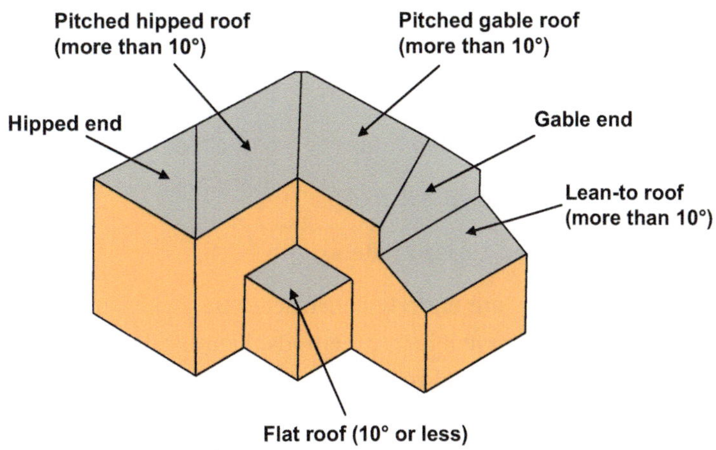

FIGURE 4.32
Roof types

Roof terminology

- Pitch – the angle of a roof's inclination to the horizontal, or the ratio to span, e. g. a one-third pitch roof with a span of 6 m will rise 2 m.
- Verge – the termination or edge of a pitched roof at the gable end or a flat roof at the sloping edge. Both often overhang the wall and are finished with a barge board and soffit.
- Eaves – the lowest part of a pitched roof slope where the ends of the rafters terminate, or the level edge of a flat roof. Both usually overhang the wall and are finished with a fascia board soffit.
- Valley – the intersection of two pitched roof surfaces at an internal corner.
- Ridge – fixed at the apex of the structure where the top of the rafters are fixed.
- Gable – the triangular portion of the end wall of the building with a pitched roof.
- Hip – part of the roof where two external sloping surfaces meet.
- Purlins – horizontal beams that support the rafters midway between the ridge and wall plate.
- Wall plate – timber beams bedded on mortar around the top of the brickwork. The wall plate supports the bottoms of the rafters and ceiling joists.
- Soffit board – horizontal board fixed to the underside of the purposely cut rafters to block in the eaves. Ventilation holes should be incorporated to ventilate the roof space.
- Fascia board – vertical board fixed to the ends of the purposely cut rafters and which support the guttering.

Construction terminology

Timber pitched roofs may be divided into two broad classifications:

- traditional framed cut roofs
- prefabricated trussed rafters.

Traditional framed cut roofs are entirely constructed in situ from loose, sawn timber sections and using simple jointing methods. Figure 4.33 shows many of the roof timbers used in traditional roofs.

Prefabricated trussed rafters are normally manufactured in a factory from prepared timbers, butt jointed and secured using various types of gussets and nail plates. A typical roof truss is shown in Figure 4.34.

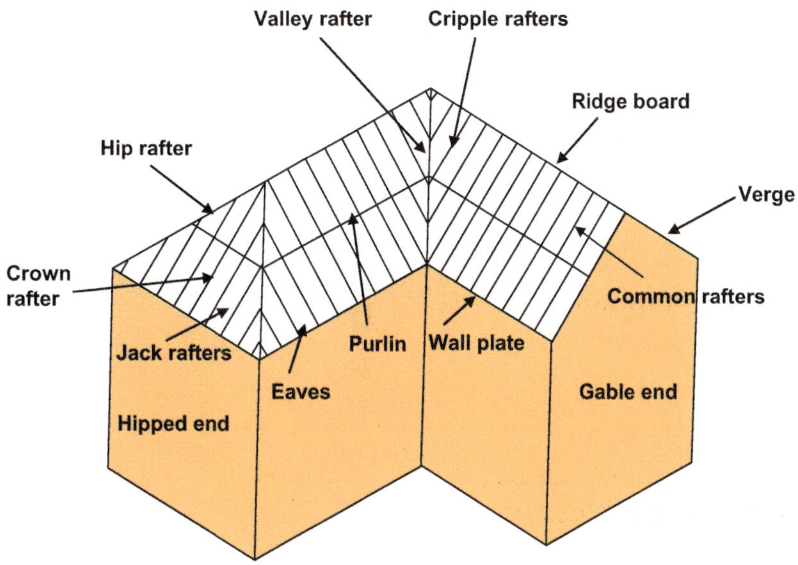

FIGURE 4.33
Roof timbers

Connectors

Members of the trusses may be butt jointed or single- or double-lap jointed. Butt joints or double-lap joints are preferred, since no eccentric loading occurs.

Butt joints are made by butting the two members together and covering the joint with one or preferably two gussets or plate connectors.

The gussets are usually of plywood, while the toothed plate is the most popular connector.

Wall plates

Wall plates transfer the load imposed on the roof uniformly over the supporting brick-work. They also provide a bearing and fixing point for the feet of the trussed rafters.

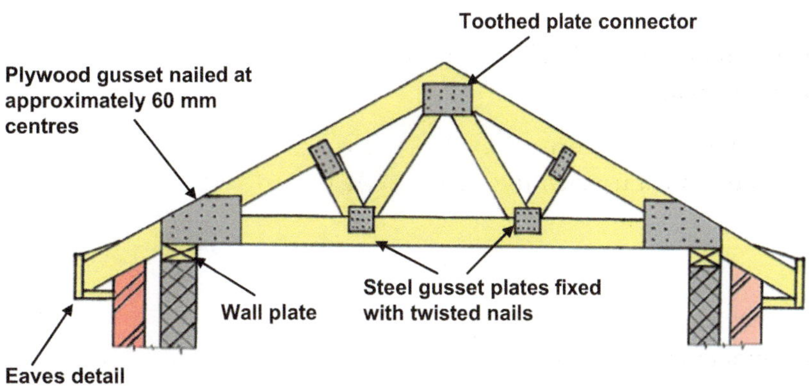

FIGURE 4.34
Typical roof truss details

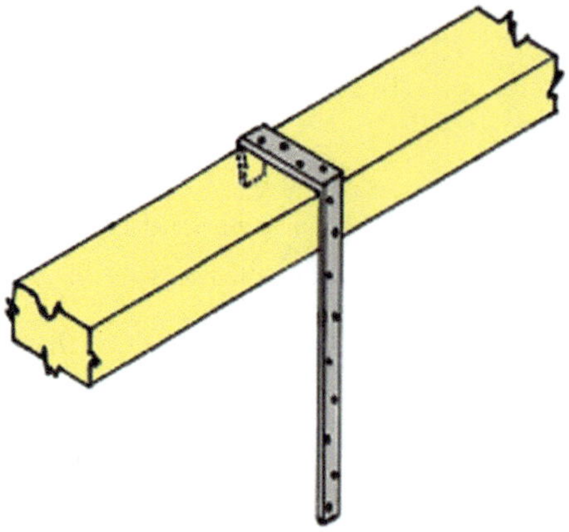

FIGURE 4.35
Wall plate strap

Wall plates are bedded onto the brickwork by the bricklayer.

The wall plate may have to be joined in length with half-lap joints, whereby the lap equals the width of the wall plate.

In order to prevent the whole of a roof being lifted off the building as a result of wind suction, it is necessary to anchor the wall plate to the wall by galvanized restraint straps at 2 m centres.

These straps are built into the inner leaf of the cavity wall approximately 1.5 m below eaves level (Figure 4.35).

Roof truss fixing

The roof trusses should be spaced along the wall plate at between 400 and 600 mm centres and fixed to the wall plate, preferably using truss clips (Figure 4.36).

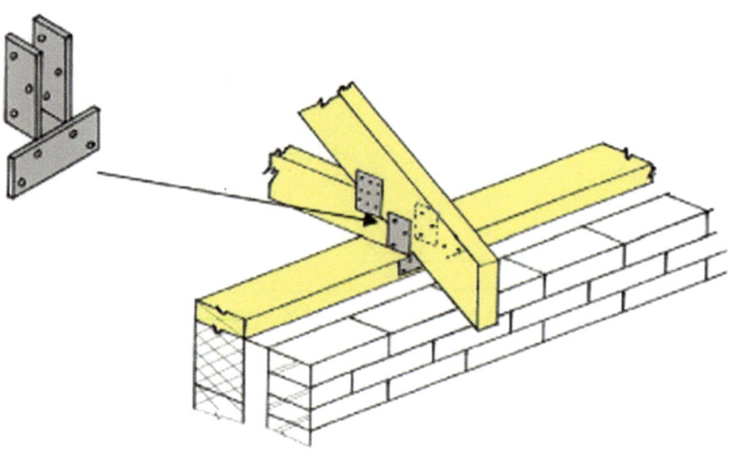

FIGURE 4.36
Galvanized restraint straps at 600 mm centres

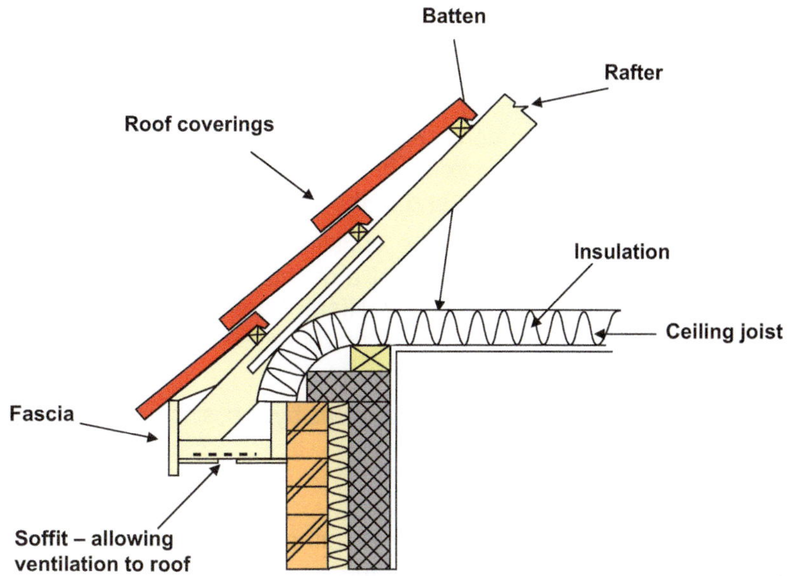

FIGURE 4.37
Typical eaves details

Eaves detail

The ends of the rafters are cut to allow the soffit and fascia to be fitted. The roof should be ventilated by providing vents in the soffit (Figure 4.37).

Flat roofs

Roofs with less than 10 degree slope are termed flat roofs. Most flat roofs have sufficient slope to prevent water standing on the roof surface. Typical details of a flat roof are shown in Figure 4.38.

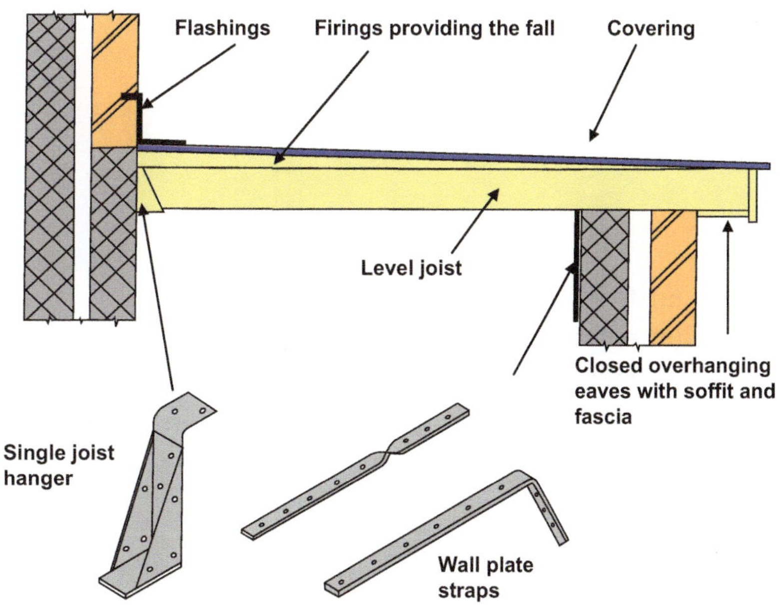

FIGURE 4.38
Types of flat roof showing support and restraint fixings

Multiple-choice questions

Self-assessment

This section of the book is designed to allow you to check your level of knowledge. The section consists of revision questions for this chapter. The questions are all multiple choice and have four possible answers. The answers are to be found at the end of the book.

The main type of multiple-choice question will be the four-option multiple-choice question. This will consist of a question or statement, known as the stem, followed by a choice of four different answers, called the responses. Only one of these responses is the correct answer; the others are incorrect and are known as distracters.

You should attempt to answer the questions by choosing either (a), (b), (c) or (d).

Example

The person employed by the local authority to ensure that the Building Regulations are observed is called the:

(a) clerk of works
(b) building control officer
(c) council inspector
(d) safety officer

The correct answer is the building control officer, and therefore (b) would be the correct response.

Construction technology

Question 1 Identify the drawing shown:

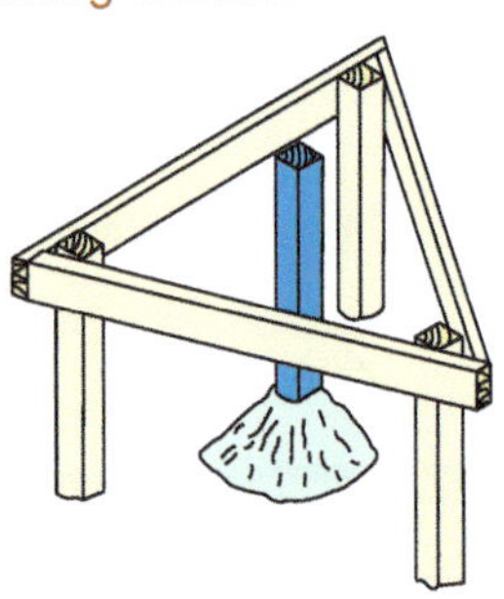

(a) temporary bench mark
(b) ordnance bench mark
(c) fixed bench mark
(d) closed bench mark

Question 2 Bagged material should not be stored more than:

(a) 2 bags high
(b) 4 bags high
(c) 6 bags high
(d) 8 bags high

Question 3 Why is it necessary to load out bricks for a wall from several different packs?

(a) to prevent banding of the wall
(b) to select the best bricks for the wall
(c) to use the bricks as quickly as possible
(d) to ensure the correct number of bricks

Question 4 Which one of the following machines is best suited for clearing a building site prior to setting out?

(a) dump truck
(b) excavator
(c) front loader
(d) backactor

Question 5 Which of the following damp-proof materials is known as rigid?

(a) bituminous felt
(b) engineering brick
(c) lead
(d) copper

Question 6 Which of the following would be classed as bulk materials?

(a) window frames
(b) aggregates
(c) plasterboards
(d) ironmongery

Question 7 Identify the following support fixing shown:

(a) double joist hanger
(b) single joist hanger
(c) wall plate strap
(d) truss clip

Question 8 Identify the roof form shown:

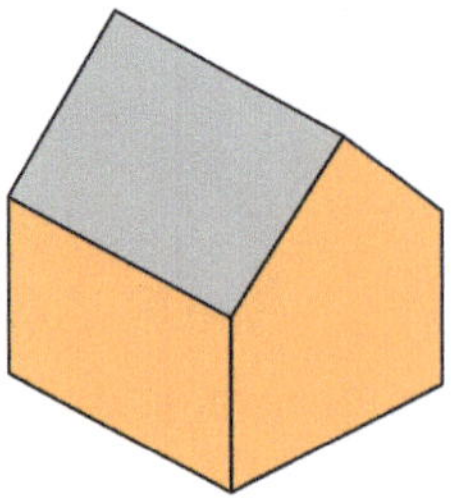

(a) flat roof
(b) lean-to roof
(c) gabled roof
(d) hipped roof

CHAPTER 5

Tools

This chapter will cover the following City and Guilds units:

- Apprenticeship   Module 12   K1.5
- Diploma   Unit 103   1.5, 3.4
    Unit 104   1.5
    Unit 105   1.7, 3.2

This chapter is about:

- Bricklaying tools and equipment

The following City and Guilds performance criteria will be covered:

- The following chapter has no comparable City and Guilds modules, but it gives the apprentice an introduction to the tools of the trade

The following Diploma outcomes will be covered:

- Select tools and equipment
- Identify correct use of hand tools and equipment

Select tools and equipment

A craftsperson is judged by his or her tools.

The saying 'clean tools, clean job' applies to all crafts, and bricklaying is no exception. The apprentice will never regret the small amount of time and trouble spent in cleaning tools at the end of a day's work. All that is required for this is an old rag and a piece of brick for scouring.

Tool kit

Bricklayers can carry their complete tool kit in a medium-sized canvas shoulder bag or heavy duty holdall in one hand, and move about site with a spirit level in the other.

A kit of tools is very personal property and the apprentice will become familiar with all the items, particularly the bricklaying trowel, and will not want other bricklayers to use them.

A tool kit must be maintained in good order so as to ensure efficient and safe working, avoiding injury to self and others.

On construction sites safety helmets must be worn at all times for brain protection - we only have one! Eye protection must be worn when cutting bricks, blocks or concrete to protect irreplaceable eyes - we only have two! Feet should be protected with stout safety shoes or boots. Soft-top track shoes give no protection against a hammer accidentally dropped from hand height, never mind anything worse. Do not throw your tool bag down, as this will damage and loosen handles.

Brick trowel

This is the most heavily worked item in a bricklayer's tool kit, used for gathering and spreading mortar, and for rough cutting some kinds of brick. Brick trowels are available in a range of shapes (Figure 5.1), sizes and thickness of steel, with length of blade from 230 mm to 330 mm. Choice of trowel has a lot to do with personal preference and what feels most comfortable for the individual.

The apprentice should avoid the temptation of thinking that biggest is best.

Narrow-blade London pattern trowels are suited to cavity work, while broad heel trowels lend themselves to solid walling. The best trowels are solid forged from a single piece of steel from tip to tang.

The majority have one side of the blade of thicker steel to withstand the wear and tear of tapping bricks and rough cutting. For this reason, left- and right-handed versions are produced. Philadelphia pattern brick trowels are not intended to be used for rough cutting, and so are not handed. Use of the trowel handle to tap bricks down to the line should be avoided as this tends to loosen the handle and burr the end, making it uncom fortable to use.

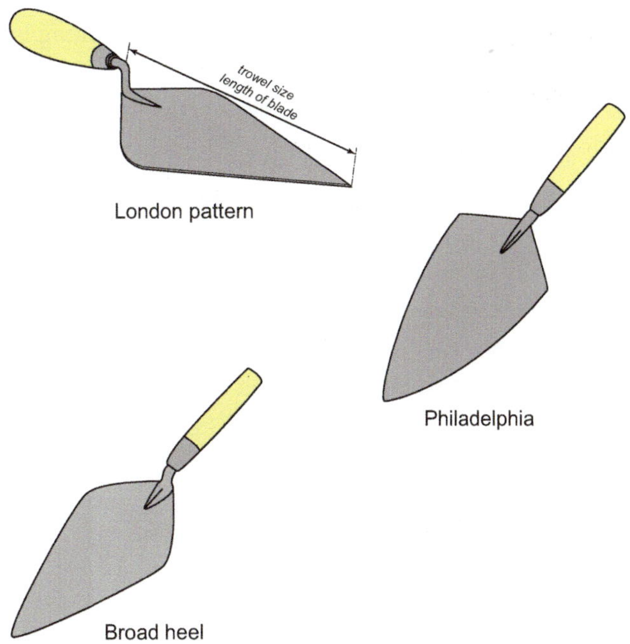

FIGURE 5.1
Trowels

Pointing trowels

These are made with blade lengths from 75 mm to 175 mm. The trowel with the shortest blade, sometimes referred to as a 'dotter', is used for filling and striking cross-joints.

The longer blade pointing trowels are used with a hand hawk for filling and striking bed joints.

Jointing tools

A variety of tools are shown for applying a permanent finish to the exposed surface of mortar joints, some purpose made, some produced by the bricklayer.

Patent wheeled jointers are ideal for raking out mortar to a constant depth in preparation for a square recessed joint finish.

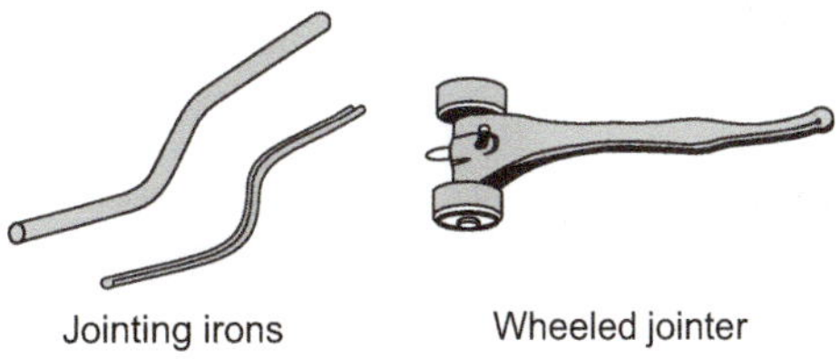

Line and pins

No tool kit is complete without a set of lines for controlling level, line, plumb and gauge of any walls over 1.2 m long. Always buy the best quality hardened steel pins you can afford so that they last.

Bricklayer's line made from traditional hemp can be spliced if accidentally broken so as to avoid knots. Cotton, polyester and various types of nylon are also sold, but it is impossible to join any break in these materials by splicing, as the strands do not separate neatly. The finer the line the better for accurate work. When renewing line, wrap insulating tape around the pins first to prevent rust staining. Always tie the line-end on to pins before winding on, in case you drop the pin when working on a high scaffold.

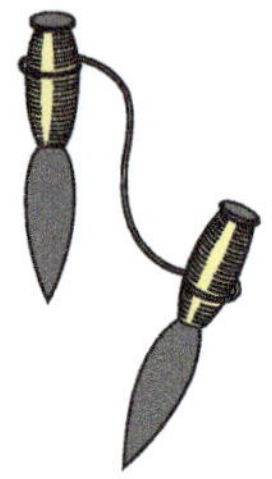

Lines and pins

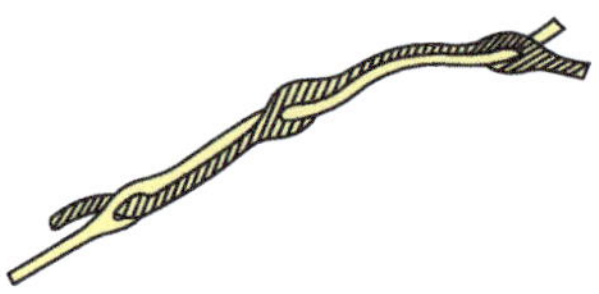

Splicing lines

Club or lump hammer

One kilogram size is ideal for use with a bolster for fair cutting and also for cutting away existing brickwork with a cold chisel.

These have ash- or hickory-wood handles or are made of steel with a rubber sleeve.

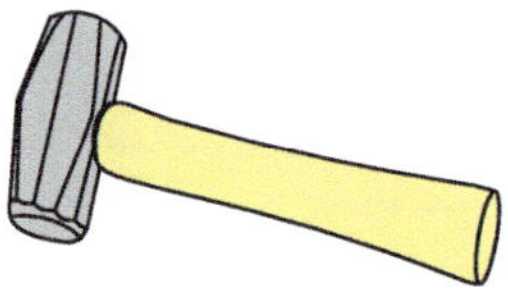

Brick hammer

This is used for rough cutting very hard bricks which would damage a trowel.

They have one hammer head and one forged chisel end which can be reground when worn.

Bolster chisels

A 100 mm width blade is most useful for fair cutting bricks, and bolsters can be supplied with or without rubber or plastic collars for hand protection.

Bolsters must be kept sharp for efficient operation.

Comb hammer

This has a hammer head on one side and is slotted to take hardened steel replacement combs on the other side.

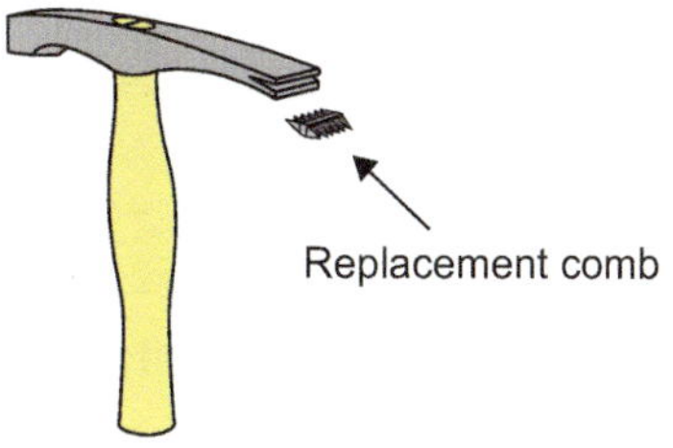

Scutch

This is slotted on both sides to receive renewable hardened steel combs or blades.

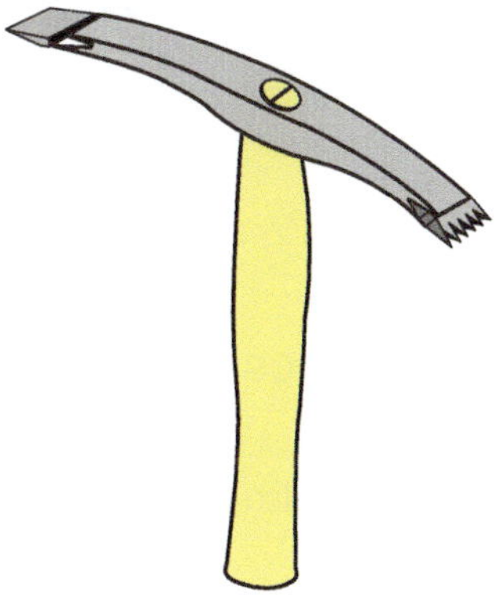

Cold chisels

These must be kept sharp if cutting away brickwork is to be as painless as possible. Grind off the first beginnings of burring over; do not wait until a fully grown and dangerous 'mushroom' has developed. Always use a steel point and wear eye protection if you are unfortunate enough to be cutting away concrete: do not ruin a cold chisel.

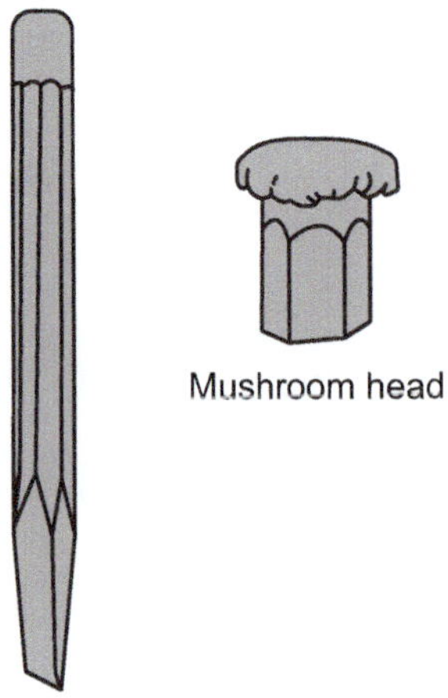

Comb chisel

A comb chisel is a very useful substitute for a cold chisel when cutting away brickwork.

Plugging chisel

This purpose-made tool is used for carefully cutting and toothing out joints in existing brickwork.

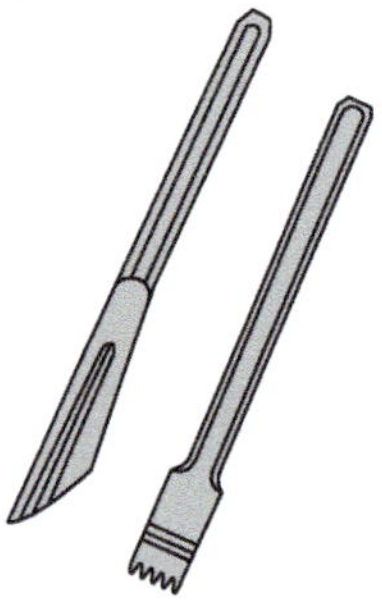

Measuring tapes and rules

Folding boxwood or plastic rules tend to break easily and have been largely replaced by steel tapes. These also have a limited life as they tend to get full of grit and do not retract fully.

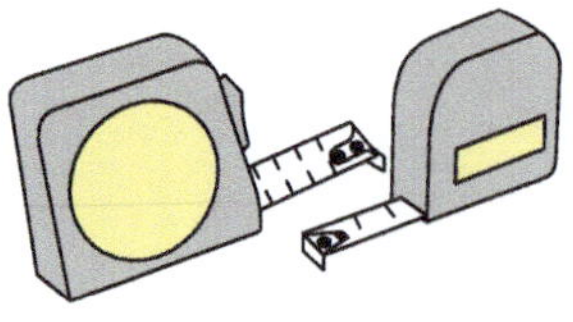

Hand hawk

This allows mortar to be picked up conveniently with a pointing trowel. The removable handle makes it easier to fit into a tool bag.

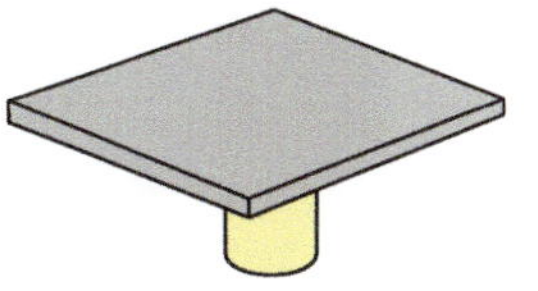

Hand brush

Medium to fine bristle is used for lightly brushing face brickwork at the end of a day's work. Great care must be taken not to leave bristle marks in any mortar that is still soft.

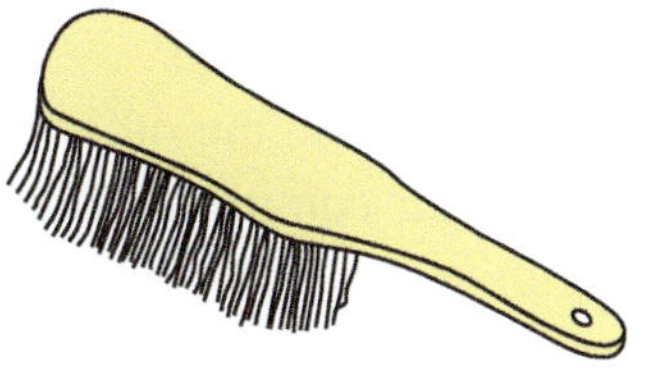

Small square

This is used for pencil marking bricks accurately before cutting.

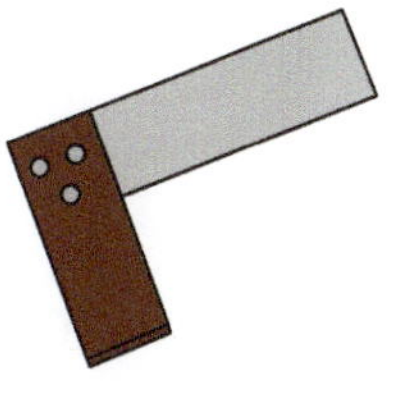

Dividers

Dividers are used for spacing out voussoir positions on an arch support centre either side of the key brick location before starting work.

Bricklayer's sliding bevel

This is used for transferring and marking the same angle of cut for all the bricks when raking cutting to a gable-end wall or when carrying out tumbling-in.

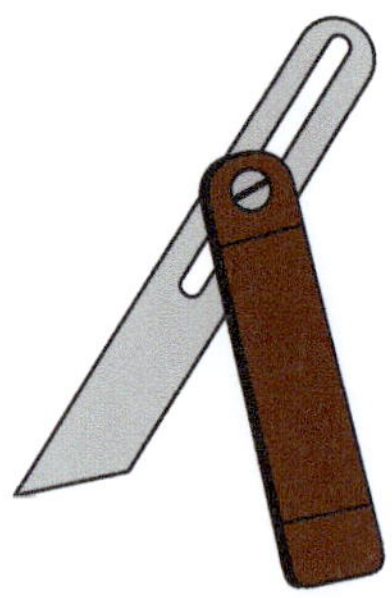

Pair of trammel heads

One is a steel compass point, while the other holds a pencil. These are used with a timber lath as a beam compass, when making a template for constructing a curved wall, drawing arches full size or striking the shape of the brick core to a bullseye.

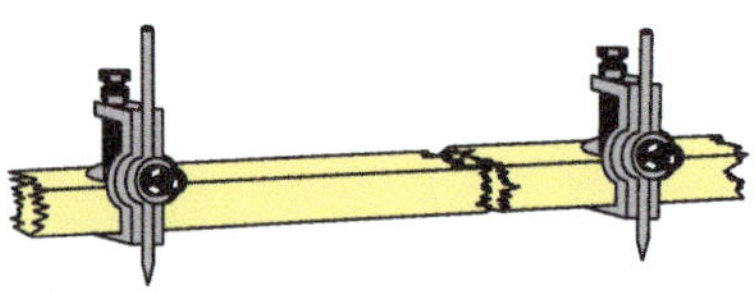

Pencil

For plumbing perpends on face brickwork, H or 2H grade pencils will last longer than HB grade.

Masonry hand saw

This saw has tungsten carbide-tipped teeth for toothing out existing brickwork or blockwork where the mortar is not too hard, and also for cutting neatly through lightweight aerated concrete blocks.

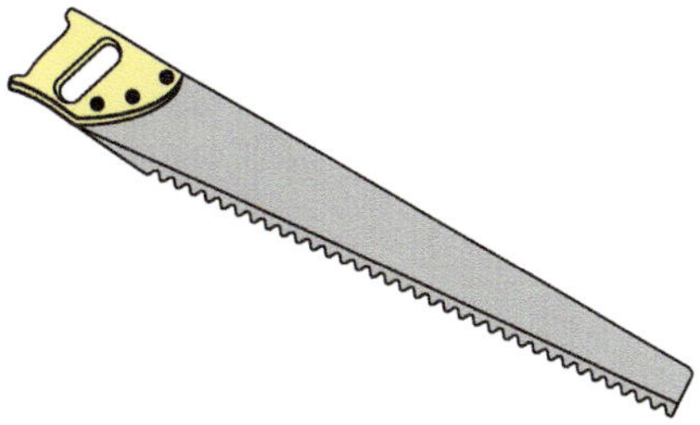

Retractable blade knife

A knife is very useful for cutting damp-proof course material and sharpening pencils.

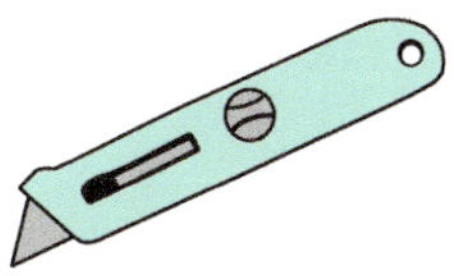

Spirit levels

These have a dual purpose and are used for checking the horizontal and vertical accuracy of brickwork (Figure 5.2).

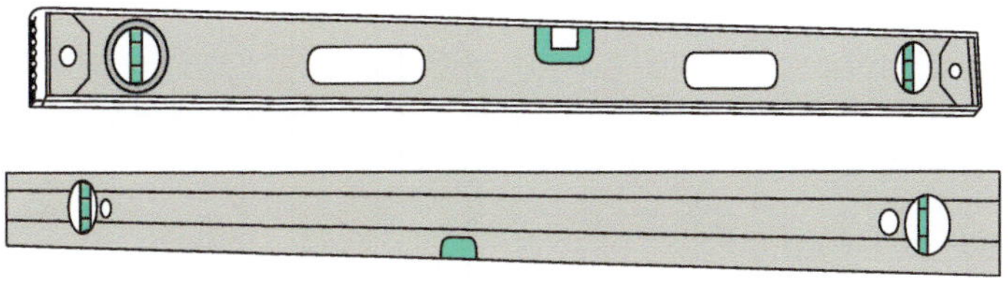

FIGURE 5.2
Spirit levels

For general-purpose use, a bricklayer's spirit level should be not less than 1 m in length, and preferably 1200 mm long.

Spirit levels are available as girder section hardened aluminium with hand-hold slots, as a hollow box section enamelled aluminium, or in hardwood.

A professional bricklayer never taps or knocks a spirit level when plumbing or levelling brickwork. Spirit levels 600 mm long are made for use in restricted places, and boat levels 200 mm long are made for plumbing soldier bricks and for decorative brickwork.

Checking the spirit level

It is important to remember that some spirit levels are adjustable but the cheaper ones are not.

Great care should be taken of the spirit level as it is expensive and if ill-used can lead to inaccurate levelling and plumbing.

Checking for level

It is important to check the spirit level occasionally to ensure its accuracy.

If a course of bricks, a sill or a lintel has been set perfectly horizontal, 'reversing the level' end-for-end should confirm this (Figure 5.3).

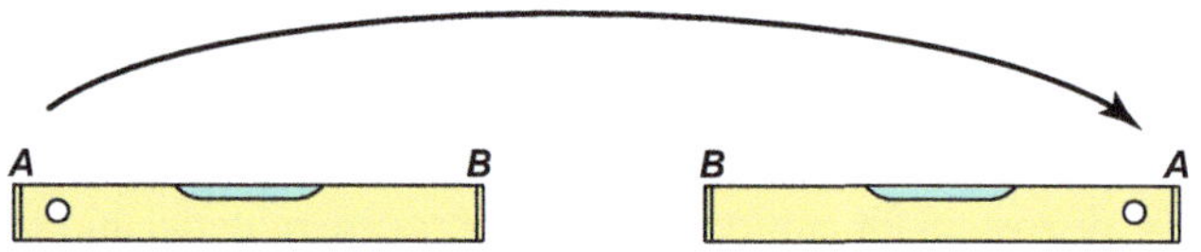

FIGURE 5.3
Daily spirit level check: horizontal level

If the spirit bubble reads truly horizontal in Figure 5.3, but is not horizontal when the level is reversed, then the spirit level is out and needs to be adjusted.

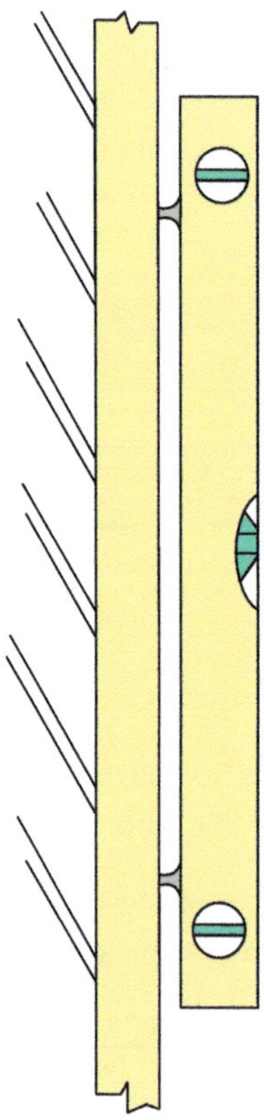

FIGURE 5.4
Checking for plumb

The clamping screws for the horizontal bubble tube must be slackened, and the necessary adjustment made.

Checking for plumb

- Set two screws equal to the length of the level apart in a vertical position on a door frame or jamb.
- Check that they are plumb with a plumb bob or with a spirit level that is known to be accurate (Figure 5.4).
- Position a faulty level on to the screws and adjust until the spirit level reads plumb.
- Reverse the level and adjust if required.

This procedure needs to be repeated if the level has double bubbles.

Once these positions have been produced they could be used for all tradespeople who need levels checking.

Items of equipment additional to the tool kit

These may be made up by the bricklayer.

Brick-cutting gauge

This is used for quick, accurate marking of standard size cut bricks required for bonding purposes (Figure 5.5). It should be made from oak or other hardwood to survive the rigours of life in a tool bag.

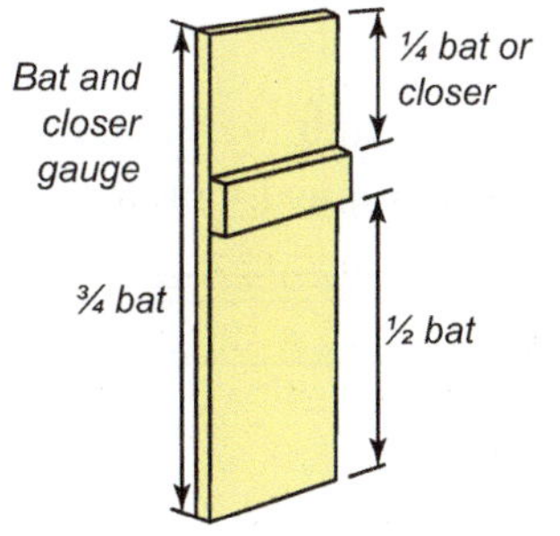

FIGURE 5.5
Brick cutting gauge

Corner blocks

These provide a simple means of supporting bricklayer's line at quoins that do not leave pinholes behind (Figure 5.6). If made from hardwood they will last longer.

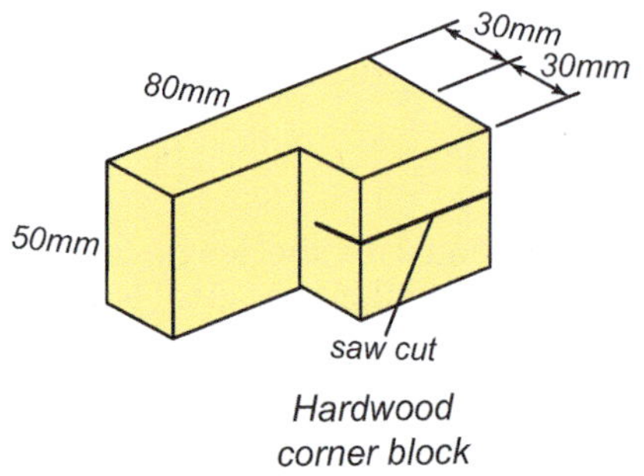

FIGURE 5.6
Corner blocks

Frenchman

This simple device is made from an old table knife, heated, bent over and filed to shape for trimming mortar bed joints when carrying out weather-struck and cut pointing, or tuck pointing.

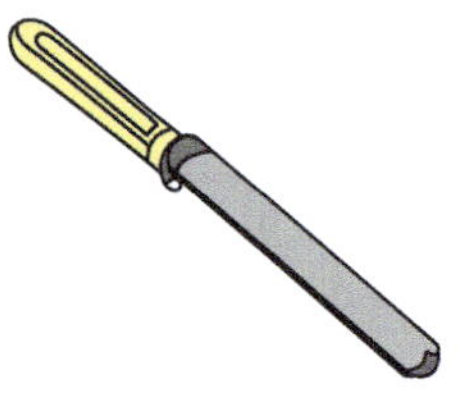

Storey rod

This is indispensable for checking courses of brickwork and blockwork, so as to ensure consistent vertical gauge above and below datum levels (Figure 5.7).

Feather edge pointing rule

This is used in conjunction with a Frenchman for trimming bed joints after pointing. Cork pads allow trimmed mortar to fall away cleanly.

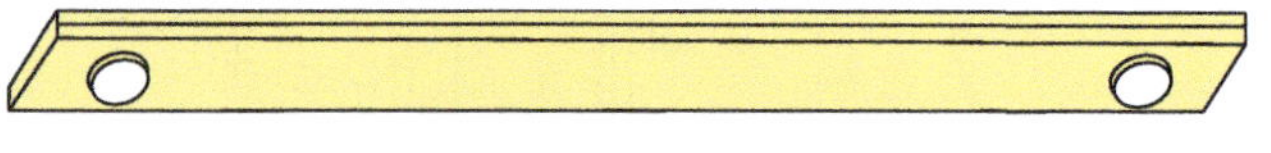

Maintenance of tools

It is important to take care of tools and equipment to ensure that they are clean and sharp when required.

Mortar should never be allowed to harden on the blade of trowels as this creates a rough surface and prevents free and easy movements when picking up and spreading mortar.

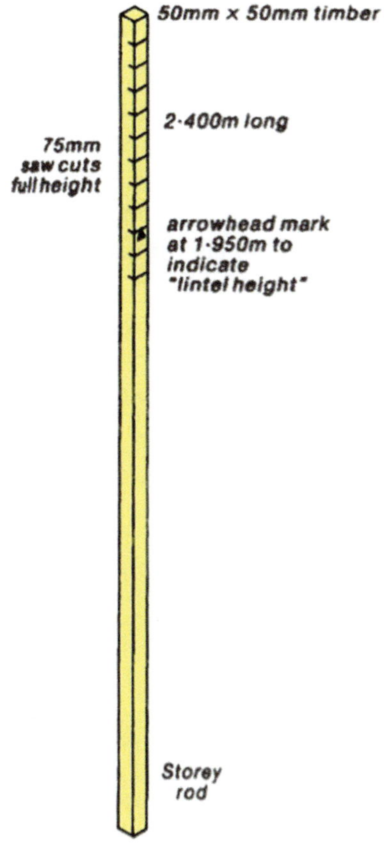

FIGURE 5.7
Storey rod

All cutting tools should be cleaned after use. Never knock bricks or blocks with the end of the handle as this will damage it. Have the chisel end sharpened and tempered regularly. Scutch hammers need cleaning after use and the combs replaced as necessary. Check all handles for cracks and splits and change where necessary.

All chisels need to be kept sharp and regular attention should be paid to the head to prevent mushrooming. Plastic mushroom sleeves are available for most chisels to reduce the risk of injury to the hand holding the chisel.

Levels should be checked regularly for level and plumb. They should be kept clean, especially the glass or plastic covering the bubbles.

Lines need constant attention to prevent twisting, fraying and stretching. Avoid knocking line pins into hardened joints as this will bend the point of the line pins.

Safety

Eyes

Remember to have personal protective equipment ready when required. Always use goggles when cutting.

Heads

Since 1990 everyone on a construction site has to wear a helmet by law. These have to be renewed when out of date.

Feet

Wear safety boots to support ankles and protect the feet. If wellingtons are required ensure that these are also protected.

Hearing

Wear ear protectors when using items of plant and equipment that emit noise over the recommended level.

Lungs

Wear masks where dust is produced.

Multiple-choice questions

Self-assessment

This section of the book is designed to allow you to check your level of knowledge. The section consists of revision questions for this chapter. The questions are all multiple choice and have four possible answers. The answers are to be found at the end of the book.

The main type of multiple-choice question will be the four-option multiple-choice question. This will consist of a question or statement, known as the stem, followed by a choice of four different answers, called the responses. Only one of these responses is the correct answer; the others are incorrect and are known as distracters.

You should attempt to answer the questions by choosing either (a), (b), (c) or (d).

Example

The person employed by the local authority to ensure that the Building Regulations are observed is called the:

(a) clerk of works
(b) building control officer
(c) council inspector
(d) safety officer

The correct answer is the building control officer, and therefore (b) would be the correct response.

Tools

Question 1 Which of the following PPE is essential when cutting bricks with a club hammer and bolster chisel?

(a) hard hat
(b) ear protector
(c) eye protector
(d) overalls

Question 2 Name the chisel used for cutting out the bed or cross-joint in an existing wall:

(a) comb chisel
(b) plugging chisel
(c) bolster chisel
(d) cold chisel

Question 3 Name the item of equipment used for supporting mortar when pointing brickwork:

(a) mortar board
(b) flat hawk
(c) hand board
(d) hand hawk

Question 4 Name the piece of timber marked off with brick courses, used for checking the height of brickwork:

(a) gauge rod
(b) timber rod
(c) pinch rod
(d) black rod

Question 5 A pair of trammel rods is used to set out:

(a) curved brickwork
(b) angled brickwork
(c) acute brickwork
(d) obtuse brickwork

Question 6 Name the item of equipment used by the bricklayer for supporting bricklaying lines on a long wall:

(a) corner block
(b) tingle plate
(c) gauge block
(d) storey plate

Question 7 Identify the tool shown:

(a) brick hammer
(b) scutch hammer
(c) club hammer
(d) claw hammer

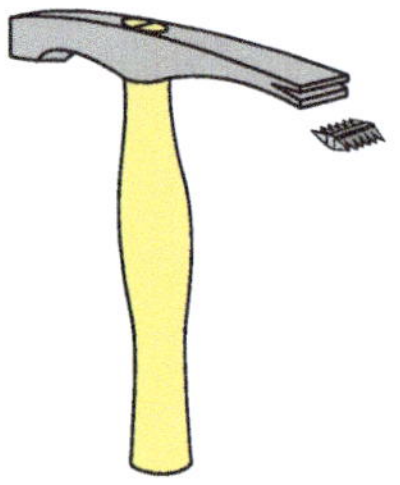

Question 8 Identify the item of equipment shown:

(a) gauge rod
(b) closer gauge
(c) corner block
(d) tingle plate

CHAPTER 6

Preparing and Mixing Concrete and Mortar

This chapter will cover the following City and Guilds units:

• Apprenticeship	Module 4	K1.4, K1.6
	Module 5	K1.4
	Module 6	S3.1, S3.2, S3.3
	Module 6	K4.1, K4.2
	Module 12	S1.4
	Module 13	S2.1, S2.2, S2.3
• Diploma	Unit 101	4.5

This chapter is about:

- Interpreting instructions
- Adopting safe and health working practices
- Selecting materials, components and equipment
- Understand how concrete is produced and placed on-site and its applications
- Mix mortar in the correct proportions

The following City and Guilds performance criteria will be covered:

- Select materials to make concrete
- Produce concrete
- Place and finish concrete
- Mix mortar in the correct proportions

The following Diploma outcomes will be covered:

- This chapter gives the student an insight into mixing both concrete and mortar.
- Transporting, placing, finishing and curing concrete.

Safe work practices

This chapter deals with the mixing of mortar and concrete. Concrete has already been briefly dealt with in Chapter 4.

Mortar is required by several trades in the construction industry. This book will deal mainly with its use for laying and jointing both bricks and blocks by the bricklayer.

Mortar can also be used for plasterwork and laying both wall and floor tiles, although there are more modern materials available. The bedding of ridge tiles by the roof tiler is still required as no modern material is available.

It is important to read and understand any instructions given to ensure that the mortar or concrete is mixed correctly.

It is a well-known saying in the industry that the craftsperson needs to understand the materials that they use, and lack of knowledge could result in materials being spoilt and work having to be taken down, both causing extra costs on the job.

Materials and methods are constantly being introduced and it is important that the users of these materials keep up to date with this ever-changing industry.

Types of instruction

There are many sources of information available, both oral and written. Oral information may come from your supervisor and written from a manufacturer's information details.

When mixing on site information is usually from the person in charge, who finds the information from contract documents such as the specification or bill of quantities.

When a new material or item of equipment is introduced to the industry the manufacturer will provide written instructions on its safe use.

Every designer and contractor must be able to call on a team of specialists and other back-up information.

It is essential that all parties have access to various sources of up-to-date information. No single person in isolation could satisfy the demands of all members of the design team and construction teams.

It is also very important that the student understands where information can be found. Information can be gathered from numerous sources, such as:

- manufacturers' literature
- organization handbooks and manuals
- legal documents
- general reference books.

Manufacturers' technical information

Many manufacturers produce technical information; which is free for the asking. These papers, brochures, leaflets, etc., should be stored in the office in an easily retrievable system.

Organization handbooks and manuals

Most bodies, including professional institutions and employers' and research organizations, publish handbooks and manuals, listing details of their members and giving additional information relevant to the function of the body.

Legal documents

Constant reference must be made throughout the duration of any contract to all relevant contractual documents. It is important that these are always available. Examples are:

- the contract document
- specifications
- bills of quantities
- form of tender
- schedules.

General reference books

There are numerous books to meet the needs of all members of the construction industry. Publishers ensure that the industry keeps abreast of technical advances that take place in construction.

Legislation

It is important that operatives understand their responsibilities regarding current legislation while working:

- in the workplace
- below ground level
- at height
- with tools and equipment
- with materials and substances
- with movement and storage of materials
- with mechanical handling and lifting equipment.

Legislation, the main form of English law, is an Act of Parliament. Most legislation only provides a basis from which to operate.

The Health and Safety at Work Act is a typical example, where the agency carrying out the details and arrangements of the Act is the Health and Safety Commission. Health and safety legislation was covered in detail in Chapter 2.

Other building legislation is outlined below.

Building Regulations

Under the control of the Secretary of State for the Environment, the main purpose of the Building Regulations is to ensure that buildings are constructed so that they are safe and do not present danger to those who occupy them. In simple terms, they describe how a building should be constructed.

The current regulations appropriate to this chapter control the materials used in the production of both concrete and mortar.

Town and Country Planning Acts

There are many Acts in existence that affect planning and the control of development. The main aim of them all is to safeguard the public regarding buildings and developments. They are mainly dealt with by local government officers

Other legislation

The following legislation also affects the construction industry:

- Building Control Act
- Clean Air Act
- Factories Act
- Guard Dogs Act
- Historic Building Act
- Housing Act
- Road Safety Act
- Water Acts, etc.

British Standards

These are issued by the British Standards Institution which give recommended minimum standards for materials and components (e.g. walling blocks and doors) used in construction and other industries. All materials and components complying with a particular British Standard are marked with the Kite Mark together with the appropriate BS number. (Figure 6.1)

FIGURE 6.1
British Standard kite mark

Standards greatly simplify the workings of construction and are quoted in:

- specifications
- Building Regulations, etc.

British Standard Codes of Practice

Codes of good practice are issued by the British Standards Institution, to cover workmanship in specific areas, e.g. building drainage, and brick and block masonry.

Agrément Certificate

These certificates are granted by an independent testing organization, called the British Board of Agrément, stating that the manufacturer's products have satisfactorily passed agreed tests.

Subsequent to the granting of the certificate strict quality control has to be continued.

The British Standards are now being overtaken by the European Standards, with the CE logo (Figure 6.2).

FIGURE 6.2
European Standard mark

European Standards

The European Committee for Standardisation (CEN) develops standards and other publications on an enormous number of subjects.

Products which satisfy the European requirements for safety, durability and energy efficiency carry the CE mark.

Respond to emergencies

When dealing with mixing mortar or concrete there could be times when operatives involved have to respond to a situation.

Your firm will have put into place a procedure to follow in such an instance. Always ensure that you follow it.

Safety

There are three main dangers when batching and mixing materials:

- stress injuries
- mechanical injuries
- chemical injuries.

Stress injuries

Lifting and shovelling heavy materials incorrectly may lead to many injuries, in particular to the back muscles and spine.

Whenever handling materials it is essential to take care of yourself and others. Always use the correct lifting techniques.

Injuries can occur through:

- poor material handling
- poor mechanical usage.

Material handling

Lifting or carrying heavy or awkward objects, such as bags of cement, can cause injury if performed incorrectly. Bad handling also increases fatigue (Figure 6.3).

The weight of cement bags has now been reduced to 25 kg to avoid strain.

Before handling any materials always read the instructions and find out how heavy the materials are. If in doubt seek the help of other workmates.

Examples of typical mixing materials are:

- cement = 25 kg (previously 50 kg)
- lime = 25 kg
- gypsum plaster = 40–50 kg
- wheelbarrow full of concrete = 100–150 kg
- buckets of adhesive = various depending on the size
- 1 litre of water = 1 kg.

Mechanical injuries

Many injuries are caused by a lack of proper maintenance and incorrect use of mixers and associated equipment.

Remember that there is a correct way of lifting and pushing a wheelbarrow.

FIGURE 6.3
Safe handling techniques

Note

NEVER, ever, put your hand inside the drum of a mixer while it is moving. This is obviously a danger.

Catching your hand or coat sleeve in the paddle fins leads to the most spectacular injuries, from tearing off a few fingers to screwing the arm round until it is wrenched off at the shoulder.

Never put the end of the shovel or anything else into a moving drum: if it snatched, it will spin round and hit you before you know what is happening.

Never use an electric mixer unless it has a reduced voltage, 110 V, and all cables, sockets and transformer are in good working condition.

Never refuel a petrol or diesel mixer while it is running.

Remember

NO SMOKING – while you are using a petrol or diesel mixer.

Note

ALWAYS make sure the mixer is on a reasonably level standing. A mixer that is leaning badly when full can easily fall on top of you.

Check that all the guards are in position and fuel tanks are full before starting work. Fuel spilling on to a hot exhaust can cause a serious fire.

Read all the manufacturer's instructions on how to operate the machine, maximum loads and essential maintenance.

Clean the machine properly after use – a clean machine is a more efficient machine.

Remember

NEVER throw mortar. ALWAYS report accidents.

Remember

Accidents do not just happen – they are caused.

A few simple dos and don'ts can make using mixers quite safe. The first is probably the most important!

Chemical injuries

Cement, lime and plaster are caustic when wet and will cause burns to the skin, especially if it is already broken.

Eyes are particularly vulnerable to these materials, even as a dry powder because the eye is always wet.

Always read the manufacturer's instructions and warning notices printed on the bag or can, until you are completely familiar with them.

Maintaining safety

If in any doubt about safety on the building site, refer to Chapter 2 on health and safety.

Remember that any type of work undertaken by the construction industry is often difficult and hazardous.

The type of work and conditions are different on each site; consequently, the hazards are also different.

It is of utmost importance that *all* trainees are capable of using machinery and equipment efficiently and safely.

Furthermore, they should be aware of the causes of accidents and be able to take actions to deal with any accidents that may occur.

Security procedures

All sites should have arrangements in place to store items of equipment and personal belongings. These can range from a small metal or wooden cabin to the property that is being worked on.

Materials suitability

During this chapter you will be dealing with the mixing of:

- concrete – cement/lime mortars
- concrete – used for foundations, oversite concrete, reinforced concrete- framed building, paths and drives.
- mortars – used for the laying of bricks and blocks, rendering walls, etc.

You will also be dealing with the placing and finishing of concrete.

Characteristics of materials

There are numerous materials with which you could come into contact with while mixing and placing and finishing concrete and while mixing mortars:

- aggregates
- sharp sand
- building sand
- lime
- cement
- plasticizer
- water
- retardcr
- colouring agents
- formwork
- reinforcement
- release agents
- damp-proof membrane

Aggregates

Aggregates are divided into two main groups:

- fine aggregates
- coarse aggregates.

Fine aggregates

Fine aggregates can pass through a 5 mm sieve. These include the sands used for both concrete and mortar.

Sand for bricklaying mortar should be 'well graded'; that is, a mixture of fine, medium and large grains (Figure 6.4).

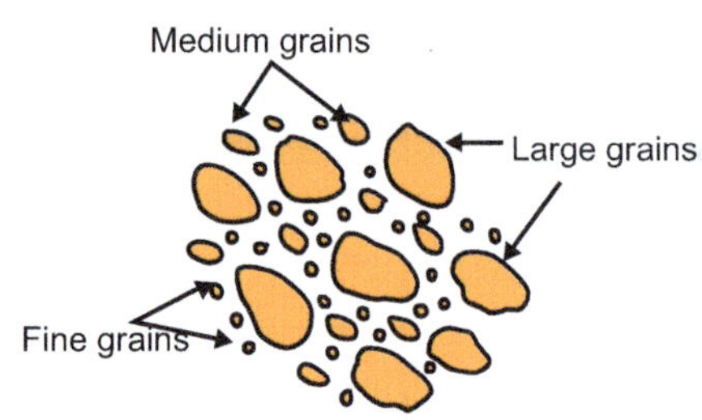

FIGURE 6.4
Well-graded sand

Sand can be excavated from the ground in pits or dredged from the sea. In both situations the sand has to be thoroughly washed before use. It is possible to test the sand for silt content on site with simple equipment (Figure 6.5).

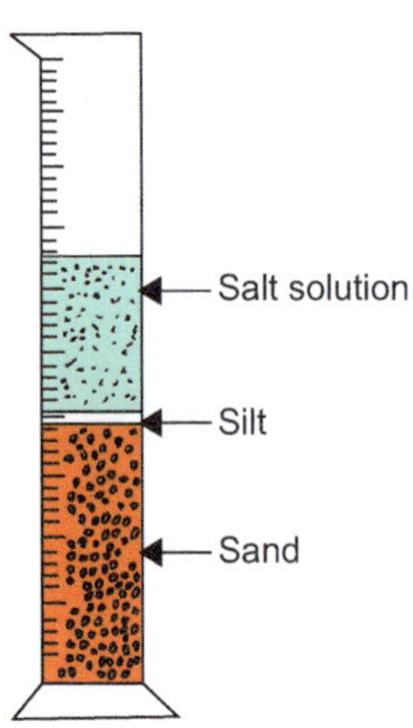

FIGURE 6.5
Silt test

Silt test for sand

A simple test which can be carried out on site to give a guide to the amount of silt in natural sand is the field settling test. It should not be used for crushed rock sands.

To carry out the test you should preferably use a 250 ml measuring cylinder.

1. Fill the cylinder up to about the 50 ml mark with a salt water solution (one teaspoonful to 750 ml).
2. Pour in the sand until the level of the sand is up to the 100 ml mark.
3. Add more salt solution until it reaches 150 ml.
4. Shake the solution well.
5. Stand the cylinder on a level surface and tap it until the top of the sand is level.
6. Leave it to stand for three hours.
7. Measure the volume of the silt layer and the volume of sand.

Silt content = Height of silt layer/Height of sand x 100

Note

The silt content should not be more than 8 per cent.

If it is more than 8 per cent, report this to the engineer as a more detailed test may have to be carried out.

If a measuring cylinder is not available a 500 g jam jar or similar bottle can be used, although this will not be quite as accurate.

Coarse aggregates

Coarse aggregates are retained by a 5 mm sieve. These are used mainly in concrete production.

To produce good concrete, the aggregates must be sound, and of the type and quality specified.

Shapes

Aggregates may be excavated from river beds or quarries or dredged from sand or shingle banks under the sea.

The three most common shapes of aggregates are:

- Rounded and regular – these aggregates have been rounded by natural wear. There is no sign of any faces and they have a smooth surface. They are obtained from river and sea beds.

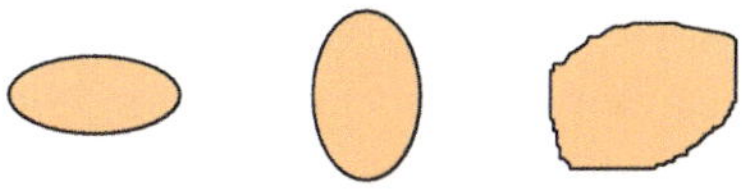

- irregular – irregularly shaped aggregates show more signs of wear. There is some evidence of faces and they have rough surfaces. Irregular aggregates are usually obtained from quarries.

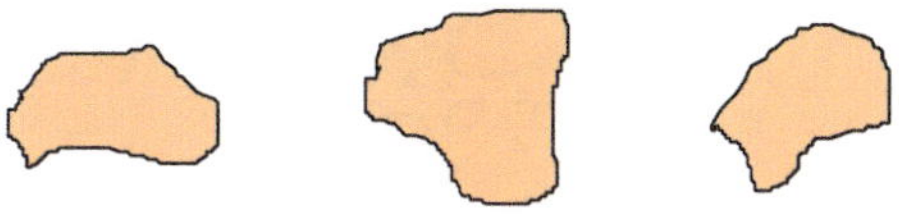

- angular - angular aggregates show little sign of natural wear. The faces have sharp edges and the surfaces are rough. These are obtained by crushing rocks.

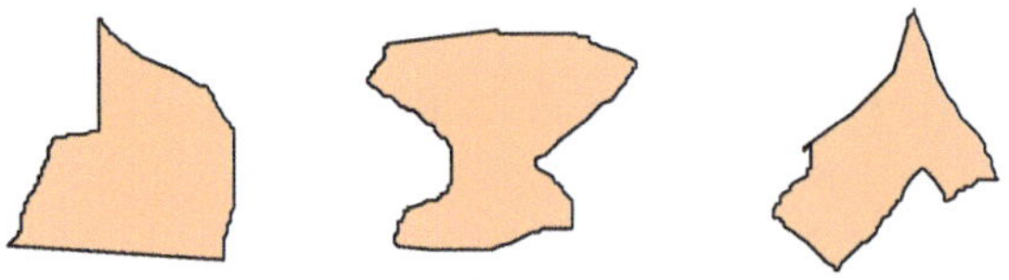

Surface area

The surface area of an aggregate has an effect on the workability of a concrete mix and the binding strength between the aggregate and the cement. An aggregate with a large surface area will decrease workability but increase the binding strength.

Rounded aggregates have less surface area than the irregularly shaped aggregates, which in turn have less surface area than the angular aggregates

Therefore, rounded aggregates give a cement mix greater workability but less binding strength than angular aggregates.

To produce a good mix the aggregates should be well graded (Figure 6.6).

Aggregates form the bulk of mortar and concrete used throughout the construction industry.

Lime

Lime has been used very successfully for thousands of years as the only cementitious ingredient in mortars.

The lime used in the past was hydraulic, meaning that it was made from a naturally occurring chalk raw material which contains clay impurities. This clay content gave the lime a slow setting action not unlike the chemical hydration-setting-hardening process of ordinary Portland cement (OPC), but the lime mortar would take years to harden.

The raw materials used to produce building lime are either chalk or limestone. When chalk or limestone is burnt at a very high temperature it turns into quicklime.

This material cannot be used for building work in this state so it undergoes treatment by the addition of water - this process is known as slaking. The end product is hydrated lime, which is available in bags (Figure 6.7).

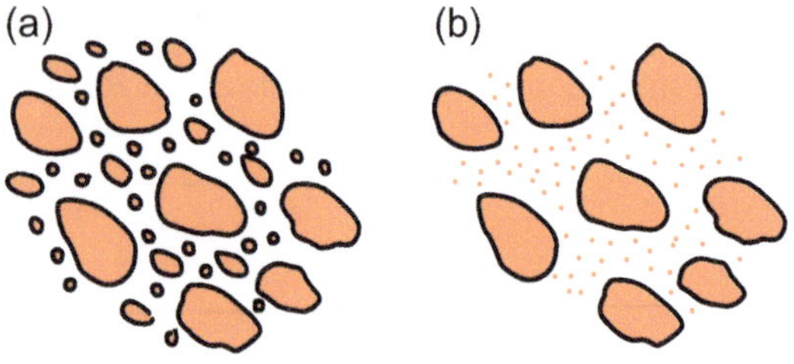

FIGURE 6.6
(a) Well-graded aggregates; (b) poorly graded aggregates.

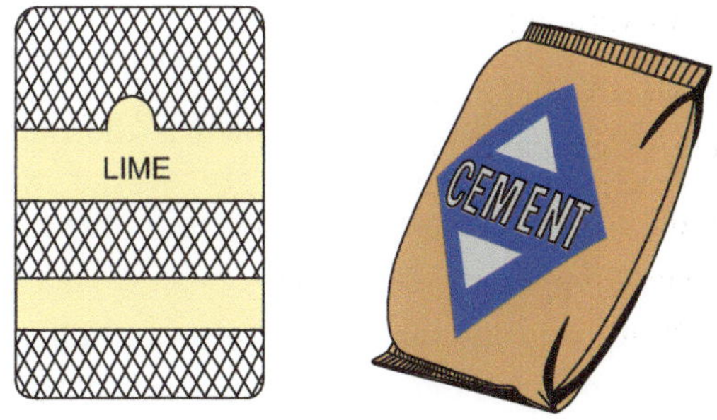

FIGURE 6.7
Bags of lime and cement

Cement

If you read the writing on a cement bag you will see that its correct name is ordinary Portland cement. This is the most commonly used cement on construction sites for making mortar and concrete and is available in bags (Figure 6.7) or loose. It is used in the manufacture of concrete and mortar.

Portland cements are so called because, when solid, they resemble natural Portland stone, which is obtained from Portland in Dorset. Portland cements are also known as hydraulic cements because when mixed with water they set to form a solid.

Cement is the all-important 'glue' in mortar which binds the grains of sand together when water is added and the setting and hardening process is completed.

Used neat on its own, cement is too sticky, sets too hard and would develop severe shrinkage cracks. It is therefore always diluted with three, four or six equal volumes of sand.

The setting and early part of the hardening processes of Portland cement involve complicated chemical reactions between the mixing water and the cement powder in a batch of mortar or concrete.

These reactions need to take place in damp conditions, called 'curing', if they are to be totally completed and give full hardened strength.

This early setting and hardening (Figure 6.8), taking place over hours and weeks, respectively, must never be hurried by early 'drying out', as this will severely reduce the final strength of OPC mortar or concrete.

Figure 6.8 shows the stages in hydration of OPC as it sets and hardens, whether used in mortar or concrete.

FIGURE 6.8
Typical setting and hardening timescale for ordinary Portland cement.

The principal raw materials of cement are:

chalk or limestone and clay or shale.

These are burnt to produce a clinker which is then ground down to a fine powder and mixed with gypsum.

Modified Portland cements

The student should be aware that there are several different types of Portland cements available and each one has a specific purpose.

While these may have different rates of strength gain, it is useful to remember that they all will eventually reach about the same strength.

Rapid-hardening cement

The main feature of this type of cement is its finer grinding than that of OPC, and its main use is in concrete work where greater early strength is required.

Sulphate-resisting cement

This is a modified Portland cement with improved resistance to chemical attack by sulphates, which are salts found in certain ground waters, and as impurities in some building materials. Sulphates in solution can cause softening, and considerable expansion of cement-based materials.

White cement

This is made from using china clay (kaolin), which is pure clay, free of iron oxides. Iron oxides give OPC its grey colour.

Coloured cements

These are made by blending inert pigment powders with white or grey cements, depending on the colour required. The pigments may be added at the works or by the user.

Water-repellent cements

This is produced by intergrinding the Portland cement clinker with a small proportion of a water-repellent agent, such as gypsum with tannic acid or certain metallic soaps.

These cements are used in water-repellent renderings and in base coats where the background is of uneven suction, especially where a coloured finish coat is to be applied, or to avoid patchiness.

Masonry cement

This is produced especially to give the high workability required in mortars for masonry, brickwork and renderings without the addition of lime or other plasticizer.

It usually consists of a blend of OPC and finely ground chalk or silica, possibly with the addition of a plasticizing chemical.

Low-heat Portland cement

This cement has a modified composition to give a low rate of heat during hydration. It is suitable when mass concrete is being used. Owing to its low heat output the mass concrete is less likely to crack.

Plasticizer

Generally speaking, bricklaying mortar made from cement and sand only is not sufficiently fatty or easily workable with the trowel. Such mortar is described as 'short' or 'harsh' and does not hold together when rolled on the spot board.

Lime added to a mix improves the workability by temporarily retaining more mixing water, and results in a denser mortar. Lime is usually added in a volume equal to that of the cement in gauged mortar; i.e. 1:1:6.

As an alternative, patent liquid plasticizers can be added, which generate millions of microbubbles of air within the mortar as it turns in the mixer. Liquid plasticizers are available in tins (Figure 6.9). These must be added to the mixer on site with great care, following the manufacturer's instructions as printed on the container. They are known as air-entrained mortars.

Water

Mixing water triggers a chemical reaction with cement, which causes the setting and hardening of mortar and concrete.

The water should be clean enough to drink, as impurities can seriously delay or prevent this setting action.

Retarder

A retarder can be used to delay the initial setting action of mortar for 36 hours. A retarder is also available when placing concrete which requires the face of the finished concrete to be exposed aggregate.

The retarder is applied to the face of formwork and it delays the setting action of the cement on the skin of the concrete. This allows the cement to be washed from the face of the concrete, exposing the aggregates.

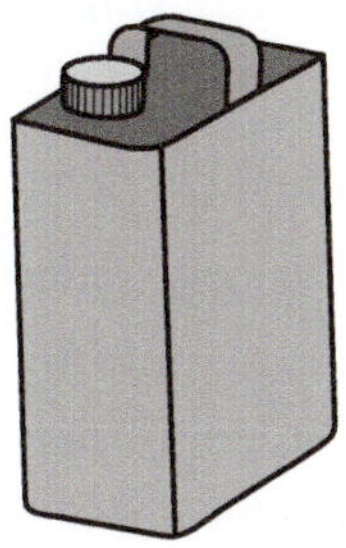

FIGURE 6.9
Plasticizer

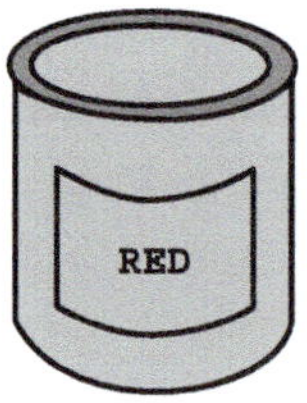

FIGURE 6.10
Colouring agent

Colouring agent

Colouring agents are applied during mixing to provide a decorative finish to the mortar or concrete. They are usually supplied in tins (Figure 6.10).

Mortars

It is not usually difficult to cause an argument between bricklayers - just ask them the following question: Does mortar stick bricks together or keep them apart? The answer is of course that it does both, but a lot more besides.

Mortar must stick firmly to bricks and blocks in external walling so as to keep the rain out. Mortar bed joints also hold bricks apart, so that the courses can be kept level and to an even vertical gauge of four courses to 300 mm, with standard metric bricks.

When walls were much thicker and before Portland cement was invented, a mortar mix of lime and sand was used. The lime was hydraulic, as described above, and lime mortar would take years to harden.

Purpose

Bricklaying mortar is the ideal material for getting bricks to rest firmly upon each other, whether these are accurately shaped class A clay engineering and calcium silicate bricks, or more irregularly shaped hand-made bricks.

The mortar must remain soft enough for each brick to be pressed down to the line, before suction causes the bed to stiffen up. Not only does mortar accommodate irregularities, but it must stick firmly to each brick so as to stop rain from penetrating exposed joints.

Types of mortar

The basic raw materials for bricklaying mortar can be prepared in a number of different ways, depending on specification and site requirements.

The apprentice or trainee should understand the definitions given in Table 6.1.

The European Standard for mortar, BS 5628:1985, has been replaced by BS EN 998-2:2003 and specifies requirements for factory-made masonry mortars (bedding, jointing and pointing) for use in masonry walls, columns and partitions (e.g. facing and rendered masonry, load-bearing or non-load-bearing masonry structures for building and civil engineering). This European Standard defines for fresh mortars the performance related to workable life,

Table 6.1 Types of bricklaying mortar

Mortar type	Definition	Advantages	Disadvantages	Remarks
Lime mortar	Mortar made from 1 volume of lime to 3 equal volumes of sand	Smooth and workable. Ideal for training purposes where materials must be recycled	Unsuited to modern construction, particularly cavity work, due to slow rate of hardening and final strength	1:3 mix using hydraulic lime, recommended by conservationists for repairs and repointing centuries-old brickwork
Cement mortar	Mortar made from 1 volume of OPC to 3 equal volumes of sand	Recommended only for use with class A engineering bricks	Stiffens and sets too rapidly. Will cause shrinkage crack damage to other bricks and blocks if used by mistake	Usually has ¼ volume of lime added to improve workability
Compo	Composition mortar using separate site deliveries of dry OPC, lime powder in bags and sand in bulk	Combines quicker strength gain of OPC with good workability of lime	Messy splitting of bags of cement and lime powder on site, giving health and safety risks	Obsolete for site use due to disadvantages given
Gauged mortar	Mortar prepared on site from bulk deliveries of wet-mixed LSM, to which 1 equal volume of OPC is added or 'gauged' each time a batch of mortar is required	Added powder pigments can be carefully measured and blended at works, before delivery to site, to ensure consistent colour of hardened mortar	Bulk delivery or stockpile of LSM, to last a week or so, must be kept covered to prevent surface drying in summer and pigment blowing away. This, and rain washing, can weaken the colour strength of mortar batches	A reliable way of producing tinted bricklaying mortar

OPC: Ordinary Portland cement; LSM: lime–sand mortar

chloride content, air content, density and correction time (for thin-layer mortars only). For hardened mortars it defines, among other things, performances related to compressive strength, bond strength and density measured according to the corresponding test methods contained in separate European Standards.

This European Standard provides for the evaluation of conformity of the product to this European Standard.

BS EN 998-2:2003 covers masonry mortars, with the exception of site-made mortars. However, this standard or part of this standard may be used in conjunction with codes of applications and national specifications covering site-made mortars (Table 6.2). The number after the letter M is the compressive strength after 28 days.

Selection of mortar

As a general guide, the hardness or eventual compressive strength of mortar should be related to the hardness of, or preferably slightly weaker than, the bricks or blocks to be laid. Therefore, if, as a result of slight foundation

Table 6.2 Mix designation, compressive strength and composition

Traditional mortar designation	BS EN 998-2 mortar class	Nominal proportions by volume		Mass of original dry materials	
		Cement:lime:sand	Cement:sand	Cement (%)	Lime (%)
i	M12	$1:\frac{1}{4}:3$	–	20.0–25.0	1.0–3.0
		–	1:3	20.5–25.0	Nil
ii	M6	1:½:4 to 4½	–	14.0–19.0	1.5–4.5
		–	1:3 to 4	16.0–25.0	Nil
iii	M4	1:1:5 to 6	–	11.0–15.5	3.0–7.0
		–	1:5 to 6	11.5–16.5	Nil
iv	M2	1:2:8 to 9	–	7.5–10.0	4.0–8.5
		–	1:7 to 8	8.5–12.5	Nil

settlement, cracks develop, these will follow the joint lines, which can easily be cut out and repointed.

Excessively hard mortar can result in the bricks becoming fractured at settlement cracks, thereby leading to a more extensive repair operation. Another reason for relating strength of mortar to the compressive strength of the bricks is to ensure that external walls weather evenly during the lifetime of the building, with any absorbed water evaporating at a similar rate from the surfaces of bricks and joints alike.

In addition to being hard enough to transfer loads evenly between irregular surfaces of bricks, the choice of mortar must resist the effects of rain and frost in the long term (Table 6.3).

Mixing mortar

Machine mixing is the most effective way of turning cement/lime and sand, i.e. dry materials, into mortar on site. A typical small tilting drum mixer is shown in Figure 6.11. The mixer driver must be instructed in the importance of using gauge boxes of the correct size for every batch of mortar produced, and given the reasons why it is important to do so.

The description of this standard site mixer as a '5/3{1/2}' indicates the volume of dry materials put in and the volume of wet mortar discharged per batch (Figure 6.11).

Table 6.3 Mortar requirements

Good workability	Smooth and easy to handle with the trowel when transferring from spot board to wall, and applying cross-joints
Water retention	Mortar should contain sufficient fine particles of sand, together with cement and lime to prevent mixing water 'bleeding out' on the spot board
Adhesion	Mortar must stick to bricks and blocks to prevent rain penetration of joints
Durability	Mortar for externally exposed brickwork must resist the combined effects of rain, frost and any soluble sulphate salts in bricks of fired clay

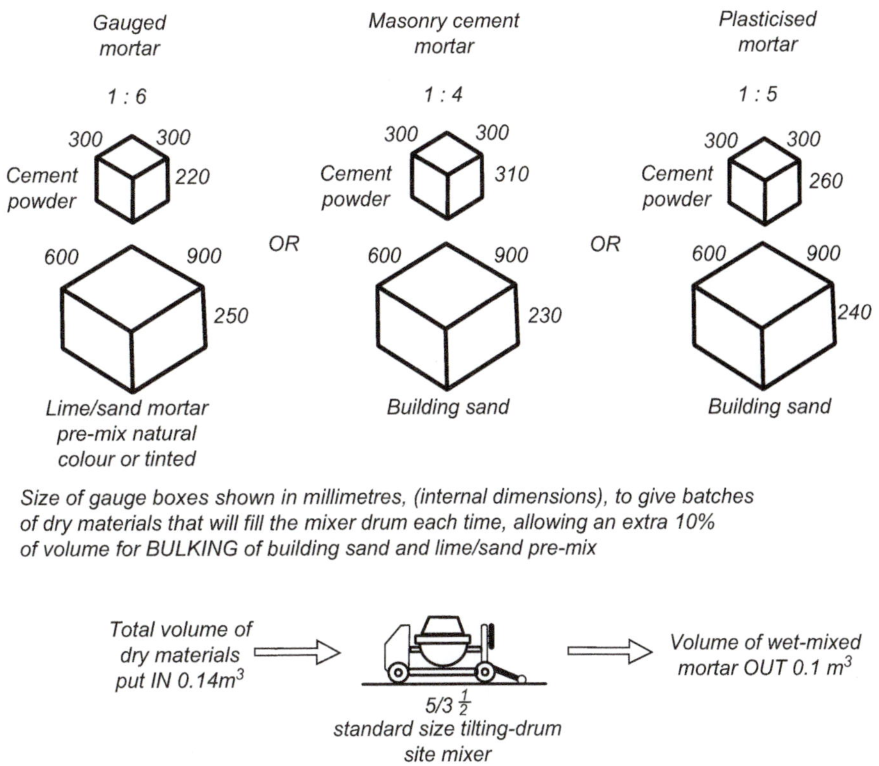

FIGURE 6.11
Three ways of producing a designation group (iii) M4 mortar

Mixing water

The water added to each batch of mortar is not measured in the same way because an exact water:cement ratio controls what is added to a batch of concrete.

A typical water:concrete ratio for concrete is 0.5. This means that the amount of mixing water allowed is 50 per cent of the weight of cement powder per batch. For example, with 50 kg cement, only 25 kg of water is permitted (25 litres).

Too much water weakens concrete.

Mixing mortar is a matter of what feels right: it must be workable with the trowel. Too dry and it will not spread into bed joints properly, too wet and it will smear the face of the bricks.

The experienced mixer driver knows the workability or consistency that bricklayers require, and he will soon be told if it is coming out wrong!

Batching

The proportions of dry materials for each mix of mortar must be carefully and regularly measured and batched to keep the strength and colour of the hardened mortar consistent. Variations can seriously affect durability of brickwork and result in conspicuous patchiness on completed facework elevations.

Bricks and mortar walls near the coast, for example, will need to be more resistant to the eroding effects of the weather than those sheltered by other buildings in towns and cities. Colour-matching or contrasting joints with the facing bricks, from the range of tinted mortar colours available, is another consideration.

Bulk deliveries of tinted lime-sand mortar, to which cement is added in carefully regulated proportions using a gauge box, to produce gauged mortar, or daily deliveries of tinted ready-to-use mortar, are both recommended methods.

Large-scale use of powder pigments added to the site mixer is not a reliable method, owing to the difficulty in ensuring the colour consistency of successive batches.

Measuring dry materials by the shovelful is totally unsatisfactory, as each will hold a different volume of cement powder or sand.

Figure 6.12 shows typical bottomless gauge boxes, the regular use of which will ensure that measurement or batching remains consistent where mortar is mixed on site. Placed on a flat surface behind the mixer, the larger box is filled with sand and struck off level. The smaller one, placed on the flat sand surface, is filled level with cement powder.

Both gauge boxes are lifted and the total contents transferred to the mixer by shovel. Figure 6.12 indicates the size of gauge boxes that will produce a designation group (iii) mortar/M4 mortar which will fill the standard size mixer used on sites.

Retarded ready-to-use mortar

Factory produced ready-to-use mortars fulfil the requirements of specifiers and users seeking factory-made materials. They are delivered to site ready to use in every respect and require no further mixing; no further constituents should be added. They have guaranteed mix proportions and overcome any potential problems relating to site mixing.

The photograph in Figure 6.13 shows a typical delivery of retarded ready-to-use bricklaying/blocklaying mortar, direct to sites.

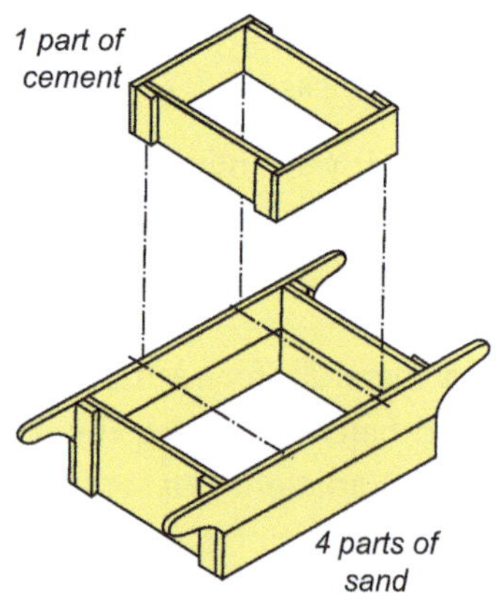

FIGURE 6.12
Gauge boxes

FIGURE 6.13
Delivery of retarded cement/lime/sand premix mortar. Note the delivery day marked on each polythene cover, to ensure use in rotation

The cement, lime and sand dry materials are mixed at a factory depot, complete with the required amount of water for optimum workability, plus a chemical retarder. This delays the onset of the initial setting action for 36 hours, while the mortar is in the delivery tubs.

Upon delivery, these 0.3 m^3 (600 kg) plastic tubs of mortar can be transported by forklift truck or tower crane directly to the point of use.

Ready-to-use mortar avoids the risk of possible carelessness with site batching that can be the cause of colour and strength variations in finished brickwork.

Bulk deliveries of lime/sand mortar plus bagged cement and a mechanical site mixer are all unnecessary if ready-to-use mortar is specified, a useful consideration on congested city centre sites.

Dry silo mortars

The modern approach on large mortar-using sites is to have a silo installed which is loaded with dry blended mortars that provide instant availability and consistent quality (Figure 6.14).

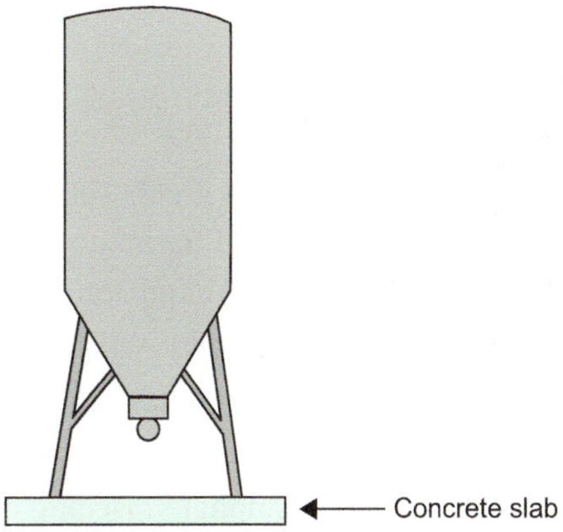

FIGURE 6.14
Dry mortar silo

This system requires space on the site to set up but reduces waste mortar as small quantities can be extracted. The silo has to be connected to water and electricity supplies.

Using a silo allows for increased consistency of the product whatever the environmental conditions, with minimal wastage and no risk of contamination of the mortar.

Any amount of mortar can be produced as required and uninterrupted by delays of any kind. The water content can be adjusted to allow for the absorption of the various bricks and blocks.

A concrete pad is required for the silo. It should measure at least 3 × 3 m and be constructed in an accessible position. The height of the silo is approximately 7.2 m and it may weigh between 33 and 35 tonnes when full.

The silo will be delivered and erected in position. It could be preloaded with up to 14 tonnes of dry mortar inside. A sensor is attached to the silo which tells the operator when there is only 10 tonnes of dry mortar remaining. The company providing the silo will provide a top-up service.

Mortar testing

Increasingly, quality control procedures call for compressive strength tests to be carried out on a regular basis, throughout the bricklaying operations on site. The same steel moulds familiar for testing concrete are used to make 100 × 100 × 100 mm mortar cubes for laboratory testing at seven- and 28-day intervals.

The Bricklayer section of a job specification will describe these mortar test requirements, and whether a copy of the mortar supplier's own test results will be acceptable, or whether an independent laboratory is to be appointed.

Typical compressive strengths of different mortar mixes were given in Table 6.2. These are results based on site tests, and will show a continuing slight increase in the weeks following a 28-day test.

It is worth noting that the typical compressive strength of ready-to-use mortar is ultimately 25–30 per cent higher than site-mixed mortar, owing to better batching control by weight rather than volume, and also because the slowed rate of initial setting and hardening improves hydration and eventual strength gain of the cement in the mortar.

Sulphate attack on Portland cement mortars

The use of sulphate-resisting cement prevents sulphate attack from developing in any brickwork that may be saturated for long periods (Table 6.4). Sulphate-resisting cement will therefore be required for brickwork below the damp-proof course level in subsoils that contain water-soluble sulphates.

Chimney stacks, parapets and free-standing boundary walls constructed of fired clay bricks, manufactured from a brick-earth that contains soluble sulphates, will also be at risk unless sulphate-resisting cement is specified.

Table 6.4 General recommendations for mortar mixes

Use	BS EN 988-2 mortar class	Types of mortar		
		Cement/ lime/sand	Masonry/ cement/sand	Plasticized cement sand
Class A and B engineering quality brickwork	M12	1:$^1/_4$:3	–	–
Work up to DPC level including extremely exposed brickwork	M6	1:$^1/_2$:4	1:3	1:4
External facing brickwork above DPC	M4	1:1:5–6	1:4	1:5
Internal walls of lightweight blockwork	M2	1:2:8	1:7	1:7

DPC: damp-proof course.

Reinforced mortars

This is a term used to describe admixtures such as styrene butadiene, which are added to cement/lime mortars to improve adhesion and water-resisting properties of brick- on-edge sills and copings. Great care must be exercised to follow the manufacturers' instructions carefully if these admixtures are to work effectively and be used safely.

Concrete

BS 5328 has been replaced by the new European Standard BS EN 206-1:2000 Concrete Part 1, which deals with producing fresh concrete.

The main changes are in the terminology. There are some new terms and some existing terms with new meanings. Table 6.5 explains some of the changes.

Concrete is an artificial rock made from a mixture of coarse aggregates, sand, a cement binder and water. It is one of the few building materials that can be produced on the building site. The appearance and properties of concrete are similar to those of limestone rock.

Other materials added to the mixer are referred to as admixtures.

The main advantage of using concrete is its versatility. It can be moulded to any required shape and its load-bearing capabilities can be increased by casting in steel reinforcing bars.

Table 6.5 New terminology

Old standards	New standards
Mix	Concrete
Strength or grade	Strength class
Slump or workability	Consistence (target or class)
PC/OCP	CEM1
20 mm aggregate	10/20
10 mm aggregate	4/10
Sand	0/2MF

There are several types of concrete:

- dense concrete
- lightweight concrete
- air-entrained concrete.

Constituents

The three main constituents used to manufacture concrete are:

- cement
- aggregates
- water.

Manufacture of concrete

The manufacture of hardened concrete involves two stages.

These are the plastic (setting) stage and the rigid (hardening) stage.

During both of these stages the chemical process of hydration occurs, where the cement reacts with the water.

The aggregate, although present, does not take part in the chemical reaction.

Cement + Aggregate + Water → Concrete + Heat

Plastic stage

Mixing

The constituents are mechanically mixed together in the correct proportions to give a homogeneous (same consistency throughout) concrete mixture.

During mixing, the cement and water produce a paste, and a film is formed around each aggregate particle. The finer aggregate particles fill the voids between the coarse aggregate particles.

Placing

The concrete mixture is placed into a mould to obtain the required shape.

Compaction

The concrete mixture may need to be vibrated to remove any air voids formed during placing. This is known as compaction of the concrete mixture.

Rigid stage

When the concrete mixture has set in the mould, the hardening process starts.

Curing

This is the process of retaining water in the concrete mix and maintaining the temperature of the concrete at about 20°C. This will ensure that the cement binds the aggregate particles together and that the concrete hardens at a favourable rate.

Curing is carried out by protecting the concrete from the weather. The exposed concrete surface is covered with a water-resistant material such as plastic or with damp canvas or hessian. This stops evaporation.

If the temperature drops below 5°C, the hardening process almost stops, and if the temperature is too high, the temperature difference between the concrete and the surroundings can cause cracking.

The concrete mix

Two essential properties of hardened concrete are durability and strength.

Both properties are affected by the voids or capillaries in the concrete which are caused by incomplete compaction or by excessive water in the mix.

It is important in the manufacture of concrete to mix the aggregates, cement and water in the correct proportions, in order to obtain the correct workability and strength required for the job.

A typical mix contains the following proportions of ingredients:

- 1 part cement
- 2 parts of fine aggregate
- 4 parts of coarse aggregate
- 0.5 part of water.

Type of mix

The proportions of the concrete may be mixed by one of two methods: by volume or by mass.

Mixing by volume

If the proportions of the ingredients are measured by volume, the mix obtained is known as a nominal mix.

Specified mixes are based on the aggregate and the cement being dry when measured. Specified examples of nominal mixes are:

- mass concrete - 1:3:6
- reinforced concrete - 1:2:4.

A minimum amount of cement is used in mass concrete to reduce costs. More cement is used in reinforced concrete to increase workability.

In nominal mixes there is always a 1:2 ratio of fine aggregate to coarse aggregate. This is to ensure that there is enough fine aggregate to fill voids between coarse aggregate particles.

The amount of water has not been specified but will be found from tests carried out on site as described.

Mixing by volume has poor control. This means that there is poor control over the proportions of ingredients because mixing by volume does not take into account the water content of the aggregate. Mixing by volume is not used very often these days.

Mixing by mass

For any concrete mix, the mass of each of the ingredients needs to be known accurately. These are called the batch quantities. An accurate knowledge of the batch quantities enables the properties of the concrete to be forecast to a good precision.

The properties of concrete are much better understood today than in the past. There are many varieties of concrete. This is because:

- the ingredients can be varied
- the relative proportions of the ingredients can be varied.

The properties of the concrete can be greatly modified by these two methods.

Consistence (workability)

Consistence is a new term introduced by EN 206-1 and covers the workability of concrete. In BS 5328, concrete was specified with a value of slump, e.g. 50 mm or 75 mm slump.

In the new standards, either a class or a target can be used to specify consistence. Table 6.6 shows the new consistence classes and the typical target values for each class.

Mixing

Concrete and mortar can be mixed by either hand or machine on site.

Mix proportions

Before concrete or mortar can be mixed together, the ingredients have to be measured in their correct proportions. This can be done either by:

Volume (hand mixing) or Weight (machine mixing)

Hand mixing

- Volume – the materials used in both concrete and mortar could be

Table 6.6 Consistence class and targets

Consistence class	Class range (mm)	Typical target value	Maximum variation allowed(mm)
S1	10 to 40	25 mm	–20 to +30
S2	50 to 90	70 mm	–20 to +30
S3	100 to 150	125 mm	–20 to +30
S4	160 to 210	185 mm	–20 to +30
S5	220	–	–

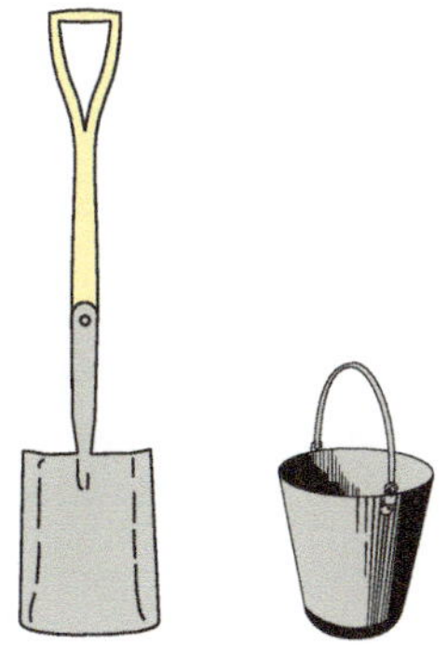

FIGURE 6.15
Shovel and bucket used for gauging

accurately 'gauged', i.e. measured into the correct quantities for each mix. There are several methods used for gauging materials:

- shovel: the method is rather crude, as the shovel size varies according to the material (Figure 6.15), e.g. you get larger shovels of sand than of gravel
- buckets: this is a much more accurate method, each bucket (Figure 6.15) being exactly the same volume
- gauge boxes: this method is similar to using buckets but two wooden boxes are made to the correct volumes for the specified mix; usually the boxes do not have bottoms (Figure 6.12).

Concrete volume batching

If you are using coarse aggregates, fine aggregates and cement then three gauge boxes are required.

- The first box is placed on the ground, filled with coarse gravel and levelled off.
- The second box is placed on the top of the first box, filled with fine sand and levelled off.
- The third and last box is placed on the top, filled with cement and levelled off.

If the volumes are calculated to work with a bag of cement then the third gauge box is not required.

The boxes are then lifted to leave the materials in a heap and ready for mixing.

Machine mixing

- Weight – by using a weigh batch mixer (Figure 6.16), the weight of the aggregates is recorded as they are shovelled into the hopper. This is much more accurate than any of the other methods.

The materials can be loaded into the hopper while the previous materials are being mixed. The hopper should be loaded in the following order:

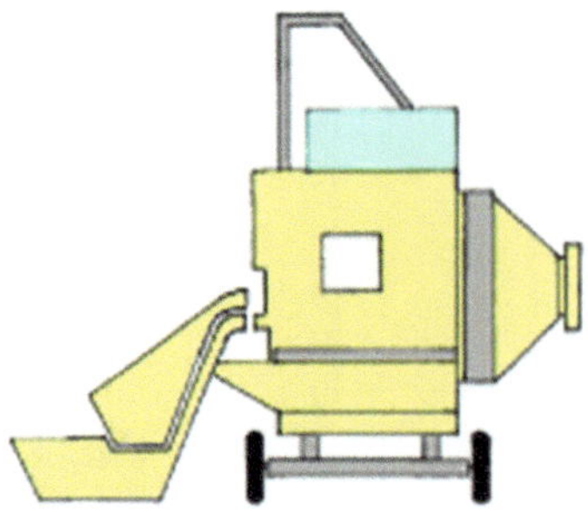

FIGURE 6.16
Weigh mixer

coarse aggregate - cement - fine aggregate.

This will keep the hopper clean and prevent the cement blowing away when the drum is loaded.

The proportions of the mix are usually adjusted to suit a bag of cement.

Example 1

A mixer with a drum capacity of 1 m^3 will produce 2240 kg of concrete.

Using a 4:2:1 mix the drum will take:

$$\text{Coarse aggregate} = \frac{2240 \times 4}{7} = 1280 \text{ kg}$$

$$\text{Fine aggregate} = \frac{2240 \times 2}{7} = 1280 \text{ kg}$$

$$\text{Cement} = \frac{2240 \times 1}{7} = 320 \text{ kg}$$

If the proportions are to be calculated using bags of cement then the following amounts could be used:

- cement = 300 kg (= 12 bags × 25 kg)
- fine aggregate = 600 kg
- coarse aggregate = 1200 kg.

The dial on the hopper would look like Figure 6.17.

Mix proportions

The mix for any job should be shown in the specification provided by the architect.

Measurements for concrete are in either cubic metres or square metres, and the mortar is measured in cubic metres:

- concrete for foundations = m^3
- concrete for paths, etc. = m^2
- mortar for brickwork = m^3.

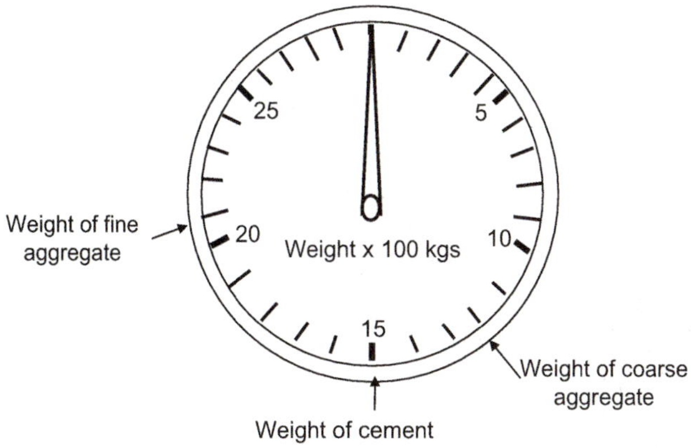

FIGURE 6.17
Dial on mixer

Formulae

$$\text{Area} = \text{Length} \times \text{Breadth}$$

$$\text{Volume} = \text{Length} \times \text{Breadth} \times \text{Depth}$$

Whenever you are dealing with quantities of materials for mortar or concrete by volume it is always taken for granted that the volume for sand is for dry sand.

Mixing time

The mixing time will vary according to whether it is by hand or machine and according to what is being mixed.

- Machine mixing – in general terms this should be between two and three minutes. On no account should the mix be allowed to stay in the machine longer as the materials will start to segregate.
- Hand mixing – this should take as long as required to ensure that all the particles have been covered and completely integrated together.

Bulking of sand

Remember that dry sand and saturated sand have the same volume. But when sand is damp it occupies a greater volume, that is, the sand swells up or bulks.

With a 5 per cent addition of water by weight to dry sand the increase in bulk can be 25–40 per cent.

A simple site test (Figure 6.18) can be carried out to find the percentage bulking of damp sand. Materials required are a glass jar, a sample of sand, water and a tape measure.

1. Fill the glass jar with 100 mm of damp sand.
2. Cover the sand with water and stir.
3. Allow to settle and measure the depth of saturated sand.
4. Check the difference and calculate the percentage bulked.

Remember

All materials should be used as soon as possible.

NEVER ADD EXTRA WATER.

Whenever you add water to a mix it will dilute the cement/lime and result in a weaker mix.

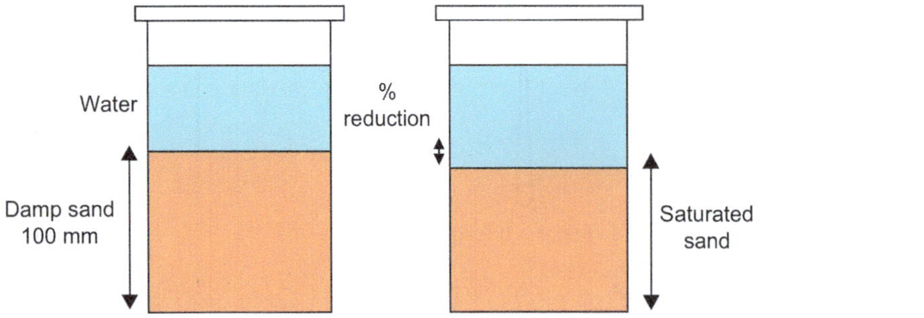

FIGURE 6.18
Bulking of sand

5. Result – this percentage must be allowed for when calculating the amount of water required in a mix.

Therefore, when using sand that is known to be damp, the estimator must increase the volume by a minimum of 25 per cent.

Water:cement ratio

1 litre of water weighs 1 kg.

The strength of concrete depends on the ratio of the weight of water in the mix to the weight of the cement:

$$\text{Water:cement ratio} = \frac{\text{Weight of water in the mix}}{\text{Weight of cement in the mix}}$$

Note

Coarse aggregate does not bul[k] at all.

Example 2

If a concrete mix has 50 kg of cement and the total water does not exceed 25 litres (i.e. 25 kg) the ratio will be:

$$\frac{25\ \text{kg}}{50\ \text{kg}} = 0.5$$

Remember

It is easy to add water to a mix bu[t] impossible to remove it.

The water content should not exceed 0.5 or the strength will suffer, but on site it quite often rises to 0. 7 through careless mixing.

Example 3

If the rate is given as 0.45 than the quantity of water per cent is:

50 kg × 0:45 = 22:50 kg or 22:50 litres

Testing concrete

The slump test

The simplest test for consistency/workability is the slump test and is one the site supervisor is most likely to use (Figure 6.19).

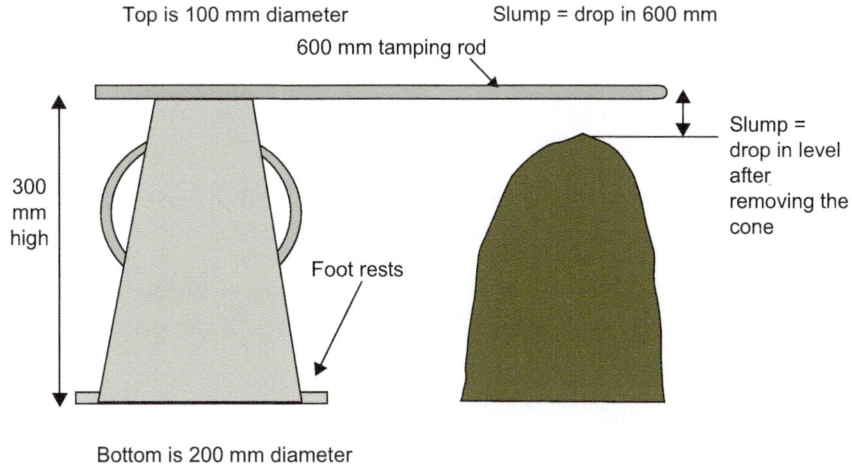

FIGURE 6.19
Slump test

The cone used is 300 mm high, 200 diameters at the base and 200 diameters at the top.

In the slump test, the distance that a cone full of concrete slumps down is measured when the cone is lifted from around it. The slump can vary from nil on dry mixes to total collapse on very wet mixes.

This highlights one drawback with the slump test, in that it is not very helpful with very dry mixes. Within this limitation, the slump test lets you compare the workability of each batch tested with earlier batches and it should stay about the same, generally within about 25 mm of the intended value.

You can check that concretes that are supposed to be the same really are the same.

You must always make sure the slump test is conducted in accordance with the following procedure:

1. Make sure the cone is clean, free from hardened concrete and dry inside. Stand it on the base plate, which also must be clean.
2. Stand with your feet on the foot rests.
3. Using the scoop fill the cone to about one-quarter of its height and rod this layer of concrete exactly 25 times using the special tamping rod. This tamping rod is 16 mm in diameter and 600 mm long with one end rounded. Add three further layers of equal height (each about 75 mm deep), rodding each one in turn exactly 25 times, allowing the rod to penetrate through into the layer below. After rodding the top layer make sure that there is a slight surcharge of concrete, i.e. that some concrete sticks out of the top.
4. Strike off the surplus concrete using a steel float.
5. Wipe the cone and base plate clean, keeping your feet still on the foot rests.

6. Take hold of the handles and pushing downwards remove your feet from the foot rests.
7. Very carefully lift the cone straight up, turn it over and place it down on the base plate next to the mound of concrete. As soon as the cone is lifted the concrete will slump to some extent.
8. Rest the tamping rod across the top of the empty inverted cone so that it reaches over the slumped concrete.
9. Using a ruler measure from the underside of the rod to the highest point of the concrete, to the nearest 5 mm. If the distance is, say, 50 mm, this represents a 50 mm slump.

There are three kinds of slump (Figure 6.20):

- true slump - where the concrete just subsides, keeping its approximate shape
- shear slump - where the top half of the cone shears off and slips sideways down an inclined plane, possibly because the mix was too dry
- collapse slump - where the concrete collapses completely, because the mix was too wet.

Both true and shear slump can happen with the same mix, but one must not be compared with the other. The only one that is of any use is the true slump.

If you get a shear slump, you should conduct a second test to obtain a shape closer to the true slump. If this also shears, it is probably due to the design of the mix and you should record this fact on the report. Similarly, collapsed slumps should be recorded as 'collapsed slumps'.

Testing of hardened concrete: cube moulds

The most common test for hardened concrete involves taking a sample of fresh concrete and putting it into special cube moulds so that, when hard, the cubes can be tested to failure in a special machine to measure the compressive strength of the concrete (Figure 6.21).

Cube moulds are available in two sizes:

- 150 × 150 mm
- 100 × 100 mm.

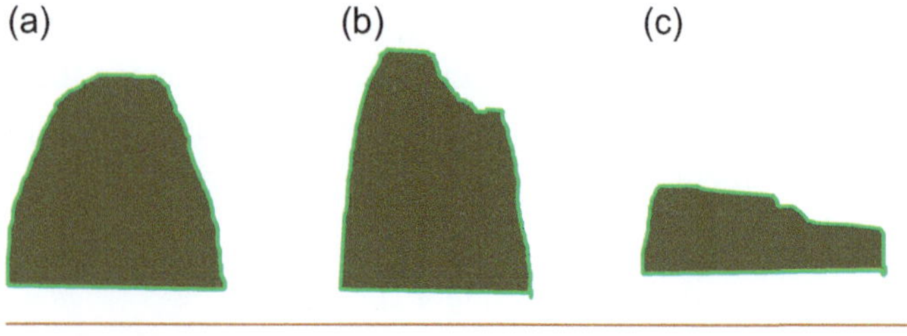

FIGURE 6.20
Types of slump: (a) true slump; (b) shear slump; (c) collapse slump

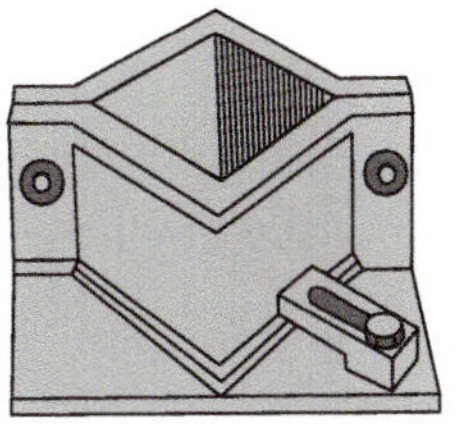

FIGURE 6.21
Concrete mould

Only use moulds made of steel or cast iron.

A cube mould is usually in two halves which bolt together, the whole thing being fastened by clamps to a separate metal plate forming the base.

The inside faces of the moulds have been machine planed to a high degree of accuracy and if they become pitted, scratched or out of plane, the concrete cube can be misshapen, giving an incorrect compressive strength when tested.

When assembling the moulds, thinly coat the internal faces with mould oil to prevent leakage during filling and to prevent the concrete sticking to the moulds.

You must make sure at all times that cube moulds are well looked after. Clean and oil them after every use and never leave them in the open to become rusty.

Sampling the concrete

Whenever possible, take the sample as the concrete is being discharged from the mixer or truck mixer. Collect the sample from various parts of the mix and thoroughly mix the sample together.

You will require approximately 10 kg of concrete to make one 150 mm cube. You should record the date and time you took the sample.

Record the following:

- the temperature and weather conditions
- the batch number and the mixer
- the place on the job where the remainder of the concrete was used.

Filling the moulds

After thoroughly remixing the sample, fill the moulds and compact them as soon as possible.

A 150 mm cube mould should be filled in three layers and a 100 mm cube in two layers.

When compacting by hand, each layer should be rammed with at least 35 strokes for the 150 mm cube and at least 25 strokes for the 100 mm cube, with a steel bar 380 mm long, weighing 1.8 kg and having a ramming face 25 mm square.

Alternatively, the concrete can be compacted by vibration, again in layers, using a suitable table vibrator.

After compaction the surface of the concrete should be trowelled as smooth as possible, level with the top of the mould.

Each cube should be clearly and indelibly marked for later identification.

Initial curing

Immediately after making, the cubes should be stored in a place free from vibration under damp matting, or similar, and wrapped completely with plastic sheeting to prevent loss of moisture.

Cubes to be tested at an age of seven days or more should be kept at a temperature of 20 ± 5°C, while cubes to be tested at earlier age should be kept at a temperature of 20 ± 2°C.

Cubes should be demoulded within 16–28 hours after the time of making. Each cube should then be immediately submerged into a tank of clean water maintained at a temperature of 20 ± 2°C until the time for testing.

Testing of the cubes

1. Cubes should be tested immediately upon removal from the water. Surface water, grit and projecting fins should be removed and the dimensions and weight recorded.
2. The bearing surface of the testing machine should be wiped clean and the cube should be placed on the machine in the centre so that the load is applied to faces other than top and bottom of the cube as cast.
3. The load must be applied without shock and increased continuously at a rate within the range of 0.2–0.4 $N/(mm^2.s)$ until no greater load can be sustained. The maximum load applied to the cube is recorded.
4. The compressive strength is recorded to the nearest 0.5 N/mm^2
5. It is important to keep the testing machine in good working order. The compressive strength is then calculated as follows:

Remember

If the load is measured in kN and the cross-section in mm^2, then the strength is given in units of kN/mm^2.

$$\text{Compressive strength} = \frac{\text{Load at failure}}{\text{Area of cross} - \text{section}}$$

Quantities of materials

The quantities of materials must now be considered and calculated. The design proportions alter according to the type of aggregate used:

- If the aggregate arrives on site as fine and coarse aggregates then three items are used.
- But most sites order aggregates which have been mixed together and are known as 'ballast'. Within these mix designs there will only be two items.

The following designs assume that the aggregates are delivered to the site as fine and coarse aggregates:

In a concrete mix of the proportions 1:2:4:

- 1 part cement
- 2 parts fine aggregate
- 4 parts coarse aggregate.

In a concrete mix of the proportions 1:1:2:

- 1 part cement
- 1 part fine aggregate
- 2 parts coarse aggregate.

In a concrete mix of the proportions $1{:}1^1/_2{:}3$:

- 1 part cement
- $1^1/_2$ parts fine aggregate
- 3 parts coarse aggregate.

If ballast was used then the designs may be as follows:

- 1 part cement
- 6 parts ballast
- 1 part cement
- 3 parts ballast
- 1 part cement
- $4^1/_2$ parts ballast.

Yield

When mixing concrete from coarse aggregate, fine aggregate, cement and water, a loss of volume will occur as a result of the fine aggregate and cement filling the spaces (voids) between the coarse aggregate particles.

A simple test can be carried out to find the percentage reduction in volume. If the proportions of a mix were 1:2:4, there would be seven parts in total.

1. Fill seven equal-sized containers with the appropriate dry materials (Figure 6.22).
2. Empty out all the containers on to a mixing board and mix 'dry'.
3. Return the 'mixed' material into the containers.
4. Check how many containers could be filled.
5. Tip the materials back onto the mixing board and mix with water.
6. Fill the containers again.
7. Check how many containers can now be filled.

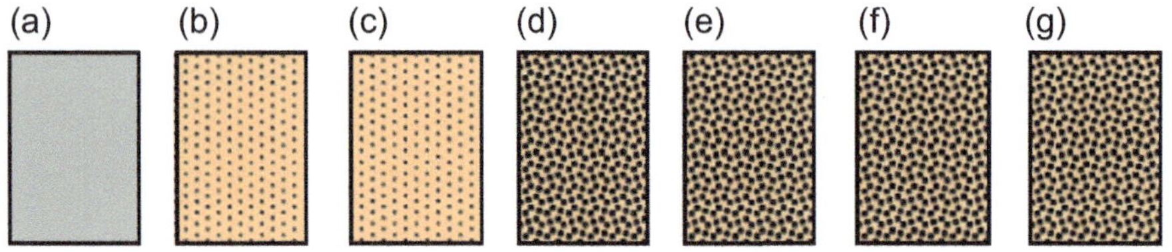

FIGURE 6.22
Yield test: (a) cement; (b) fine aggregate; (c) fine aggregate; (d) coarse aggregate; (e) coarse aggregate; (f) coarse aggregate; (g) coarse aggregate

Results:

Mixed dry there were ... containers.

Mixed wet there were ... containers.

The approximate reduction was ... %.

When calculating the volume of concrete required for a given task the percentage reduction must be added to the quantities of dry material required.

The yield has been found by practice to be 50 per cent of the original dry volume.

Typical concrete mixes

Working out the quantities of materials in volume

Before working out proportions of cement, sand and aggregate for a mix it is important to remember the following:

- 1 litre of water = 1 kg
- 1 bag of cement = 25 kg
- 50 kg cement = 0.035 m^3
- 1 m^3 of concrete = 2400 kg
- 1 m^3 ballast = 1441 kg
- 1 m^3 cement = 1441 kg
- 1 m^3 sand = 1600 kg.

Remember to add 50 per cent to allow for the voids.

Using the above information the quantities of materials can now be considered and calculated.

Basing the quantities on 50 per cent voids, a cubic metre must be increased by 50 per cent:

$$1.00\ m^3 + 50\% = 0.50\ m^3$$

$$50\%\ \text{increase} = 1.00\ m^3 + 0.50\ m^3 = 1.50\ m^3$$

or

$$50\%\ \text{increase} = 1.00\ m^3 \times 150\% = 1.50\ m^3$$

Example 4

Consider a concrete mix of the following proportions:

- 1 cement
- 1 fine aggregate
- 2 coarse aggregate

= 4 parts in total.

- The cement per m^3 will be 1/4.
- The fine aggregate per m^3 will be 1/4.
- The coarse aggregate per m^3 will be 1/2.

Proof: $1/4 + 1/4 + 1/2 = 1\ m^3$

The gross quantities will now be:

Cement $1/4 \times 1.50\ m^3 = 0.375\ m^3$

Fine aggregate $1/4 \times 1.50\ m^3 = 0.375\ m^3$

Coarse aggregate $1/2 \times 1.50\ m^3 = 0.750\ m^3$

Total = $1.5\ m^3$

Example 5

Calculate the raw materials required to produce 1 m^3 of concrete (1:2:4) mix.

Answer 5

Basing quantities on 50 per cent voids the cubic metre must be increased by 50 per cent.

Therefore $1\ m^3 + 150\% = 1.50\ m^3$

In concrete (1:2:4) the parts will be:

- 1 cement
- 2 fine aggregates
- 4 coarse aggregates

= 7 parts in total.

- The cement per m^3 will be 1/7.
- The fine aggregates per m^3 will be 2/7.
- The coarse aggregates per m^3 will be 4/7.

Proof: $1/7 + 2/7 + 4/7 = 1\ m^3$

The gross quantities will be as follows:

Cement $1/7 \times 1.50\ m^3 = 0.214\ m^3$

Fine aggregate $2/7 \times 1.50\ m^3 = 0.428\ m^3$

Coarse aggregate 4/7 × 1.50 m^3 = 0.858 m^3

Total = 1.50 m^3

BUT: Since materials are bought in kg not in m^3 it is necessary to give the answer in kg.

Cement = 0.214 m^3 × 1441 kg per m^3 = 308 kg

Fine aggregates = 0.428 m^3 × 1600 kg per m^3 = 684 kg

Coarse aggregates = 0.858 m^3 × 1441 kg per m^3 = 1236 kg

Example 6

Calculate the materials required for the foundation shown below, using a 1:6 mix.

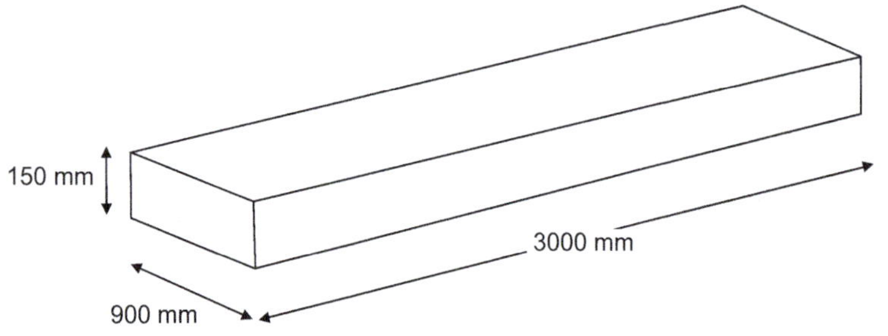

Remember

Concrete calculations – always calculate the volume required first then add 50 per cent to allow for the voids to find the total volume required.

Answer 6

Volume of the concrete foundation = Length × Width × Depth = m^3

= 3 × 0.9 × 0.15 m^3 = 0:405 m^3

Add 50 per cent for voids = 0.405 + 50% = 0.6075 m^3

Using a mix of 1:6 the parts will be as follows:

- 1 cement
- 6 aggregates

= 7 parts in total

- The cement per m^3 will be 1/7.
- The aggregates per m^3 will be 6/7.

The gross quantities will be as follows

Cement 1/7 × 0.6075 m^3 = 0.0868 m^3

Aggregates 6/7th × 0.6075 m^3 = 0.5207 m^3

Total = 0.6075 m^3

Changing to kg:

Cement = 0.0868 m^3 × 1441 kg per m^3 = 125 kg

Table 6.7 Volume batching

Mix	Cement	Fine aggregate	Coarse aggregate	Uses
1:3:6	50 kg	0.10 m^3	0.20 m^3	Foundations
1:2:4	50 kg	0.07 m^3	0.14 m^3	Reinforced concrete
1:1½:3	50 kg	0.05 m^3	0.10 m^3	Waterproof concrete

Table 6.8 Weight batching

Mix	Cement	Fine aggregate	Coarse aggregate	Uses
1:3:6	50 kg	150 kg	300 kg	Foundations
1:2:4	50 kg	100 kg	200 kg	Reinforced concrete
1:1½:3	50 kg	75 kg	150 kg	Waterproof concrete

$$\text{Aggregates} = 0.5207\ \text{m}^3 \times 1441\ \text{kg per m}^3 = 750\ \text{kg}$$

Volume batching and weight batching data are shown in Tables 6.7 and 6.8, respectively

Typical mortar mixes

Mortar is a mixture of the following materials:

- sand, cement and water
- sand, lime and water
- sand, lime, cement and water
- sand, cement, plasticizer and water.

It is used for the bedding and jointing of all masonry, including bricks, blocks and stone. In site practice terms it is possibly the least understood and most abused material on the building site.

The final strength of the mortar is determined by the strength of the brick or block to be bedded in it. The mortar strength should roughly match that of the brick or block and in no case should it be stronger than it.

From a bricklayer's point of view a mortar should be 'fatty', i.e. mortar that will handle well without being sticky, spreads easily and sets at the right pace to allow time to treat the joints in all weather conditions. If a mortar satisfies all these requirements it is said to have 'workability'. In achieving workability, other factors must not be overlooked:

- adequate compressive strength
- adequate bond between the mortar and the brick

Table 6.9 Recommended uses of mortar mixes

Class A and class B engineering quality brickwork	(1) M12
Work up to DPC level, including extremely exposed brickwork, e.g. chimney stacks, parapets, free-standing boundary walls, brick- on-edge sills and copings	(2) M6
External facing and common brickwork above DPC level and internal walls of medium-density blockwork	(3) M4
Internal walls of lightweight blockwork	(4) M2

- durability - resistance to frost and chemical attack
- joints sealed against the driving rain and wind
- an appearance that is complementary to the bricks.

You cannot expect to find all these properties in the highest degree in a single mortar mix. For every job you do, you will have to make up your mind which of them are the most important and then decide upon the most suitable mix.

Mortar will work easily if it has lime in it: the more lime it contains, the more workable it will be.

Mortar will stiffen quickly if there is cement in the mix: the more cement the more quickly it will stiffen.

When you require the highest possible strength you should use cement and sand mortar.

Mortar mixes are designed as shown in Table 6.4. As they progress from M12 to M2 they become progressively weaker, but with a higher lime content they become more tolerant to movement within the structure. Table 6.9 shows the recommended mortar mixes for certain areas of brick and blockwork.

Workability of the mortars listed here can be improved by adding plasticizers. These should never be used on the designated 1 mix or very weak mortars. Plasticizers should not be used in structural work without the consent of the structural engineer.

Mortar calculations

The nominal size of a brick is 215 × 102.5 × 65 mm:

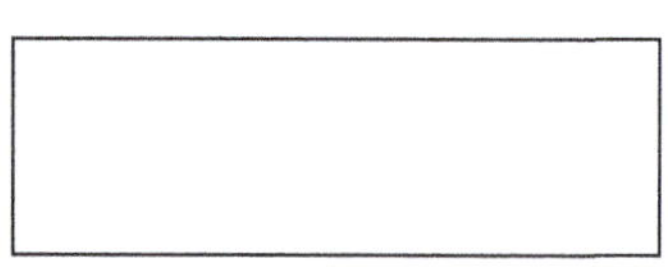

The nominal size of a joint is 10 mm:

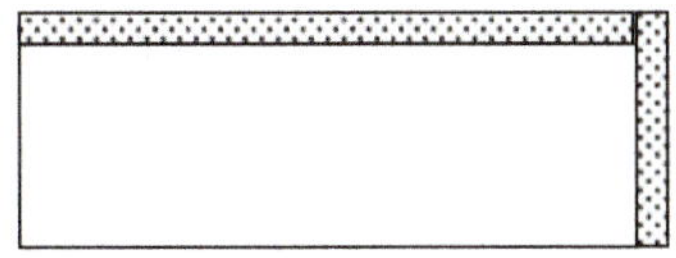

If brick size = 215 × 102.5 × 65 mm

Then

Bed joint = Length × Width × Thickness of bed joint

$= 0.125 \times 0.1025 \times 0.01 = 0.0002\ m^3$

Cross-joint = Width × Depth × Thickness of cross-joint

$= 0.1025 \times 0.075 \times 0.01 = 0.00007\ m^3$

Therefore, the total for one bed joint and one cross-joint

$= 0.0002 + 0.00007 = 0.00027\ m^3$

If the calculation was for 1000 bricks, the mortar required

$= 0.00027 \times 1000 = 0.27\ m^3$

Note

The normal allowance is around 0.6 m^3 per 1000 bricks, which allows for the frogs and waste.

Another method to simplify the calculation if to allow 1 kg of mortar for each brick.

It follows that 1000 bricks would require 1000 kg of mortar, which is 1 tonne.

Brick calculations

The size of a brick including the joints = 225 × 112.5 × 75 mm.

If we find the face area of one brick and divide that into 1 m^2 it will result in the brick requirement.

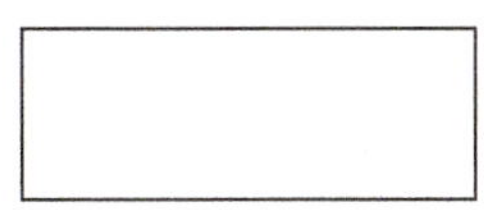

Area of the stretcher face of a brick

$= 0.225 \times 0.075 = 0.01687\ m^2$

Area of a half-brick wall

$= 1\text{ m} \times 1\text{ m} = 1\text{ m}^2$

Therefore,

$$\frac{1}{0.01687} = 59.27 \text{ bricks}$$

Allowing for waste = 60 bricks per metre for a half-brick thick wall. The other brick requirements are:

- 1-brick wall = 120 bricks
- 1 1/2-brick wall = 180 bricks.

Equipment required for mixing

Hand mixing

Whenever materials are to be prepared for mixing, it is essential to have available the correct tools and equipment to carry out the operation safely and efficiently.

You could be asked to prepare and mix the following materials:

- mortars
- concretes.

To mix these the following tools and equipment (Figure 6.23) may be required:

- Barrows - there are various types available but the main difference is often the type of wheels, i.e. solid or pneumatic. Barrows are used for transporting both the raw and mixed materials around the building site.
- Shovels - again, there are many different types, with the weight being the main difference.
- Buckets - buckets are available in various materials from plastic to steel and are mainly used for carrying water around the building site.

FIGURE 6.23
Equipment required for mixing by hand

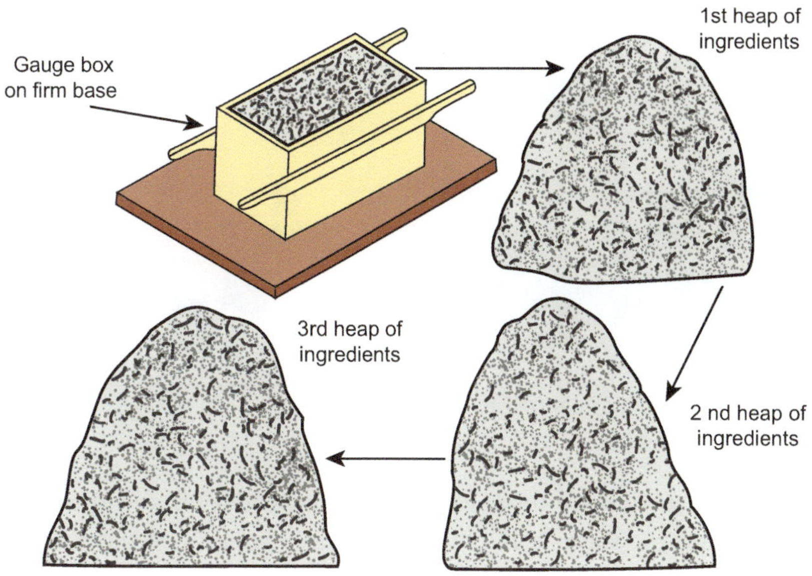

FIGURE 6.24
Mixing the materials dry

- Gauge boxes - used to produce the correct volume of raw materials for each mix. Several sizes should be available for each item of raw material.

Area

An area should be selected which is suitable for hand mixing both mortar and concrete.

The area selected for mortar or concrete mixing should be hard, level and clean. The most suitable would be a concrete slab which will not matter if it becomes stained.

The area should be swept clean of all rubbish so that there is no contamination of the mortar or concrete mix.

Mixing concrete or mortar by hand

Small quantities of mortar and concrete will often have to be mixed by hand. A clean, hard surface should always be used.

After the ingredients have been correctly gauged out they should be turned over with a shovel into a pile, ensuring good integration of the materials (Figure 6.24).

Repeat the process THREE times.

The main object of this is to distribute evenly the cement or lime with the particles of aggregate. At this point the mix should be fully integrated

Assume that a mix of 1:4 is required. The gauge box is filled three times with sand and once with cement, then once with sand. These should be turned out into one heap.

(a) (b)

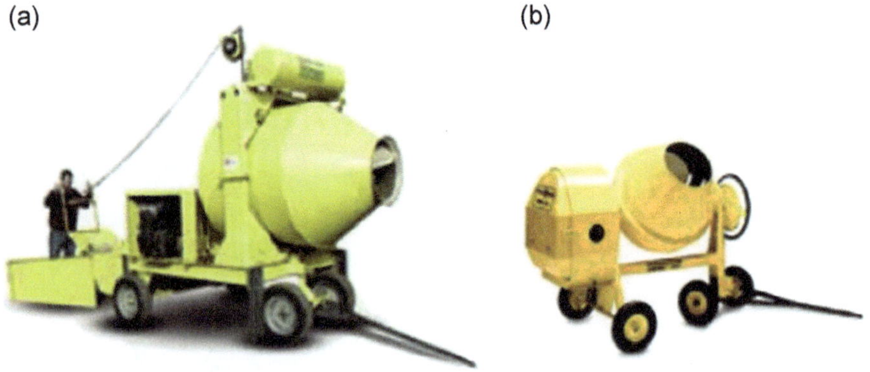

FIGURE 6.25
Types of mixers: (a) weigh batch mixer; (b) small portable mixer

The dry mix should be turned again while water is added from a sprinkler rose. This prevents the water from pooling and washing all the cement or lime off the aggregate.

Continue turning the mix until sufficient water has been added to produce a good, well-integrated and workable mix. It is vital to ensure that the water is not allowed to escape, taking cement with it and reducing the strength of the mix.

Remember that the amount of water to be added should be about 0.6 by weight of the amount of cement. Adding extra water will make the concrete easier to lay but will reduce the strength by up to 30 per cent.

Remember

All concretes and mortars shou be used as soon as possible aft mixing, for maximum strength.

The amount of time for mixing depends on the amount being mixed. Hand mixing is often more economical for small than for large mixes.

Mixing materials by mechanical mixer

There are many different types of mixer, from small portable mixers suitable for the small builder up to very large static mixers suitable for very large sites where the concrete is being mixed on site.

Mixers can be powered by petrol or diesel, and the smaller portable ones by electricity. Typical mixers are shown in Figure 6.25.

Note

NEVER add water to a mix that ha started to set.

Whenever using a mixer always ensure that safety measures are taken:

1. Set the mixer up according to the instructions of the manufacturer or your site supervisor.
2. Ensure that you have sufficient materials and small tools.
3. Start the mixer.
4. Add the required water.
5. Add the pregauged materials or separate materials in the following order:
   Concrete: coarse aggregate - cement - fine aggregate
   Mortar: half the sand - cement - half the sand.

6. Add more water if required.
7. Allow the materials to mix for three minutes - never any longer.
8. Turn out the mix into a wheelbarrow.
9. Repeat steps 2-6 until work is completed.
10. When completed clean out the mixer.

Premixed mortars

Many sites now have ready-mixed mortars delivered each day according to daily consumption.

These have the advantage of not having to provide the raw materials and a mixing area. It can also save on storage and the problem of material deliveries and handling.

There is a downside in that it is more expensive than mixing on site, so it is important to order the correct amount to avoid wastage. In addition, site mixing can be stopped for inclement weather, whereas premixed mortars cannot.

Advantages

- No space is required for the mixer and storage of materials.
- A smaller amount of transportation plant is required.
- A smaller labour force is required.
- Quality control of the materials is the responsibility of the ready-mix supplier.
- Economical for medium-volume requirements.

Disadvantages

- Good access is required.
- Delivery may be difficult to obtain because more than one day's notice may be required.
- There is no check on the quantities other than the delivery ticket.
- Vehicles delivering part loads are uneconomical.

FIGURE 6.26
Small mixer

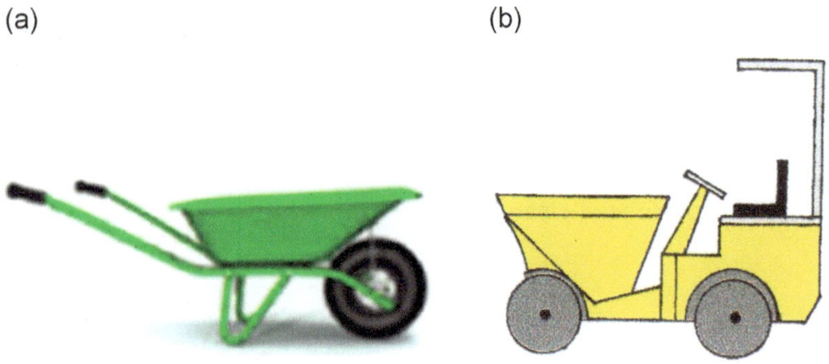

FIGURE 6.27
(a) Barrow; (b) dumper

FIGURE 6.28
Front loader

Transporting mortar

When mortar has been adequately mixed it requires discharging from the mixer (Figure 6.26). and then transporting around the site to the various positions of work.

Mortar may be transported around the site in a number of ways.

- Barrow - the most common method on small building sites is the wheelbar- row (Figure 6.27).
- Dumper - there are many sizes and designs of dumper, which can be used for transporting mortar and concrete from mixers up to 600 litre capacity (Figure 6.27).
- Cranes - these are available in all makes and sizes according to the job in hand. Cranage can be either mobile or fixed. Cranes can span the whole building area so are excellent for the placing of concrete and mortar on large multistorey buildings. There are also several types of machine available to move mortar and concrete around construction sites, such as front loaders (Figure 6.28).

Concrete

Concrete can be either mixed on site or premixed off site.

BS 5328 has been replaced by the new European Standard BS EN 206-1:2000 Concrete Part 1, which deals with producing fresh concrete.

The main changes are in the terminology. There are some new terms and some existing terms with new meanings. Table 6.5 explains some of the changes.

Concrete is one of the few building materials that can be produced on the building site.

The main advantage of using concrete is its versatility. It can be moulded to any required shape and its load-bearing capabilities can be increased by casting in steel reinforcing bars.

There are several types of concrete.

- dense concrete
- lightweight concrete
- air-entrained concrete.

Constituents

The three main constituents used to manufacture concrete are:

- cement
- aggregates
- water.

Manufacture of concrete

The manufacture of hardened concrete involves two stages: the plastic (setting) stage and the rigid (hardening) stage. During both of these stages the chemical process of hydration occurs, where the cement reacts with the water.

The aggregate, although present, does not take part in the chemical reaction.

Cement + Aggregate + Water → Concrete + Heat

Formwork

Apart from when concrete is used for foundations, where the ground supports the side of the concrete, formwork is required to hold the concrete in place while it sets.

The formwork and the way it is made and used play a great part in the finished appearance. Apart from its appearance, formwork usually needs to be used many times and this can only be done if it is carefully and properly handled, cleaned and stored.

The materials that can be used include timber, steel and fibreglass.

Formwork should:

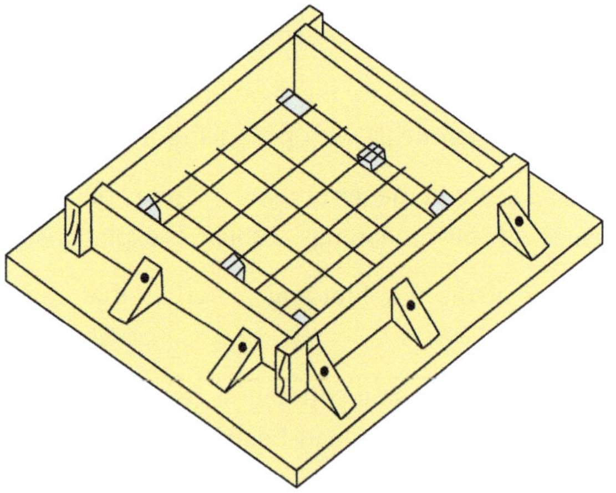

FIGURE 6.29
Formwork for small slab

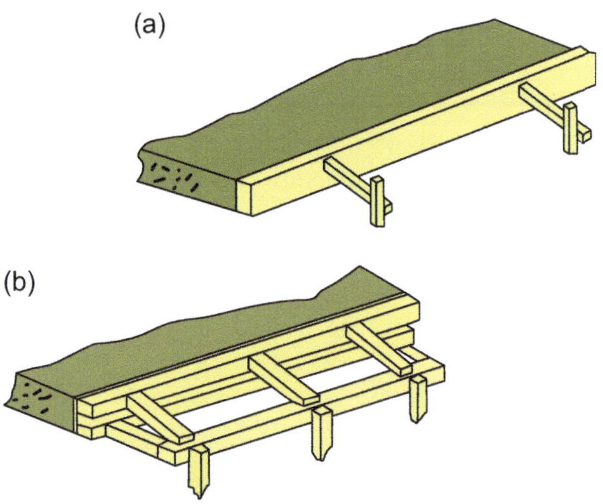

FIGURE 6.30
Timber formwork: (a) simple timber formwork; (b) timber formwork to support more pressure

- be strong enough to resist any pressure from the wet concrete
- be designed to be easily erected and dismantled without damage
- be of reasonable shape and size to be easily placed and removed
- have reasonable joints to prevent water leakage and water loss.

Formwork for small slabs, paving slabs, etc., can be constructed as shown in Figure 6.29. Formwork for the sides of large slabs may consist of timber beams supported with timber pegs (Figure 6.30a). An alternative method is used where there is a great deal of pressure from the concrete (Figure 6.30b).

Steel formwork is available but it is not as versatile as timber. It is purchased in off-the-shelf sizes. Steel formwork offers a simple means of dealing with repetitive work and can be reused quickly (Figure 6.31). All shapes and sizes are available, especially for curved work.

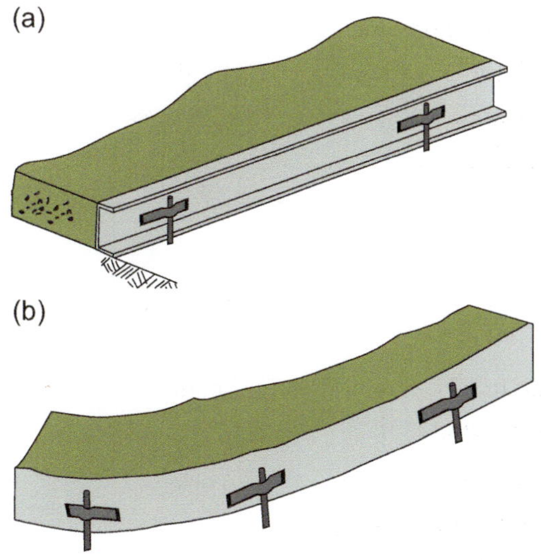

FIGURE 6.31
Steel formwork: (a) simple steel formwork; (b) curved steel formwork

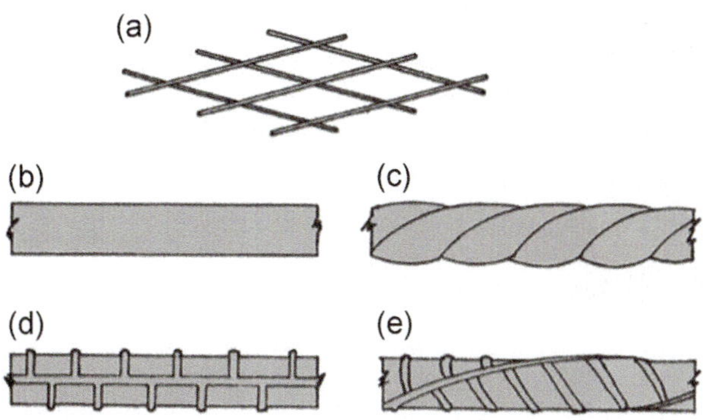

FIGURE 6.32
Types of steel reinforcement: (a) mesh reinforcement; (b) plain; (c) twisted; (d) ribbed; (e) ribbed and twisted

Reinforcement

The position of reinforcing within a floor slab or lintel should be specified.

All reinforcing requires cover from the concrete to protect it from rusting. Simple floor slabs are constructed with a reinforcing mat which is placed in position on spacers before the concrete is placed. It is important to ensure that the steel used for reinforcing concrete is free from mill scale, loose rust, grease or site mud, before placing it.

Various types of reinforcing bars are shown in Figure 6.32.

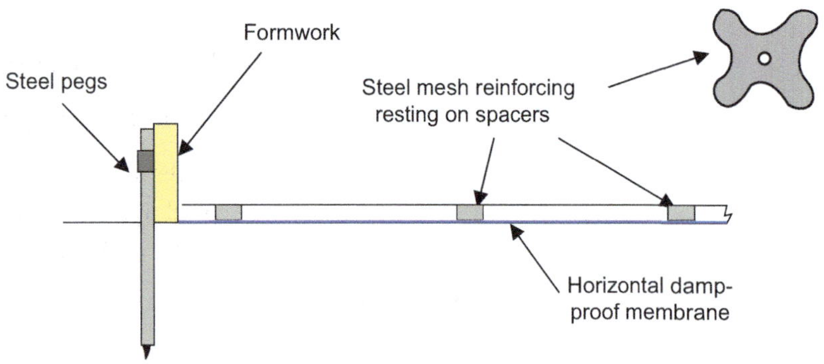

FIGURE 6.33
Simple formwork for a concrete ground floor slab

Release agent

This is applied to formwork to allow the removal of the formwork without damage to the face of the concrete. It is applied to the inside face of the formwork immediately before placing the concrete.

It is available as a liquid or a cream. The type used depends on the required finish. Care must be taken not to stain the finished face of the concrete.

Damp-proof membrane

When concrete is used for ground floor slabs it may be necessary to lay a horizontal membrane to protect the floor from rising damp (Figure 6.33).

Striking formwork

This is the term used when removing the formwork after the concrete has set.

Vertical sides of concrete slabs, etc., can be removed after 12 hours, but any soffits should be left for 7–14 days depending on the type of concrete.

Remember to clean and stack away all formwork.

Transporting concrete

Concrete may be transported from the mixer or ready-mixed concrete lorry to the point of placement in a number of ways, some specifically developed for concrete, others being multipurpose.

The chosen method will depend on the size of the site and the position where the concrete is required. In all cases the concrete must be transported so that it does not segregate and so that it does not become contaminated with any other material after it has left the mixer.

Where concrete is to be placed below the level of supply, a chute can be used because gravity is the cheapest method of transport.

Concrete should generally be placed within half an hour of mixing.

FIGURE 6.34

Transporting machinery: (a) dumper; (b) ready-mixed concrete lorry; (c) barrow; (d) lorry-mounted concrete pump; (e) crane; (f) front loader

Note

Never add too much water, or it will dry out and leave voids which will weaken the concrete.

Before concrete is placed into any mould or formwork certain preparations may have to take place.

Methods of transport range from hand wheelbarrows to concrete pumps (Figure 6.34).

- Barrow – The most common method on small building sites is the wheelbarrow.
- Dumper – There are many sizes and designs of dumper, which can be used for transporting concrete from mixers up to 600 litre capacity.
- Ready-mixed concrete lorry – When ready-mixed concrete is prescribed, large amounts can be delivered either ready mixed or site mixed to most site positions.
- Pumps – Concrete pumps are used to transport large volumes of concrete (up to 100 m^3) in a short period, in both vertical and horizontal directions.
- Cranes – These are available in many makes and sizes according to the job in hand. Cranage can be either mobile or fixed. Cranes can span the whole building area so are excellent for the placing of concrete on large multistorey buildings. There are also several types of machine

Remember

There is a danger of segregation, so concrete should not be dropped more than 1 m.

available to move concrete around construction sites, such as front loaders.

Placing concrete

The concrete mixture is usually placed into formwork to obtain the required shape.

The workability of fresh concrete is important in assessing the practicability of placing and compacting the mix, and also in maintaining consistency throughout the job.

The inside of forms should be inspected before concrete is placed to make sure that they are clean and have been treated with a release agent. Rubbish such as sawdust, shavings and wire should be blown out before the release agent is applied and the concrete is placed.

The formwork into which the concrete is to be placed should have been treated with any release agent necessary to allow its removal from the hardened concrete.

Concrete should be placed into its final position as quickly as possible.

Remember

On no account add water to a m after it has left the mixer or read mixed concrete lorry.

Compacting concrete

The concrete mixture may need to be vibrated to remove any air voids formed during placing. This is known as compaction of the concrete mixture.

If concrete is to reach its maximum strength, it must be compacted so that it contains the minimum amount of unwanted air. This is easy with wet concrete but the extra water contributes to weakness. The strongest and most durable

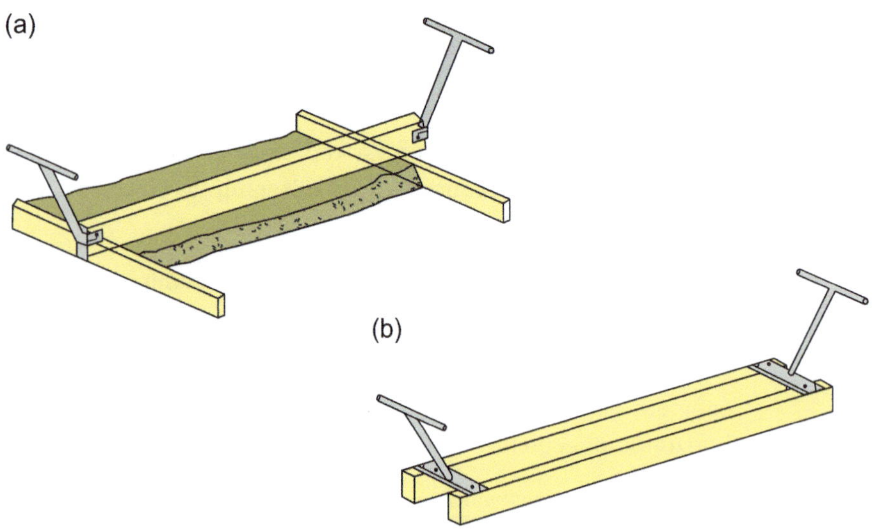

FIGURE 6.35
Tamping beams: (a) single beam; (b) double beam

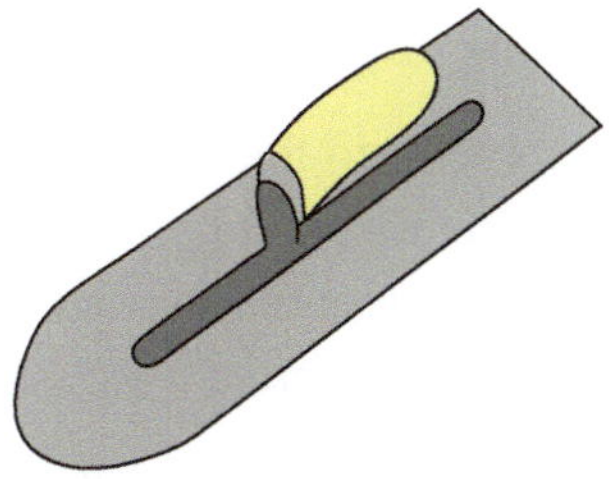

FIGURE 6.36
Concreting trowel

concrete is as dry as can be fully compacted by the means available. Vibration leads to better concrete.

Slabs can be compacted by tamping beams (Figure 6.35). These combine the action of a screed and a vibrator, but they are only effective for a limited depth.

In general, a slab more than 150 mm thick should be compacted with a poker vibrator and finished with a vibrating beam. Pokers vary in size, usually from 25 to 75 mm in diameter. They can be either internal or external.

An external vibrator can be fixed to tamping beams to give extra vibration of the concrete. An internal poker vibrator should not be dragged through the concrete to spread it out. It should be placed vertically in the concrete, held in position until the air bubbles cease to come to the surface, then slowly withdrawn so that concrete can flow into the space it produced. This should be repeated at about 450 mm centres.

Finishing concrete

In many instances concrete will not require any further treatment after the formwork has been struck. However, if the concrete will be exposed a surface finish may be required.

A good-quality concrete, well laid and finished, makes a satisfactory floor for many purposes. The concrete surface can be smooth, easily cleaned and have good resistance to abrasion at a low maintenance cost.

Most concrete finishes are applied with a trowel (Figure 6.36). A power float

FIGURE 6.37
Power float

can be used for larger areas (Figure 6.37). The finish is usually applied after the concrete has set for at least three hours.

Many finishes can be applied with the aid of the formwork. The formwork may have a pattern on the inside that will be reproduced on the concrete. One of the most common patterns is grooves formed into the face of the concrete to relieve the monotony of plain concrete.

A retarder can be applied to the inner surface of the formwork to all the concrete finish to be brushed after the formwork has been removed.

Note

Overvibration should be avoided at all costs.

Overvibration may cause segregation of the materials and form a weak layer of cement paste on the top.

Curing concrete

When concrete has been placed it needs protecting until it has sufficiently hardened.

Concrete requires the presence of water for the chemical reaction of hardening to take place. It is therefore essential that the water used in the mixing remains until full hydration has taken place. The temperature of the concrete is kept at about 20C. This ensures that the cement binds the aggregate particles together and that the concrete hardens at a favourable rate.

There are various methods of curing available, depending on where the concrete has been placed:

- The concrete surface can be sprayed with water, thereby replacing any water loss.
- Hessian or straw blankets spread over the concrete surface and suitably damped provide insulation as well as a moisture-holding medium.
- Damp sand provides similar protection to hessian but does not give the same degree of insulation.
- Spray coatings form an impervious coating and may also act as a discolouring agent or a solar reflector.

If the temperature drops below 5C, the hardening process almost stops. If the temperature is too high, the temperature difference between the concrete and the surroundings can cause cracking.

Maintenance of tools and equipment

On completion of the day's work it is essential that all items of tools and equipment are cleaned and any maintenance work is carried out. Poor tools and equipment can delay the job and cause delays.

Concrete must be removed before it hardens and forms a coating on the tools.

Washing with water is the preferred method of cleaning tools and equipment. This also applies to transporting plant and equipment such as dumpers, barrows and skips.

Maintain a clean and tidy work station:

IT SHOULD BE THE AIM OF EVERYONE TO PREVENT ACCIDENTS.

Remember, you are required by law to be aware of and fulfil your duties under the Health and Safety at Work Act.

The main contribution you as a trainee can make towards the prevention of accidents is to work in the safest possible manner at all times, thus ensuring that your actions do not put you, your workmates and the general public at risk.

Team working

The building industry relies a great deal on the co-ordination and teamwork of relatively small groups.

If any project is to be successful, not only in making a profit but also in finishing by the proposed date, all those concerned will have to work together in goodwill and harmony.

It is essential that the whole process of batching, mixing, transporting, placing and curing concrete is carried out as a team operation.

Each member of the team should be conversant with all aspects of the work and be able to fit in on any part of the construction.

Multiple-choice questions

Self-assessment

This section of the book is designed to allow you to check your level of knowledge. The section consists of revision questions for this chapter. The questions are all multiple choice and have four possible answers. The answers are to be found at the end of the book.

The main type of multiple-choice question will be the four-option multiple-choice question. This will consist of a question or statement, known as the stem, followed by a choice of four different answers, called the responses. Only one of these responses is the correct answer; the others are incorrect and are known as distracters.

You should attempt to answer the questions by choosing either (a), (b), (c) or (d).

Example

The person employed by the local authority to ensure that the Building Regulations are observed is called the:

(a) clerk of works

(b) building control officer

(c) council inspector

(d) safety officer

The correct answer is the building control officer, and therefore (b) would be the correct response.

Preparing and mixing concrete and mortar

Question 1 Identify the following logo:

(a) British Standard kite mark

(b) European Standard mark

(c) British Specification mark

(d) European Standard kite mark

Question 2 Which of the following is the correct weight of a bag of lime?

(a) 20 kg

(b) 25 kg

(c) 30 kg

(d) 50 kg

Question 3 When mixing concrete, which of the following shaped aggregates would give the greatest binding strength?

(a) rounded
(b) regular
(c) irregular
(d) square

Question 4 When mixing mortar by volume, which of the following items of equipment would give the most accurate mix?

(a) box
(b) shovel
(c) barrow
(d) gauge box

Question 5 Which of the following would be the most accurate test for workability in a concrete mix?

(a) slump test
(b) silt test
(c) bulk test
(d) cube test

Question 6 When mixing by hand, how many times should the ingredients be turned over before adding water?

(a) 1
(b) 2
(c) 3
(d) 4

Question 7 When mixing mortar by machine on site, which of the following is the correct order for adding the ingredients?

(a) water – half the sand – cement – half the sand
(b) cement – sand – water
(c) half the sand – water cement – half the sand
(d) sand – cement – water

Question 8 Name the item of plant shown for transporting materials on site:

(a) electric barrow
(b) dumper
(c) fork-lift truck
(d) dump truck

CHAPTER 7

Setting Out Basic Masonry Structures

This chapter will cover the following City and Guilds units:

• Apprenticeship	Module 4	S1.1, S1.2
	Module 4	S2.1, S2.2, S2.3, S2.3, S2.4, S2.5, S2.6
	Module 4	K1.1, K1.2, K1.3, K1.5, K1.6
	Module 11	S1.1, S1.2, S2.1, S2.2, S2.3
	Module 12	S1.1, S1.3
• Diploma	Unit 102	1.4, 1.5, 1.6, 1.7, 1.8, 1.9, 1.10, 1.11, 1.12, 1.13, 1.14, 1.15, 1.16, 1.17 2.2, 2.3, 2.4, 2.5, 2.6

This chapter is about:

- Interpreting instructions
- Adopting safe and healthy working practices
- Selecting materials, components and equipment
- Assisting in setting out basic building structures

The following City and Guild performance criteria will be covered:

- Read and follow working drawings
- Use measurements to establish walling positions
- Establish levels from temporary benchmarks
- Set out a building to square
- Transfer building line positions to the ground concrete
- Check and correct measurements

The following Diploma outcomes will be covered:

- Know how to interpret given instructions to establish setting out work to be carried out
- Know how to select required resources when assisting in the setting out
- Know how to assist in the setting out from working drawings

Types of instruction

This chapter will deal with the setting out of small buildings. The information gained from this chapter will assist the student with the practical requirements of the following chapters.

Information can be gathered from numerous sources such as drawings, specifications, schedules, etc.

Information from drawings

The student should be able to extract information required for setting out from drawings. Drawings have been dealt with in Chapter 3. Please refer back if necessary.

When drawings are received on site they should be carefully studied so that the work to be done is fully understood.

Groups of individual measurements should be added up and checked against overall dimensions.

It is also very important to find out how the measurements have been taken, e.g.

- over all the walls
- centre to centre
- between the walls.

Drawings which show the work to be carried out are drawn to scale, in one of the general scales used in the building industry. See Table 3.1 in Chapter 3. The larger the scale the more detail can be shown on a drawing.

Application of scales

It is impracticable to draw most parts of buildings, construction sites and components, etc., to full life size. It is therefore necessary to present them at a smaller size which bears a known ratio to the real thing. This is known as drawing to scale.

For example, if it is decided to reproduce an object at half life size, this can be described as drawing to a scale of 1:2 or half size.

Identifying and taking off dimensions

Taking off

This is a term used in the construction industry meaning to identify from drawings the type and amount of materials required to carry out the task.

In this unit we discuss taking off dimensions, datum positions and levels from the drawings.

Very often the site supervisor will be expected to be able to understand drawings and extract setting out information. On a larger site a site engineer will carry out all the setting out and levelling.

Site location

When planning to erect a new building, one of the first considerations is: 'where are we going to build on the plot of land'?

To determine this, a site location plan is drawn (see Figures 3.6 and 3.7 in Chapter 3). This drawing, along with other plans and documentation, is submitted to the local authority for approval. You should be able to extract sufficient information from this drawing to be able to set out the building.

There may also be information regarding the drainage runs.

Examples of various setting-out drawings

The design team will be required to produce working drawings for the builder to use on the site.

Drawings, schedules and specifications will have to be prepared explaining how the design team requires the building to be constructed. To be able to read these drawings it is essential that the trainee is able to understand them.

Drawings should be produced according to British Standard Recommendations for Drawing Office Practice BS 1192. These recommendations apply to the sizes of drawings, thickness of lines, dimension of lettering, scales, various projections, graphical symbols, etc.

The person carrying out a task should be able to read drawings and extract the required information.

Information concerning a project is normally given on drawings and written on printed sheets. Drawings should contain only information that is appropriate to the reader; other information should be produced on schedules, specifications or information sheets.

Working drawings

These have been explained in Chapter 3 and this chapter will deal with the setting out of the building shown in the site plan (see Figure 3.6).

Using the information given previously, the building has a front dimension of 12 m and a depth of 10 m.

This is calculated as:

Plot width = 25 m

8 m + 5 m = 13 m

25 m - 13 m = 12 m width

Plot depth = 25 m

8 m + 7 m = 15 m

25 m - 15 m = 10 m depth

Information from specifications

Except in the case of very small building works the drawings cannot contain all the information required by the contractor, particularly the required standard of materials and workmanship.

Therefore, the architect will prepare a document known as the *specification* to supplement the working drawings.

The specification is a precise description of all the essential information and job requirements that will affect the price of the work but cannot be shown on the drawings.

Typical items included in the specification are:

- description of materials, quality, size and tolerance
- description of workmanship, quality, fixing and jointing
- other requirements, site clearance, making good on completion, nominated suppliers and subcontractors
- inspection of the work.

Bills of quantities

The bills of quantities are produced by the quantity surveyor working for the architect.

These documents give a complete description and measure of the quantities of labour, material and other items required to carry out the works, based on the working drawings, specifications and schedules.

Schedules

These are used to record repetitive design information about a range of similar components.

The main areas where schedules are used include:

- doors
- windows
- ironmongery
- sanitary ware
- radiators
- finishes
- floor and wall tiling.

The information that schedules contain is essential when preparing estimates and tenders.

Schedules are also extremely useful when measuring quantities, locating work and checking deliveries of materials and components.

Manufacturer's information

Technical information can be produced in several formats. When items of equipment are purchased there will always be manufacturer's information sheets with them.

The information may be:

- operating instructions – how to use the item
- safety guidelines – power supply, personal protective equipment (PPE) to be worn and recommended checks
- technical information – mechanical details and possible outputs.

Normal hand tools for setting out are not usually provided with manufacturer's instructions.

Information on setting out equipment is provided by the manufacturer to ensure that the item of equipment is used correctly.

For example, manufacturer's instructions for a Cowley level will explain how to set up the instrument correctly and various types of readings which will give accurate and inaccurate levels.

Selection of resources

Setting out equipment

Before setting out any work the equipment should be carefully checked for accuracy.

Tapes

Linen tapes tend to stretch after they have been used for some time, so it is always better to use a steel tape (Figure 7.1).

There are several other types of measuring equipment such as hand tapes, wooden rules and fibreglass tapes (Figure 7.2).

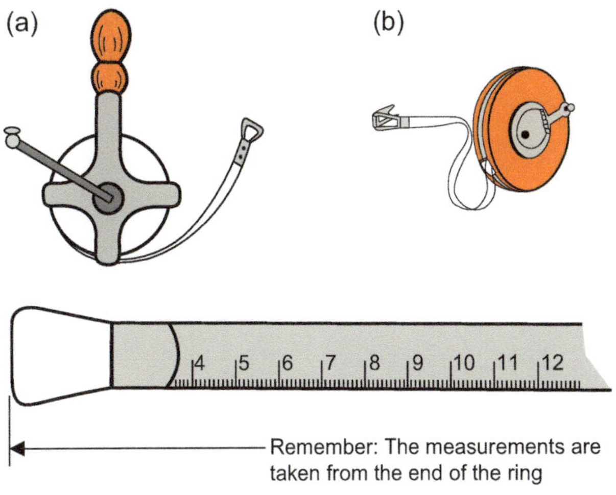

FIGURE 7.1
Types of tape: (a) steel tape; (b) metallic linen tape

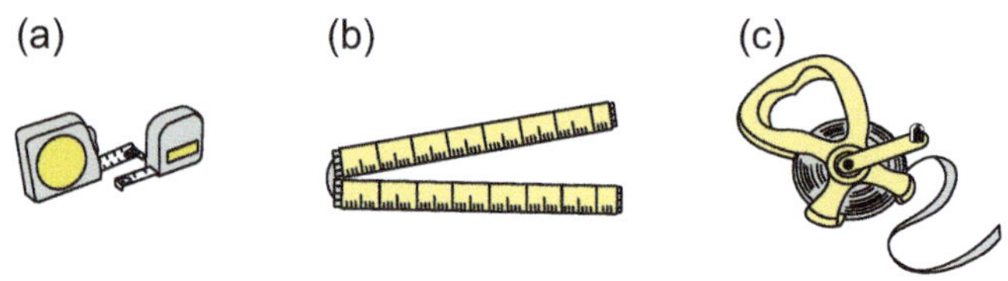

FIGURE 7.2
Other types of measuring equipment: (a) hand tapes; (b) wooden rule; (c) fibreglass tape

Spirit levels

Spirit levels have a dual purpose as they are used for checking the horizontal and vertical accuracy (Figure 7.3). It is important to take great care of spirit levels and to check them for accuracy.

Straight edge

A straight edge is normally made from timber but aluminium ones are also available.

They are usually 2–3 m long and tapered at either end (Figure 7.4).

The straight edge has to be checked from time to time for straightness and accuracy.

Remember that when transferring levels over a distance the straight edge and level should be reversed each time. This will counteract any discrepancies in either level or straight edge.

Ranging lines

There are numerous types of lines available for setting out the building, ranging from cotton and hemp to nylon. It is most important to avoid ravelling the lines.

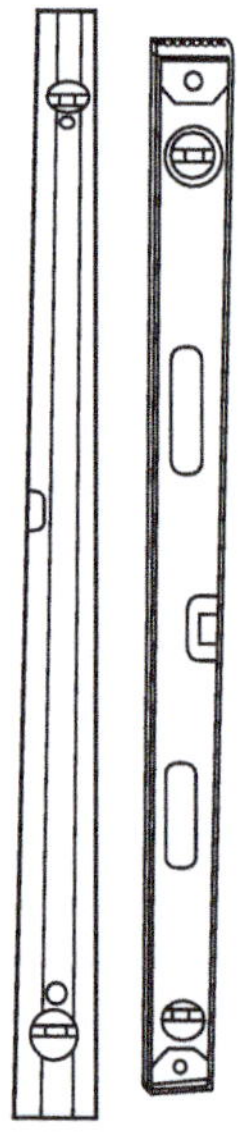

FIGURE 7.3
Spirit levels

FIGURE 7.4
Straight edge

Hammers

Various types of hammers are required, from large ones for driving in the setting-out pegs to small ones for knocking nails into the pegs and profile boards.

Builder's square

Most buildings have square corners; that is, corners set at 90 degrees (a right angle).

There are two methods of setting out a right angle which the trainee should know at this stage of training:

- the builder's square
- the 3:4:5 method.

Builder's square

The builder's square (Figure 7.5) is the most commonly used method of setting out a right-angled corner. It can be constructed of 75 mm × 25 mm timbers, half-jointed at the 90-degree angle with a diagonal brace, tenoned or dove-tailed into side lengths.

The builder's square is laid to the previously fixed front line and the end wall line is placed to the square to produce a right angle (Figure 7.6).

3:4:5 method

A right angle can also be set out using the 3:4:5 method (Figure 7.7).

A peg should be fixed exactly 3 m from the corner peg on the fixed frontage line. A tape is then hooked on to the nail on the corner peg and another tape hooked on to the peg on the front line.

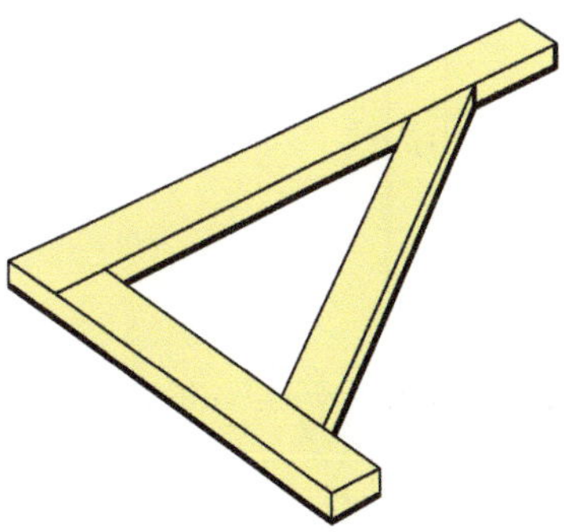

FIGURE 7.5
Builder's square

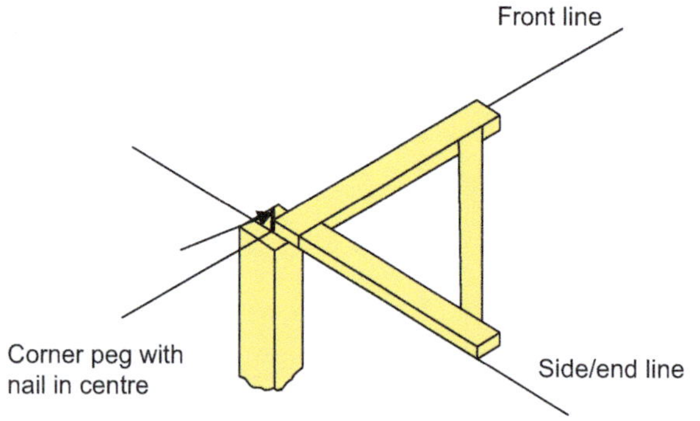

FIGURE 7.6
Setting out a right angle with a builder's square

Stretch both tapes towards the side wall line. At the crossing point of 4 m on one tape and 5 m on the other, a third peg should be fixed.

This will establish the end line at 90 degrees to the front line.

Unless using a builder's square with at least 2 m long sides, the 3:4:5 method is the most accurate method of setting out right angles on the building site.

Maintenance of setting-out equipment

There are several hidden dangers when setting out on the building site or in the workshop.

Tools used for setting out should always be maintained in a safe manner, kept clean and stored away safely until required again.

Wooden-handled hammers should have their heads fastened on correctly and never have split shafts.

Take care when using knives and saws as they are very sharp. Saws should always be sharp and stored correctly when not being used. Never cut directly

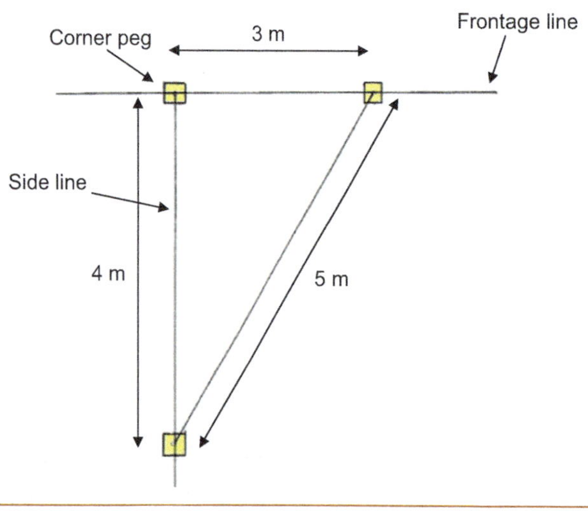

FIGURE 7.7
Setting out a right angle with the 3:4:5 method

towards the body: stand at the side of the cut and keep the hands behind the blade.

Always wear the correct PPE.

Workers should be instructed to report any faults immediately and stop using the item of equipment or cable as soon as any damage is seen.

Faulty equipment should be taken out of service as soon as the damage is noticed.

DO NOT carry out makeshift repairs.

Tapes

These should always be cleaned and oiled immediately after use and stored correctly.

Builder's squares, etc

Any home made item of equipment such as boning rods or builder's squares should be checked before use.

The dimensions of builder's squares should be checked regularly to ensure that they are true.

Automatic level

It is essential that any type of automatic level is checked regularly to ensure that all setting out on site is accurate. Inaccurate setting out on site can cause problems and delays which could be very expensive.

The camera section should be sent away to the makers periodically for calibration. This cannot be done on site!

Checking spirit levels

Great care should be taken of spirit levels as they are expensive and if ill-used can lead to inaccurate levelling and plumbing.

It is therefore important to check the spirit level occasionally to ensure its accuracy (Figure 7.8).

Checking for level

1. Set two screws into a bench, equal to the length of the level apart.
2. Turn one of the screws until the bubble reads level.
3. Reverse the level and replace it on the screws. If the bubble is between the lines the level is accurate.
4. If the bubble is not in the centre of the lines, adjustment must be made to the bubble tube.

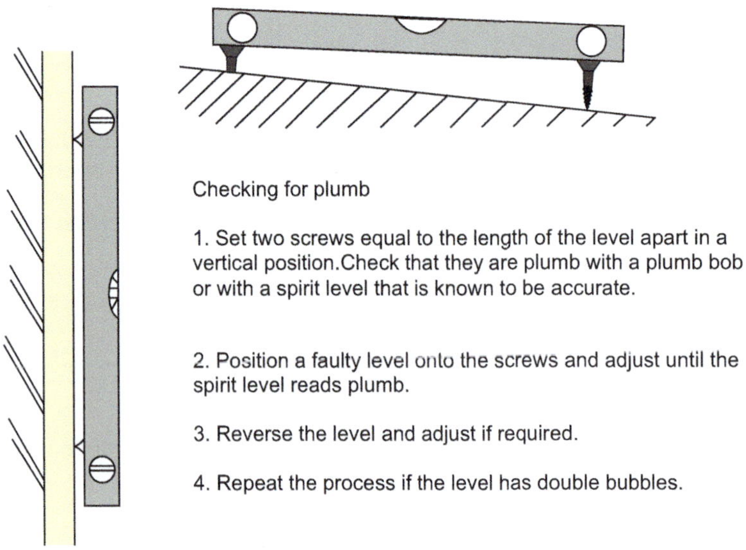

FIGURE 7.8
Checking the accuracy of the spirit level

Remember that some spirit levels cannot be adjusted.

Site clearance

Before setting out can begin it is necessary to clear the site of all rubbish and top soil.

There could be several hidden dangers in the rubbish to be cleared and care must be taken.

Services

If the site has been developed before there could be underground services such as water, gas and electricity.

Care should be taken when driving pegs into the ground that these pipes are not penetrated.

Remember

If in doubt ask your supervisor.

Setting-out calculations

Calculations required for setting out involve diagonals and the 3:4:5 method of producing a right angle.

3:4:5 Right angle

Pythagoras states that the square on the hypotenuse (longest side) is equal to the sum of the squares on the other two sides of a right-angled triangle.

Written as a formula, this is:

$$A^2 = B^2 + C^2$$

This is the basis of the 3:4:5 triangle.

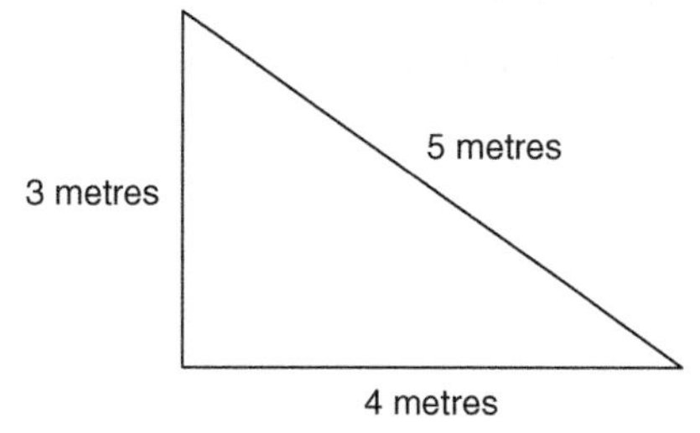

e.g. $5^2 = 4^2 + 3^2$

$25 = 16 + 9$

$25 = 25$

By transposing this formula it is possible to find the length of any side of a right- angled triangle provided we know the length of the other two sides.

Example 1

To find the length of side A:

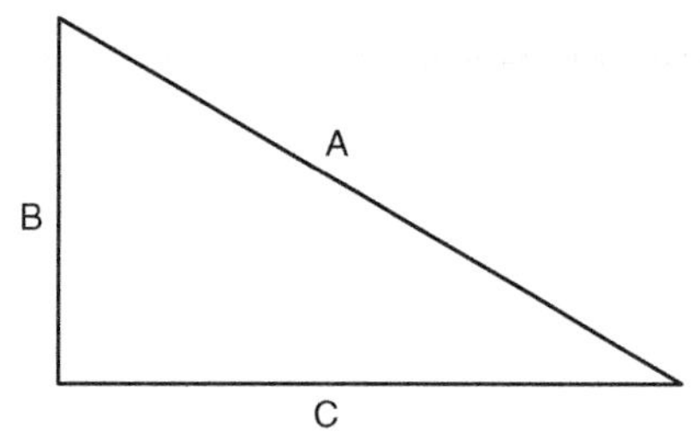

$$A^2 = B^2 + C^2$$

$$A = \sqrt{(B^2 + C^2)}$$

To find the length of side B:

$$A^2 = B^2 + C^2$$

$$A^2 - C^2 = B^2$$

$$B^2 = A^2 - C^2$$

$$B = \sqrt{(A^2 - C^2)}$$

To find the length of side C:

$$A^2 = B^2 + C^2$$

$$A^2 - B^2 = C^2$$

$$C^2 = A^2 - B^2$$

$$C = \sqrt{(A^2 + B^2)}$$

Example 2

To find the length of the sides of a triangle given the base and the perpendicular height:

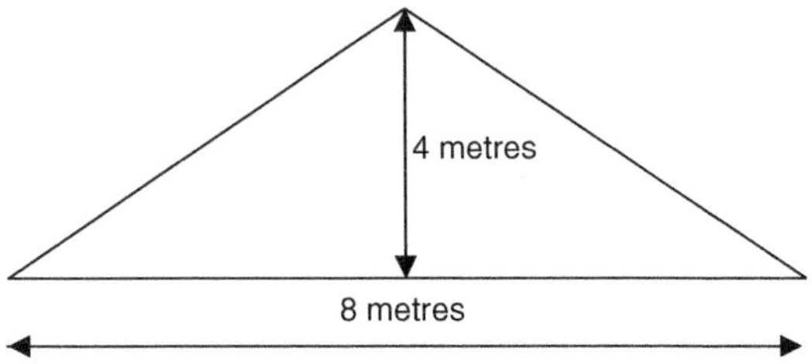

$$A = \sqrt{(B^2 + C^2)}$$

$$A = \sqrt{(4^2 + 4^2)}$$

$$A = \sqrt{(16 + 16)}$$

$$A = \sqrt{32}$$

$$A = 5.657 \text{ m each side}$$

Example 3

To find the length of the diagonal of a rectangle given the length of the two sides:

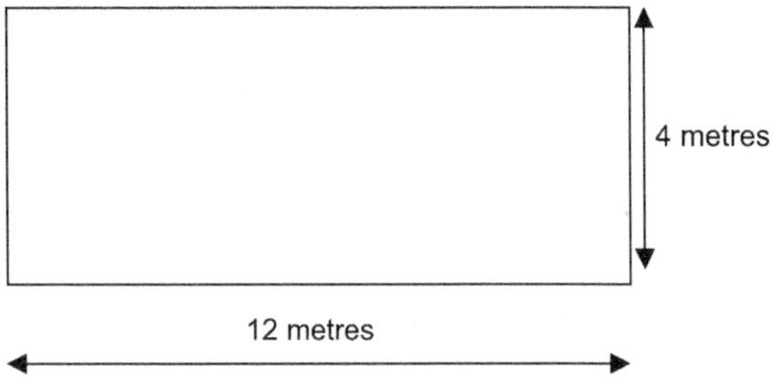

$$\text{D agonal} = \sqrt{(12m^2 + 4m^2)}$$
$$= \sqrt{(144 + 16)}$$
$$= \sqrt{160}$$
$$\text{D agonal} = 12.65$$

Minimize the risk of damage

As stated before, there are several hidden dangers when setting out on the building site or in the workshop.

Setting out is usually the first item of work to take place on a building site.

It is usual for the top soil to be removed first. At this point the site is usually clear of all obstacles and dangers. If the setting out takes place before removal of the top soil then there could be hidden dangers on the site.

It is therefore essential to be aware and to ensure that correct PPE is worn, especially safety boots.

Accident reporting

Every accident should be reported – there should be an accident report book on every site and in every workshop, usually with the general supervisor or whoever is in charge of the site or workshop.

Make sure that you report any accident in which you are involved as soon as possible.

Site security

It is the responsibility of everyone on the work site to ensure that security of that site is maintained.

Security can take many forms and they are all equally important.

Visual security

- alarms – positioned in an accessible place within view of the general public
- bars, mesh, locks – fitted to glass-panelled doors and windows
- padlocks and padlock and chains – fitted to compound gates, pieces of plant and machinery
- lighting – flood lights and movement-activated lights
- security firms.

Individual security

It is the responsibility of all employees to contribute to the overall security of the firm. For example:

- Tidiness – do not invite crime by leaving tools and equipment where they may be easily seen. If there is a secure store, lock them away.
- If necessary immobilize large plant.
- Unauthorized access – it is the responsibility of all employees on site to

challenge anyone who they feel has no authorization to be within a particular area. (Politeness is the best approach.)

If, despite the security measures taken, your site is breached, there are certain procedures you should follow. These should be given to you by your line manager, and may include:

- reporting the incident to the site supervisor
- reporting the incident to the police
- checking the inventory to find out what has been taken
- recording damage done to the premises and/or equipment.

Remember

Vandalism and theft do not alway happen at night.

Always take care of tools an equipment with which you ar working and report any suspiciou people or events to your lir manager.

Safety on site and in the workshop

Maintain a clean and tidy work station:

It should be the aim of everyone to prevent accidents.

Remember, you are required by *law* to be aware and fulfil your duties under the Health and Safety at Work Act.

The main contribution you as a trainee can make towards the prevention of accidents is to work in the safest possible manner at all times, thus ensuring that your actions do not put you, your workmates and the general public at risk.

A safe working area is a tidy working area.

All unnecessary obstructions that may create a hazard should be removed, e.g. offcuts of material, unwanted materials, disused items of plant, bricks, etc., and nails should be extracted from discarded pieces of timber, or flattened.

Clean your work area periodically as offcuts of any materials are potential tripping hazards.

Stacking and storing reusable materials and components

When clearing the work area there may be many materials that can be salvaged and used again.

Materials used for setting out consist mainly of timber. Any materials left should be stored away safely and securely. These materials should be properly cleaned before storing.

Any item that is damaged and cannot be reused is classed as waste and should be correctly disposed of.

Hazards

It is important that all materials are checked for any potential hazards, such as nails in timber or loose tin lids.

Always report any potential hazards to your supervisor.

Remedial actions

Any actual or potential hazards should be correctly dealt with. Remedial action may involve:

- correcting – removing a nail from a piece of wood
- reporting – telling your line manager about the hazard
- warning others – placing a sign to warn others about the hazard.

Disposal of waste

There should be very little waste to be disposed of after completing setting out.

Most substances and materials are safe provided they are handled or worked on sensibly and with proper precautions.

Almost anything can be dangerous if handled or used irresponsibly. Some materials require extra care, and a few need extreme caution.

It is important to know what the hazards are, when they occur and how they can be prevented.

Usually prevention will involve the use of protective clothing and equipment.

Removal of waste

Since 1 April 1992, the law on waste has included a duty of care that means you must take all reasonable steps to look after any waste you have and prevent its illegal disposal by others.

If you break the law, you could be fined an unlimited amount.

'Controlled waste' is any household, commercial or industrial waste, such as waste from a house, shop, office, factory, building site or any other business premises.

Protection of the surrounding area

It is necessary to protect not only the site you are working on but also the neighbouring land or property. The surrounding properties could be affected by noise, smoke or dust.

Try to arrange parking so as not to upset the surrounding properties and cause traffic problems throughout the duration of the work.

A little understanding can prevent problems during the works.

Setting out

The first task in setting out a building is to establish a base line to which all the setting out can be related.

The base line is very often the building line; this is an imaginary line that is established by the local authority. You may build behind the building line and even up to the building line.

Note

NEVER build in front of the building line.

It is usual for the building line to be given as a distance from one of the following:

- the centre line of the road
- the kerb line
- existing buildings.

The frontage line of the building must then be on or behind the building line, never in front of it.

Degree of accuracy

British Standard BS 5964-1: 1990 states that the permissible deviation for horizontal brick walls up to 40 m in length is ± 40 mm.

This required accuracy can be achieved if a steel tape is used and supported to avoid sag.

Procedure for setting out

For this example we will use the site plan (Figure 3.6) in Chapter 3.

In this example, the building line has been set by the local authority 7 m from the front of the plot.

Two square lines are set from the kerb to the building line, at a distance apart larger than the required building, and two pegs are knocked in.

A line is stretched between the two pegs to establish the building line.

Stage 1

The building line is fixed to two pegs on either side of the plot (Figure 7.9).

Stage 2

In this example the frontage line of the building is on the building line (Figure 7.10).

After taking off the dimensions from the drawing the frontage line is calculated at 12 m.

The distance from the boundaries should be read from the drawing. In this case the left-hand dimension is 5 m and 8 m from the right-hand boundary. Pegs should be knocked in to represent both points.

This will determine the front corners of the building.

In our example the two front corner pegs should be on the building line

Remember

The building cannot be in front of the building line, but it can be behind it.

Stage 3

End wall lines can now be set out at right angles to the frontage line (Figure 7.11).

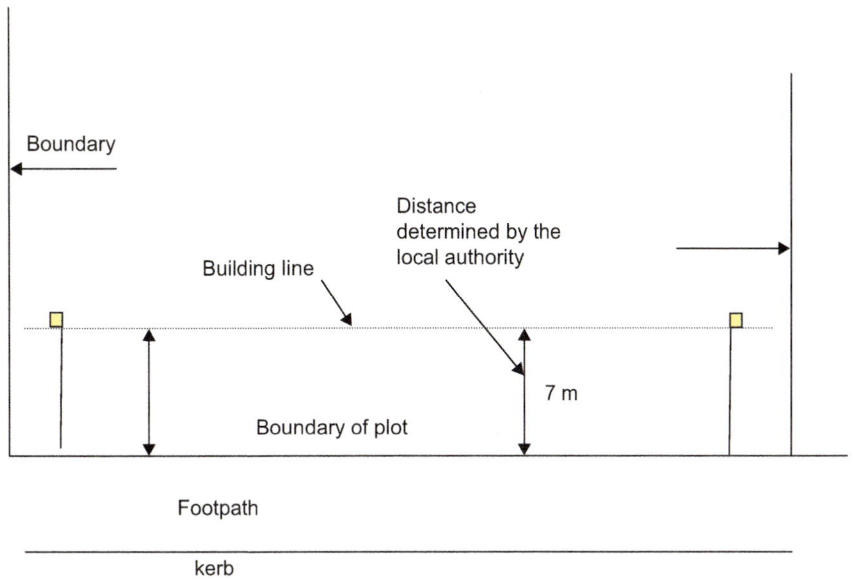

FIGURE 7.9
Stage 1

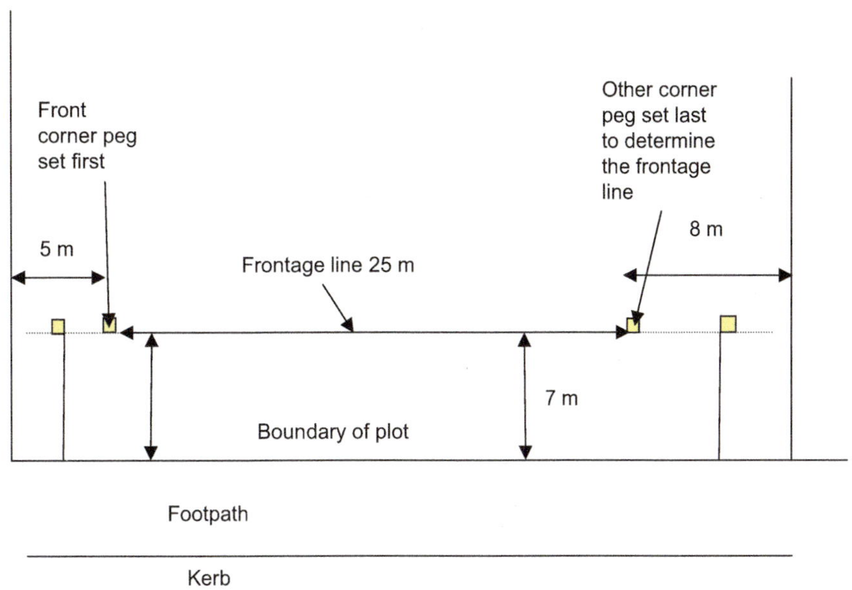

FIGURE 7.10
Stage 2

The depth of the building is calculated from the drawing and is 10 m.

The lengths of the end walls are pegged out.

Produce a right angle on both front corners of the building using a builder's square, the 3:4:5 method or an optical square.

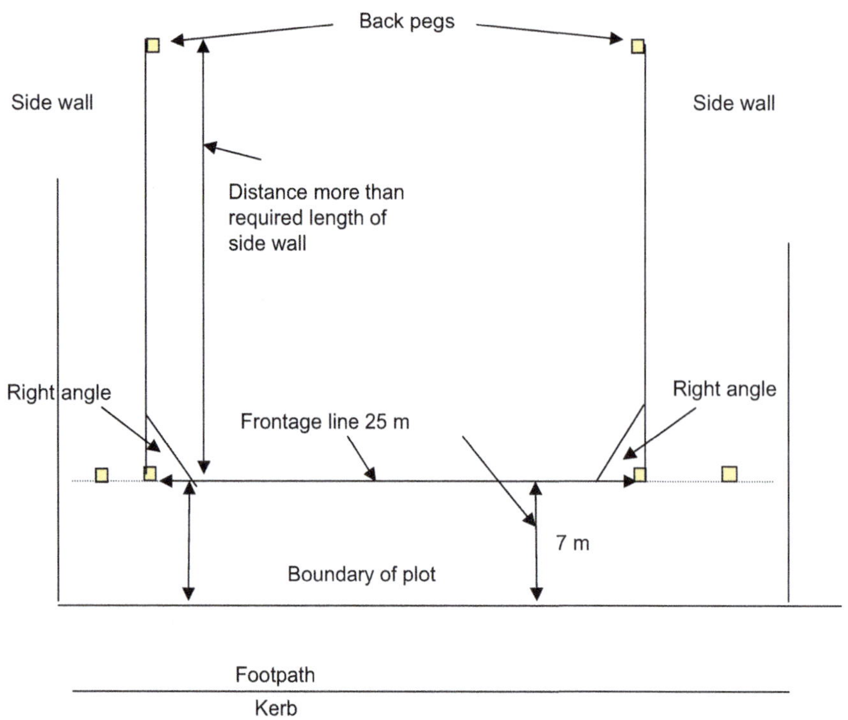

FIGURE 7.11
Stage 3

Extend the lines back farther than the side wall dimensions of 10 m and knock two pegs in to hold the side wall lines.

Stage 4

Knock two pegs in on each of the side walls at the required distance, i.e. 10 m.

Connect a line to the two back pegs to give the back wall line (Figure 7.12).

Stage 5

The building is square if the diagonals are equal.

When completed check all dimensions and then check the diagonals.

Measure each diagonal and if they are the same then the building is square (Figure 7.13). If not, then an adjustment has to be made.

Depending on which diagonal is out adjustments have to be made to make the diagonals equal.

Always check the dimensions of the walls after any adjustments.

Remember

Do not alter the frontage line.

Profiles

When the building has been set out and proved by checking the diagonals, profiles can be erected.

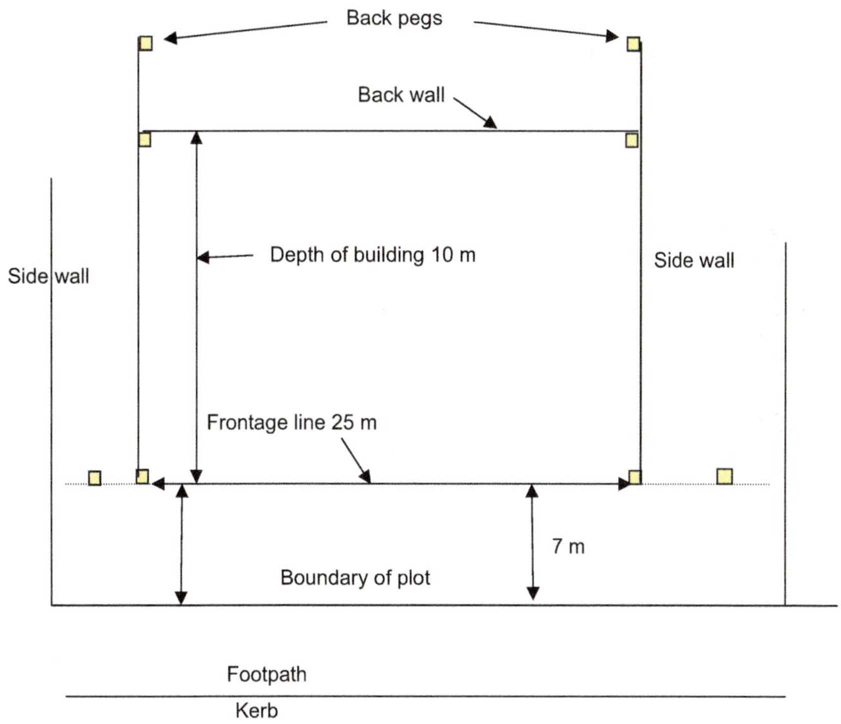

FIGURE 7.12
Stage 4

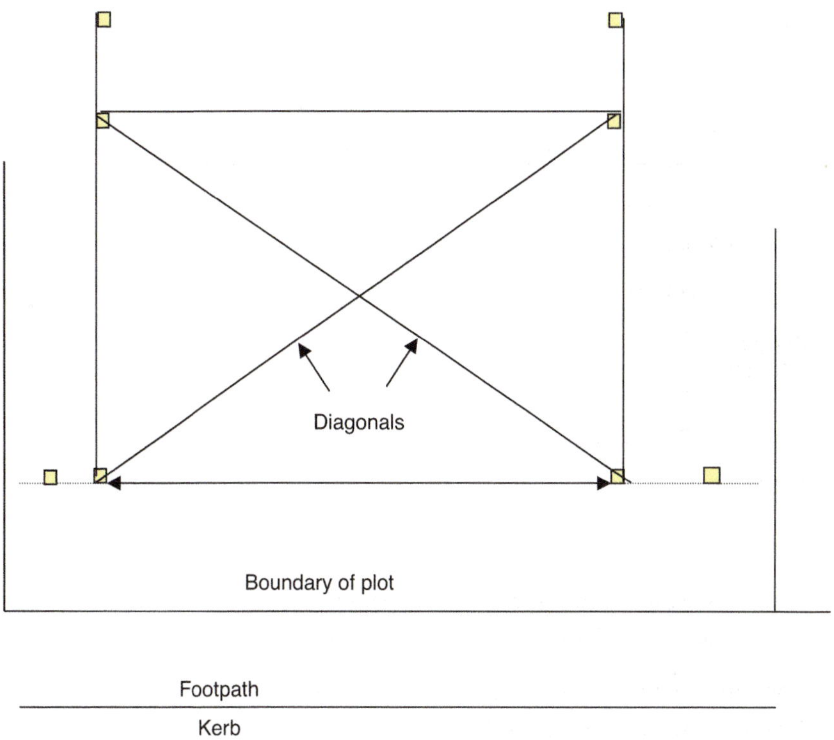

FIGURE 7.13
Stage 5

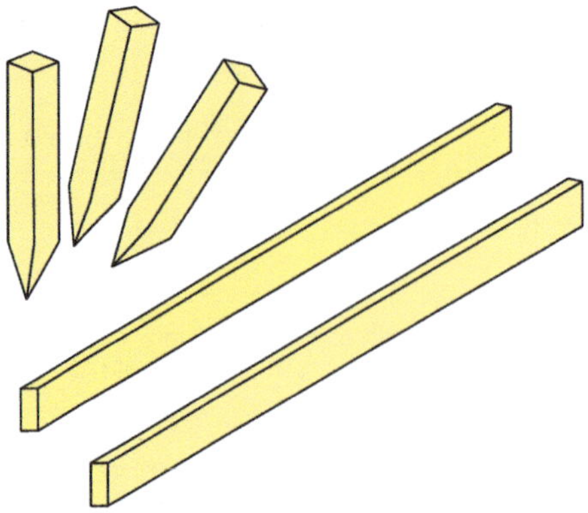

FIGURE 7.14
Material for making profiles

At the moment, the setting-out pegs are in the foundation trench, so they have to be repositioned approximately 1 m away from all the wall faces.

Profiles consist of 75 mm × 75 mm pegs with 150 mm × 25 mm boarding nailed to them (Figure 7.14). They can be either single (Figure 7.15) or corner type (Figure 7.16).

Single profile type

Profiles are erected to enable the corner setting-out pegs to be removed to allow excavation to take place without disturbing the pegs.

One of these corner profiles could be set to a given level which is known as the 'datum', and may relate to the finished floor level or damp-proof course (DPC). This datum peg should be protected with concrete.

The datum peg could also be positioned away from the setting pegs but close enough to be accessible, and should also be protected with concrete and a small barrier of pegs and rails (Figure 7.17).

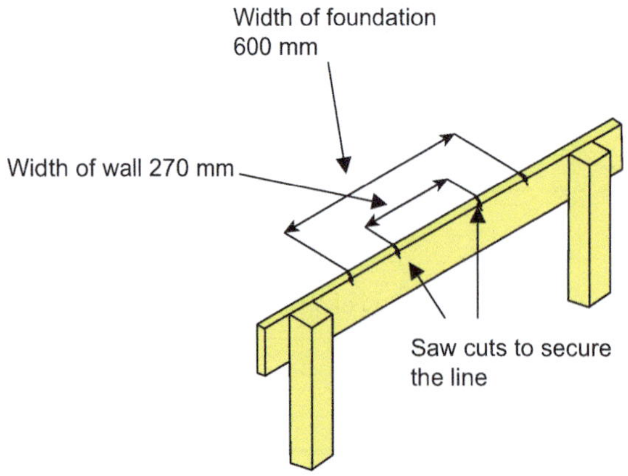

FIGURE 7.15
Single profile

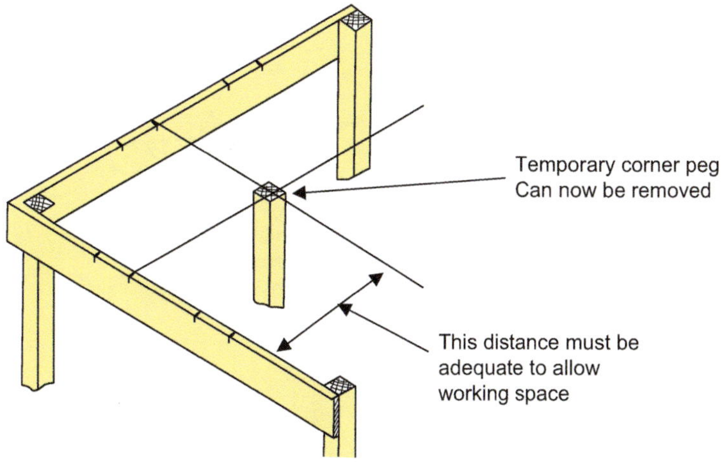

FIGURE 7.16
Corner profile

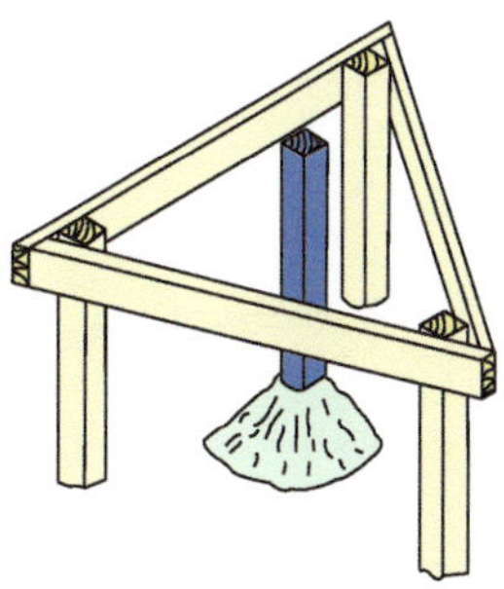

FIGURE 7.17
Datum peg protected

The profiles should be positioned approximately 1 m away from the face of the building to allow working space for the excavation.

The completed profiles can have foundation and wall widths marked on them in one of various methods. Saw cuts are best, as nails could be accidentally removed (Figure 7.18).

Once the profiles have been accurately constructed the dimensions should all be checked again, as should the diagonals.

The original setting-out pegs can now be removed.

Building lines can now be fastened to the profiles and the trench be marked out ready for excavation (Figure 7.19).

Lime or sand can be used to mark out the trenches ready for excavation (Figure 7.20).

Transferring levels

There are many methods of transferring levels on the building site, such as:

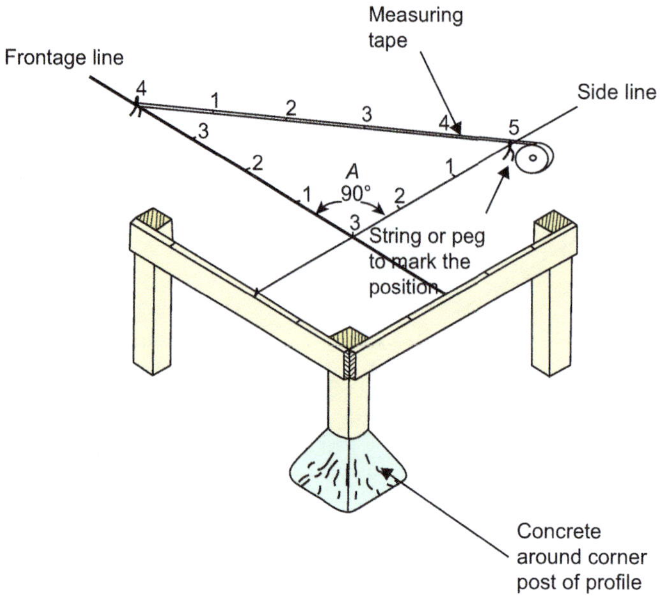

FIGURE 7.18
Checking a right angle by the 3:4:5 method

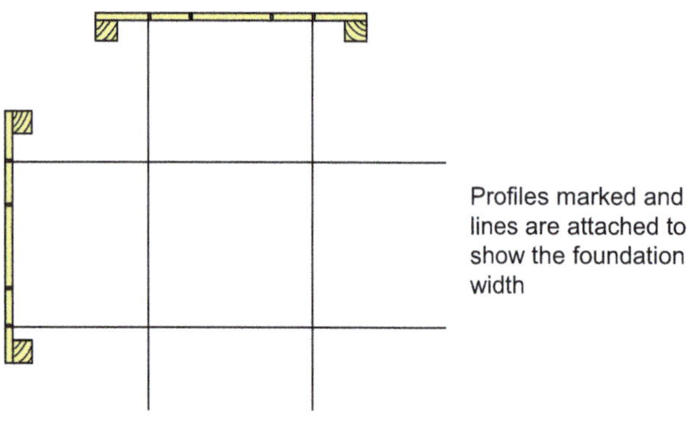

FIGURE 7.19
Fixing lines to the profiles to mark out the trench

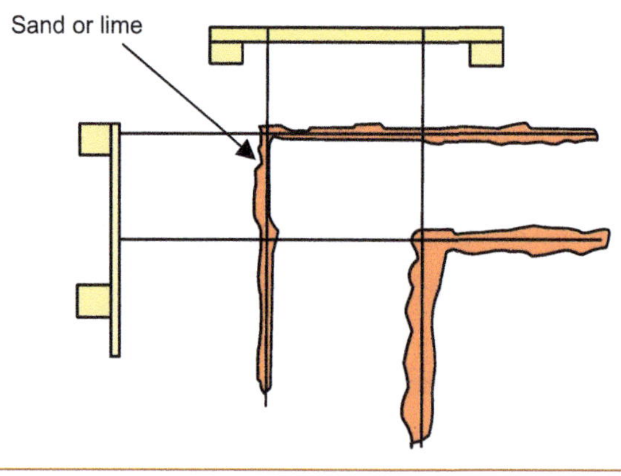

FIGURE 7.20
Marking out the ground ready for excavation

- straight edge and spirit level
- boning rods
- water level
- Cowley level
- laser level.

Straight edge and spirit level

Remember to reverse both level and straight edge for each reading (Figure 7.21).

Boning rods

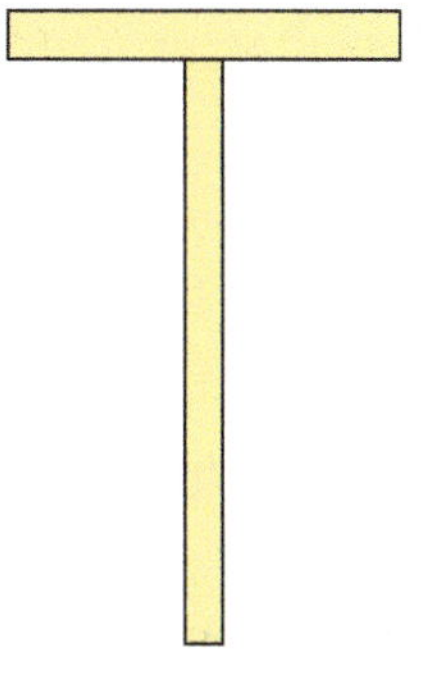

Boning rods can also be used to transfer levels between two known points.

The operation can be speeded up with the use of boning rods, as spirit levels are not used.

After the main level pegs have been set out at either end of a foundation, the extra levels between these can be set out with the use of boning rods.

Eye sight is used, instead of the spirit level, to sight in the top of the rods.

Boning rods are also used to transfer level points on large areas of hardcore, paving, etc., which are not necessarily level points.

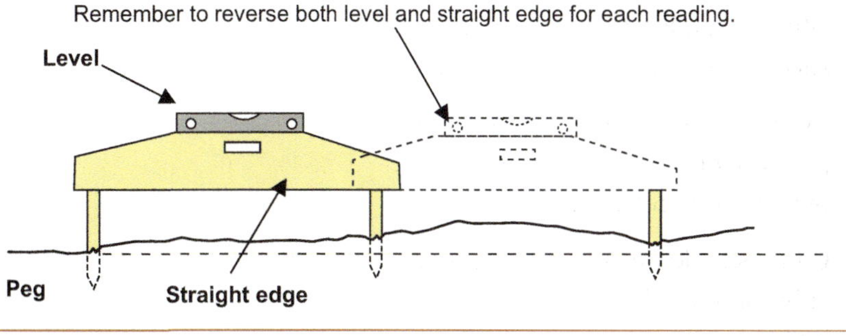

FIGURE 7.21
Use of straight edge

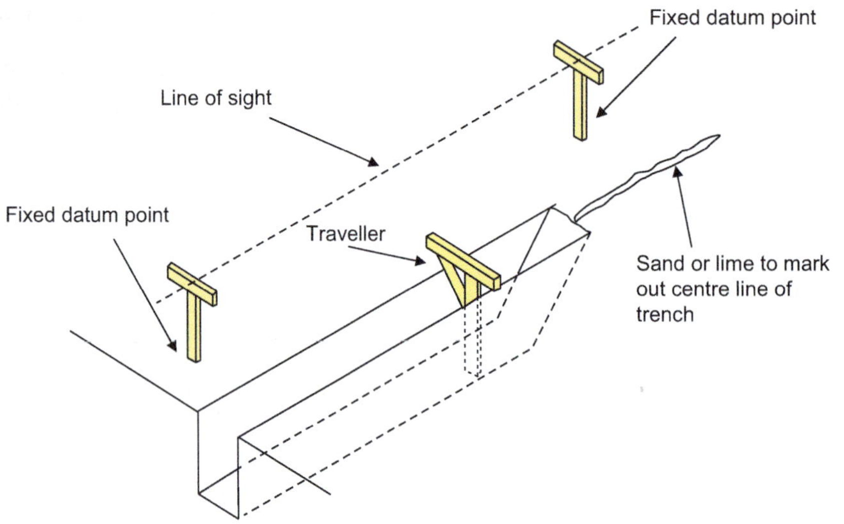

FIGURE 7.22
Use of boning rods and traveller

The boning rod that is moved around the area or trench to produce intermediate level points is called the 'traveller' (Figure 7.22).

Water level

The water level is simply a length of rubber tubing with a glass tube attached at each end (Figure 7.23).

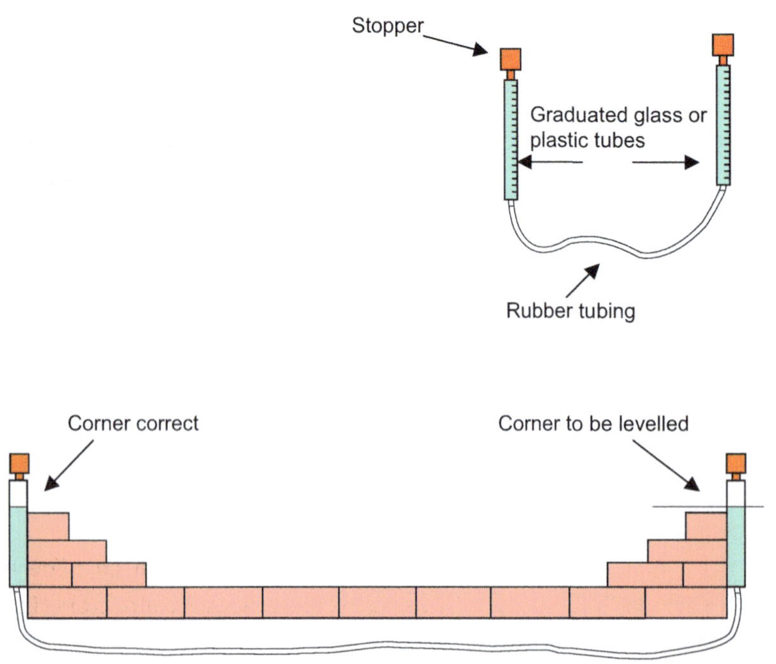

FIGURE 7.23
Use of water level

The tube is carefully filled with water at one end only, to ensure that no air bubbles are trapped in the tube.

When the two glass tubes are held at the same level, the height of the water will be the same level in each tube.

If one tube is significantly lower than the other, then water will pour out of the lower tube, until once again the water finds its own level. To prevent this happening, both tubes are fitted with stoppers.

The water level is particularly useful for transferring levels from room to room or around corners.

To transfer a level, hold one of the glass tubes so that the middle is about level with the required mark. Take the other glass tube to the position where you require the level to be transferred to.

Unplug both ends and raise or lower the glass tube nearer the level to be transferred until the water comes to rest level with the mark. The new level mark can then be marked off equal to the water level in the other glass tube.

Cowley level

When levels have to be transferred over a long distance the method of straight edge and level can be very time consuming and rather inaccurate. A more accurate method is the Cowley level.

It can be used to transfer the site datum peg to all corners of the building very quickly and very accurately.

It is an automatic device working a system of mirrors, some of which are controlled by pendulums so that the sighting is always along the same line.

The target is 450 mm long and 50 mm wide and slides up and down a staff, which is marked out on the back in metre and millimetre graduations. The equipment is shown in Figure 7.24.

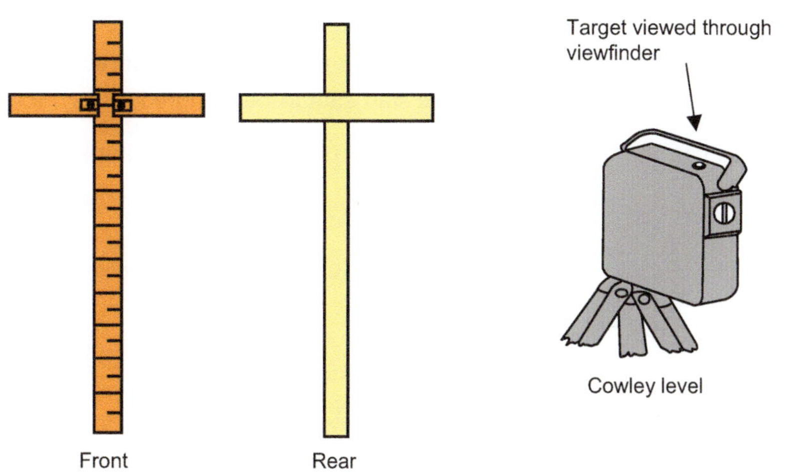

FIGURE 7.24
Cowley level equipment

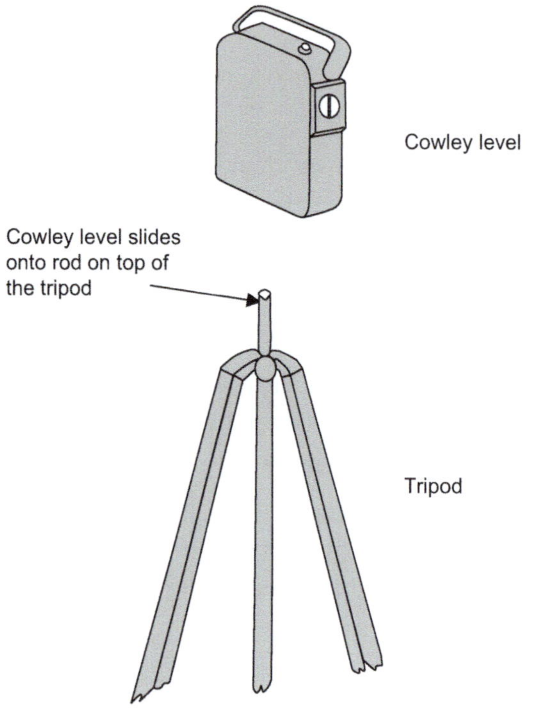

FIGURE 7.25
Setting up the Cowley level

Set up the tripod as level as possible.

There is a rod on the top of the tripod which slides into the base of the Cowley level. This releases the two mirrors.

Place the instrument over the rods on the tripod, noting the release of the glass mirrors (Figure 7.25).

Rotate the camera around until you bring the target into vision.

Sight through the eyepiece: the view is of two mirrors, one the correct way up and the other upside down. Various views are shown in Figure 7.26.

Signal the assistant to slide the target up or down as required. As soon as both sides of the target are level the staff should be fastened off.

This can then be transferred to other pegs. Each peg being knocked in to the

Note

The camera should never b moved while attached to the tripo for fear of damaging the mirrors.

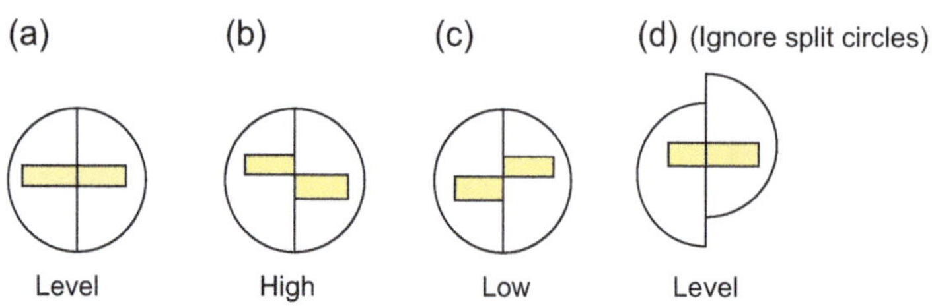

FIGURE 7.26
Various views of the staff

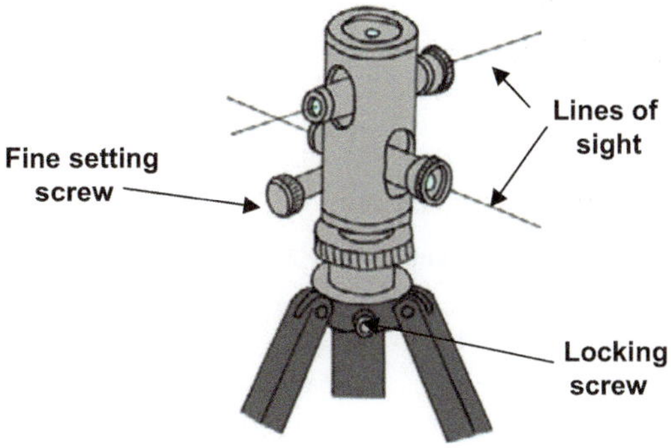

FIGURE 7.27
Site square

required depth so that the view through the eyepiece reads level at every new peg.

Laser level

Laser levels are the latest method for transferring levels around a construction site. As long as they are looked after correctly they are very accurate and easy to set up.

The level is fixed to a tripod. Press the set button and it automatically finds its own level.

The laser sends out a red dot that can be located on the staff, giving off a reading.

The laser level is a very versatile item of equipment and can be used for all kinds of levelling, both internal and external.

The optical site square

The site square is an accurate setting-out instrument (Figure 7.27).

The optical square consists of two telescopes mounted one above the other and with their lines of sight set at 90 degrees to each other. The instrument is fixed on a tripod (Figure 7.28), which is set up over a known corner peg.

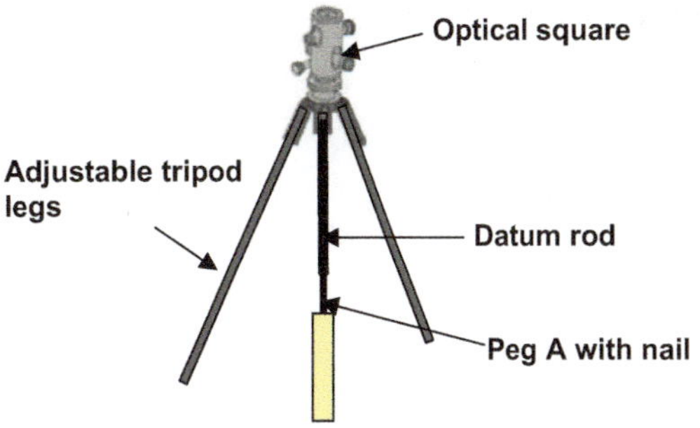

FIGURE 7.28
Site square set on tripod over corner peg

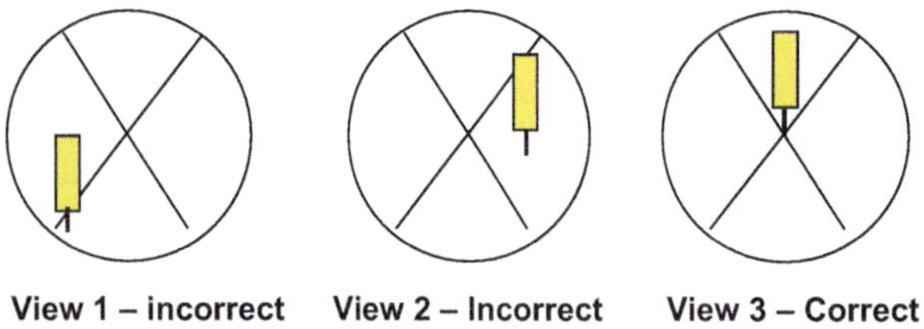

FIGURE 7.29
Views through the bottom telescope

Once the site square has been set up on a known corner peg, look through the bottom telescope and a view along one side of the building is shown until a view of the nail on the peg is in sight. Three views are shown in Figure 7.29. The correct view is when the nail in the top of the peg is in the centre of the hair lines.

When looking through the top telescope, without moving the equipment, it will trace out a line at right angles to the original line and then a further site peg can be positioned along the new line (Figure 7.30).

Completing the work on time

It is important that all work on site, even the setting out programme, is kept on time.

If one part of the programme is delayed then it has a knock-on effect on following programmes. Any problems with keeping to the programme should be reported to the site manager so that allowances can be made if possible.

One of the most common reasons for delays is poor drawings. When drawings are difficult to understand or there is a lack of clear information it can cause a time delay.

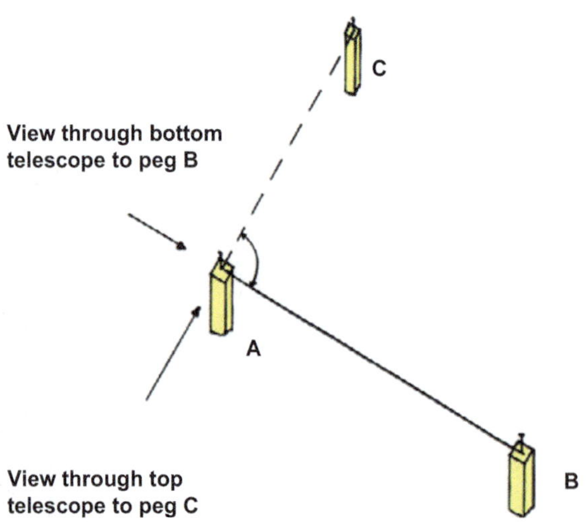

FIGURE 7.30
Producing a right angle

Another cause is the weather. Poor weather during winter months can affect the duration of setting out.

Setting out brickwork and blockwork

Once the initial setting out has been completed and the foundations have been excavated and concreted the brickwork can commence.

Brickwork up to DPC level is laid first and the correct bonding must be sorted out before any brickwork is exposed.

The next three chapters will deal with the setting out of brickwork and blockwork.

Multiple-choice questions

Self-assessment

This section of the book is designed to allow you to check your level of knowledge. The section consists of revision questions for this chapter. The questions are all multiple choice and have four possible answers. The answers are to be found at the end of the book.

The main type of multiple-choice question will be the four-option multiple-choice question. This will consist of a question or statement, known as the stem, followed by a choice of four different answers, called the responses. Only one of these responses is the correct answer; the others are incorrect and are known as distracters.

You should attempt to answer the questions by choosing either (a), (b), (c) or (d).

Example

The person employed by the local authority to ensure that the Building Regulations are observed is called the:

(a) clerk of works
(b) building control officer
(c) council inspector
(d) safety officer

The correct answer is the building control officer, and therefore (b) would be the correct response.

Setting Out Basic Masonry Structures

Question 1 Identify the item of setting out equipment shown:

(a) builder's square
(b) builder's level
(c) site square
(d) site level

Question 2 Identify the item of level-ling out equipment shown:

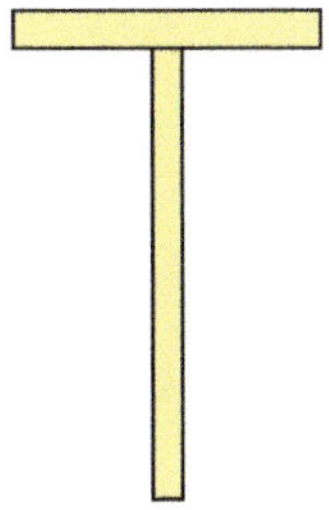

(a) profile
(b) ranging rod
(c) level
(d) boning rod

Question 3 Which of the following is a correct 3:4:5 ratio?

(a) 2:4:6
(b) 6:8:10
(c) 8:10:12
(d) 9:10:15

Question 4 Which of the following is the correct dimension for the diagonal x?

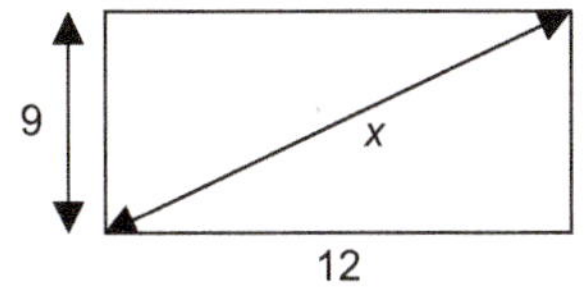

(a) 12
(b) 14
(c) 15
(d) 21

Question 5 Who is responsible for setting the building line?

(a) the local authority
(b) the local council
(c) the site manager
(d) the client

Question 6 If, when checking the diagonals after setting out, they are found to be wrong, which line should not be altered?

(a) the back line

(b) the side line

(c) the frontage line

(d) the diagonal line

Question 7 Why is it necessary to reverse both the level and straight edge when transferring levels?

(a) to save using two levels

(b) to correct any deficiency in the equipment

(c) to achieve the correct fall

(d) to be able to read the level correctly

Question 8 Name the item of equipment used for setting intermediate level points when transferring levels using boning rods:

(a) datum point

(b) straight edge

(c) range rods

(d) traveller

CHAPTER 8

Laying Bricks to Line

This chapter will cover the following City and Guilds units:

- Apprenticeship Module 5 S1.1, S1.2, S1.3
  Module 5 S3.1, S3.2
  Module 5 K3.1, K3.2, K3.3
- Diploma Unit 104 1.1, 1.4
  2.1, 2.2, 2.4, 2.5
  3.1, 3.2, 3.3, 3.4
  4.1, 4.2, 4.3, 4.5, 4.6

This chapter is about:

- Interpreting instructions
- Adopting safe and healthy working practices
- Laying bricks to line

The following City and Guild performance criteria will be covered:

- Build single skin walls
- Set out and build solid walls
- Set out and build solid walls at returns and junctions

The following Diploma outcomes will be covered:

- Know how to set out brickwork to comply with working drawings
- Know how to build straight walls in half-brick stretcher bond
- Know how to build return corners in half-brick stretcher bond
- Know how to build straight walls in one-brick walling
- Know how to build return corners in one-brick walling
- Know how to form junctions in brick walling

Bonding of brickwork

The brick is a relatively small unit and it can therefore be manipulated to fit most dimensions.

To achieve economy in both materials and labour the architect will give thought to the planning of these dimensions, while the bricklayer will exercise skill in setting out.

Whatever the system of measurement adopted, bonding principles must still apply.

Brick proportions

The apprentice should note that in the following bonding examples the standard brick format is used, namely 225 × 112.5 × 75 mm.

Using a 10 mm mortar joint the actual size of the standard brick becomes 215 × 102.5 × 65 mm. Figure 8.1 indicates the basic relationship between header and

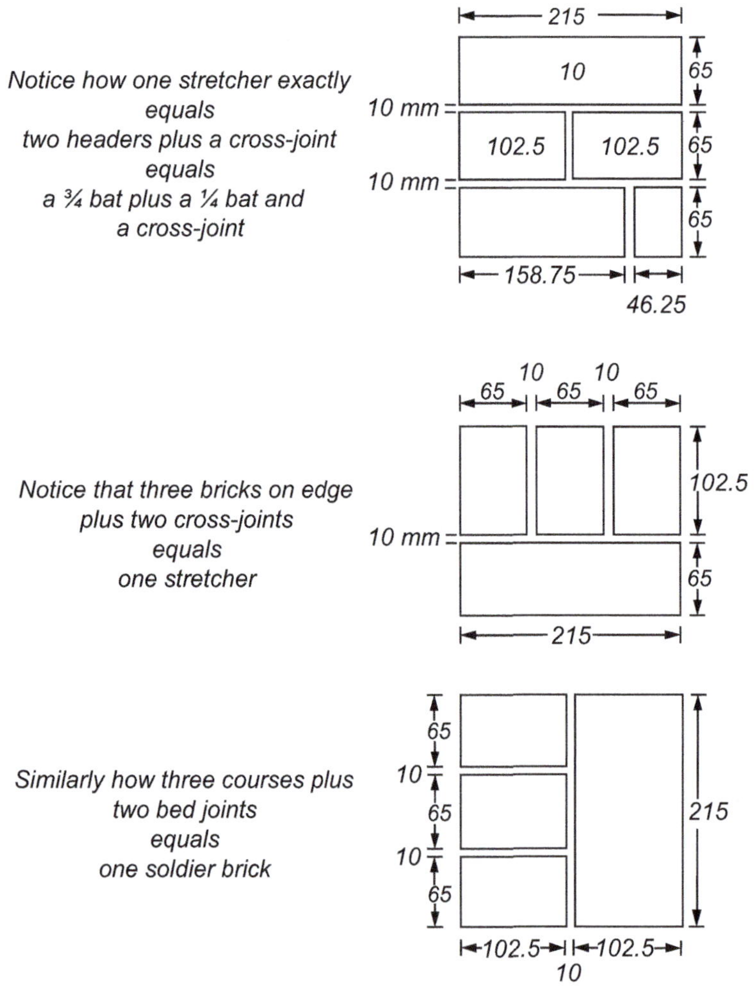

FIGURE 8.1
Brick proportions

stretcher faces with joint allowances for standard bricks. The cut pieces of whole bricks are used in bonding.

Understanding brick dimensions

The co-ordinating size of clay bricks, inclusive of mortar joints, is given as 225 mm long by 112.5 mm wide by 75 mm deep (Figure 8.2).

The work size, for which brick manufacturers aim (also given in the British Standard) is 10 mm less than each of those dimensions, to provide a 'joint allowance'. The work size for each standard metric brick is therefore 215 mm long by 102.5 mm wide by 65 mm deep.

The actual sizes of bricks as delivered vary slightly, and these variations must be absorbed by adjusting the thickness of mortar joints.

The individual differences in brick dimensions (larger or smaller than the work size) are mainly due to differing rates of drying and shrinkage of the naturally occurring clay or shale - the raw material. This is where the skill of the bricklayer is important. Opening or tightening up cross-joints must be carried out carefully and evenly, so that:

- it is not noticeable in the finished wall
- plumb perpends are maintained at three- or four-brick intervals for the full height of the wall
- the co-ordinating size of the brickwork is maintained.

Dimensional deviations

BSEN 771-1:2003 is the European clay masonry unit's product specification or standard.

The manufacturer should declare the work size in order of length, width and height in millimetres. The deviation category for the mean values must also be stated.

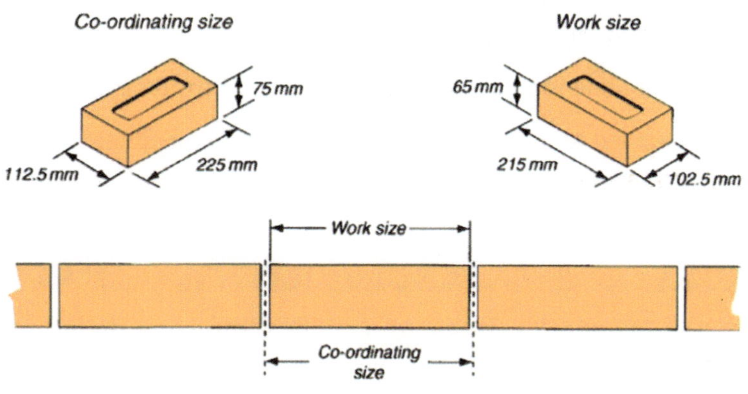

FIGURE 8.2
Sizes of standard bricks

Table 8.1 Tolerance categories

Work size (mm)	T2 (mm)	T1 (mm)	TM
Length 215	±4	±6	Deviation declared by the manufacturer
Width 102.5	±3	±4	
Height 65	±2	±3	

Different categories take into account the wide range of brick types and variation in acceptable appearance to different manufacturing methods.

All bricks will be sold with two categories, known as dimensional tolerance range, which is defined as T for tolerance and R for range classification. The tolerance categories are shown in Table 8.1.

The maximum range (difference between largest and smallest measurement) for any given dimension on the batch of 10 sampled should be within one of the three categories shown in Table 8.2.

There are a number of classifications; in most cases T2 R1 is the most common one.

To determine compliance the use of specially calibrated instruments to measure accurately the length, width and height of a brick must be used. It should be noted that the use of such instruments on site is not practicable and the British Standards Institution has published a document suitable for on-site appraisal of brick units.

This publicly available specification (PAS 70:2003) provides both guidance on appearance of facing bricks and practical on-site testing methods for size, using a conventional steel tape.

On-site test

The PAS has two procedures for measurement. Procedure A, the simpler procedure, is used to determine the size tolerance of the brick consignment delivered to site.

In procedure A, 10 bricks, taken at random from a consignment, should be set out on suitable level ground (Figure 8.3), and the overall measurement taken for the 10 bricks.

This overall dimension is then divided by 10 to provide a mean figure.

This is rounded to the nearest whole number and the difference between this figure and the work size is compared with the tolerance categories in Table 8.3.

Table 8.2 Range categories

Work size (mm)	R2 (mm)	R1 (mm)	RM
Length 215	4	9	Deviation declared by the manufacturer
Width 102.5	3	6	
Height 65	2	5	

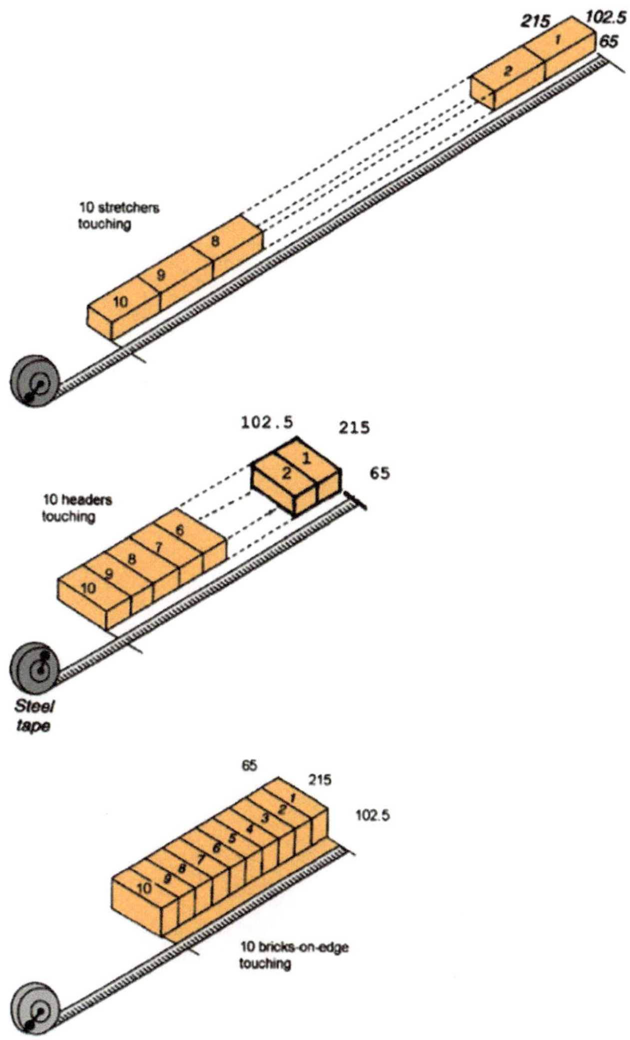

FIGURE 8.3
Carrying out an on-site brick test for dimensional deviations

Table 8.3 Tolerance categories

Work size (mm)	T2 (mm)	T1 (mm)	TM
Length 215 Width 102.5 Height 65	3 2 1	5 3 2	Option for manufacturers to declare own limits

Bonding

Purposes of bonding

The reasons for bonding brickwork are:

- to strengthen a wall
- to ensure that any loads are distributed (Figure 8.4)

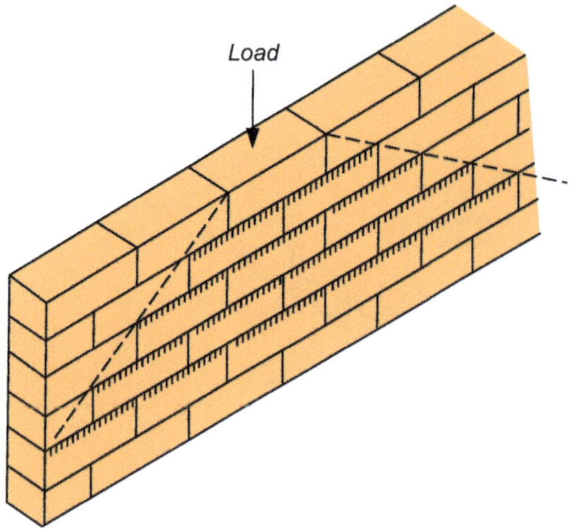

FIGURE 8.4
Bonding the bricks distributes the loads evenly

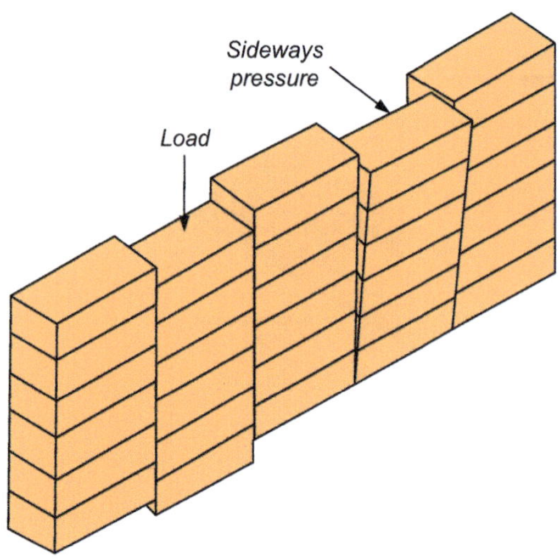

FIGURE 8.5
Tendency to failure of unbonded walls

- to make sure that it is able to resist sideways or lateral pressure (Figure 8.5).

The straight joints of an unbonded wall make it weak and liable to failure (Figure 8.5).

Principles of bonding

To maintain strength, bricks must be lapped one over the other in successive courses along the wall and in its thickness.

There are two practical methods, using either a half-brick lap or a quarter-brick lap, called half-bond and quarter-bond (Figure 8.6).

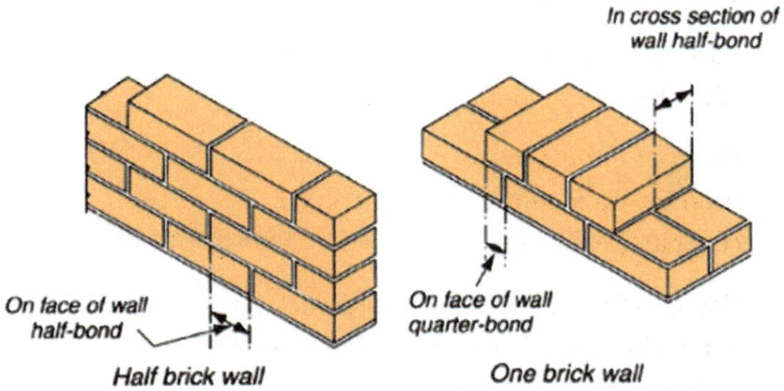

FIGURE 8.6
No straight joints

If the lap is greater or smaller than these, then both appearance and strength are affected. If bricks are so placed that no lap occurs, then the cross-joints or perpends are directly over each other (Figure 8.7). This is termed a 'straight joint', being either 'external' for those appearing on the face of the wall, or 'internal' for those occurring inside the wall, and should be avoided whenever possible.

The apprentice should note that internal straight joints will occur in some bonding problems; excessive cutting would perhaps solve a particular problem, but this wastes labour and materials and tends to weaken the wall. However, by introducing one or two straight internal joints, whole bricks can be used. This is a case where practice and theory must compromise.

The pattern in a brick wall is purposely arranged, has its particular use, and is called a 'bond'.

To summarize, the two main principles of the bonding of brickwork are:

- to maintain half- or quarter-bond, avoiding at all times external straight joints and internal straight joints wherever possible
- to show the maximum amount of specified face bond pattern possible.

To assist in maintaining these principles, rules should be remembered and applied.

The apprentice should never try to remember all the problems shown as examples. Problems must be solved as they occur by the logical application of the rules. Eventually, the bonding of brickwork becomes automatic to the bricklayer.

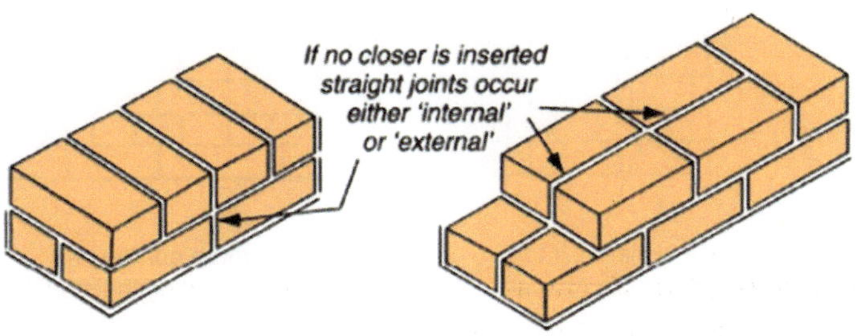

FIGURE 8.7
Straight joints

Several bonds are in general use, but for the purpose of beginning the apprentice's bonding education, stretcher, English, and Flemish bonds will be explained. Problems in other bonds can be solved by the application of the same rules.

Rules of bonding

Bonding is the term given to the various recognized arrangements of brickwork in walling.

These bond patterns are essential for any wall which is intended to carry heavy loads, and they prevent, as far as possible, structural failure.

For this to be effective the bonding must distribute the loading evenly throughout the length of the wall, so that each part of the wall carries a small amount of the load.

In addition to the even distribution of loads throughout the wall, stability is achieved by correct bonding at corners, attached piers, junctions and separating walls, and ensuring that they are well tied in together.

The bonding of brickwork, however, is not confined wholly to strength requirements. Very often a certain bond is introduced for its pleasing appearance, or another bond to make decorative patterns is incorporated in walling facework. In this way a huge flank wall which may otherwise be just a drab piece of walling may be transformed into an interesting architectural feature.

Certain general principles may be applied to bonding:

- The correct lap should be set out and maintained at quoins and stopped ends (Figure 8.8a, b).
- The perpends or cross-joints in alternate courses should be kept vertical (Figure 8.9).
- There should be no straight joints in a wall; that is, no vertical joints should coincide in consecutive courses or, if they are unavoidable, they should be kept to a minimum (Figure 8.7).

Note

Wall thicknesses are usually stated in brick sizes, e.g. the width of a brick is known as a half-brick wall; the length of a brick as a one-brick wall; the width, plus length of a brick, as a one-and-a-half-brick wall, and so on.

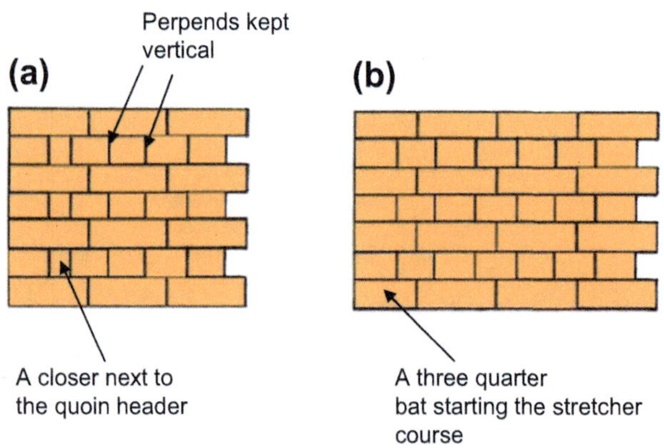

FIGURE 8.8
(a) A closer next to the quoin header; (b) a three-quarter-bat starting the stretcher course

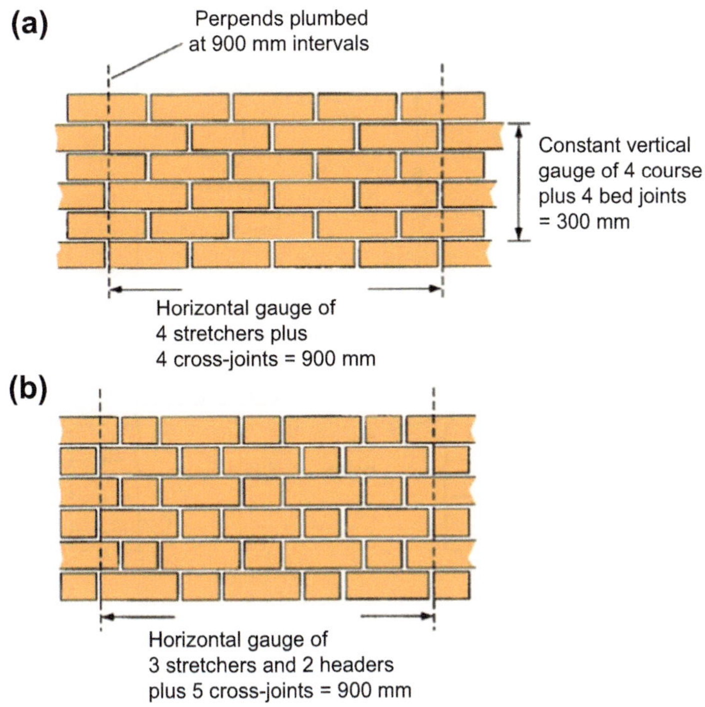

FIGURE 8.9
Plumbing perpends

- Closers should never be built in the face of the wall except next to a quoin header (Figure 8.10).
- The tie bricks at junctions or quoins should be well bonded to secure the walls together (Figure 8.11).
- In English bond, when a wall changes direction, the face bond changes from headers to stretchers or vice versa in the same course (Figure 8.12).

These co-ordinating dimensions for brickwork take into account variations in the size of bricks. The same principles shown in Figure 8.9(a) on an elevation of stretcher bond also apply to other face bonds.

To ensure maximum strength in a wall, all the joints should be flushed up with mortar on every course.

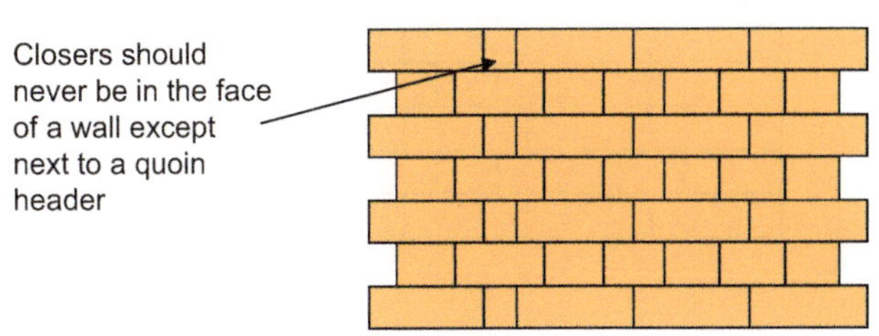

FIGURE 8.10
Wrong position of closers

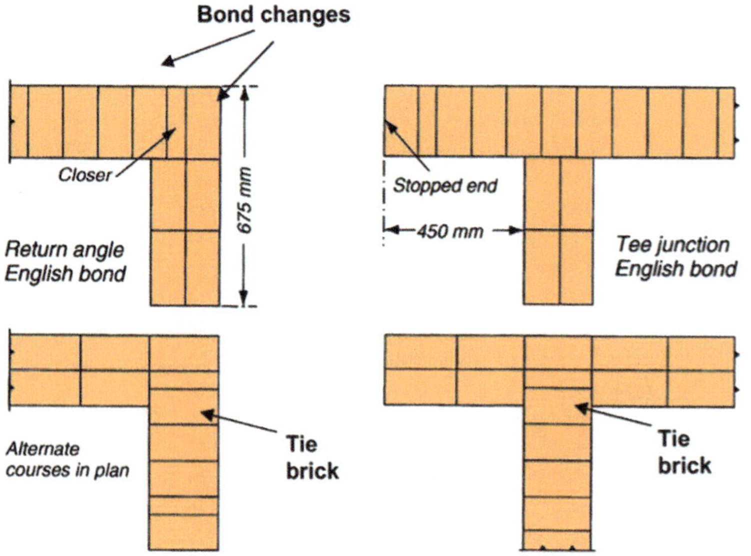

FIGURE 8.11
Tie bricks and bond changes

Types of bond

English bond

Walls built in English bond are very strong as no straight joints occur in any part of the wall (Figure 8.12).

Alternate courses of headers and stretchers produce quarter-bond, and because of its somewhat monotonous appearance it is used where strength is preferable to appearance. To achieve and maintain quarter-bond a queen closer must be laid next to the quoin header.

One-and-a-half-brick walls have two faces and the line should be used on both sides.

Attention to the following points helps to maintain a good standard of work.

- Cut queen closers neatly and keep them regular in size.

Remember

There are sometimes exceptions to rules.

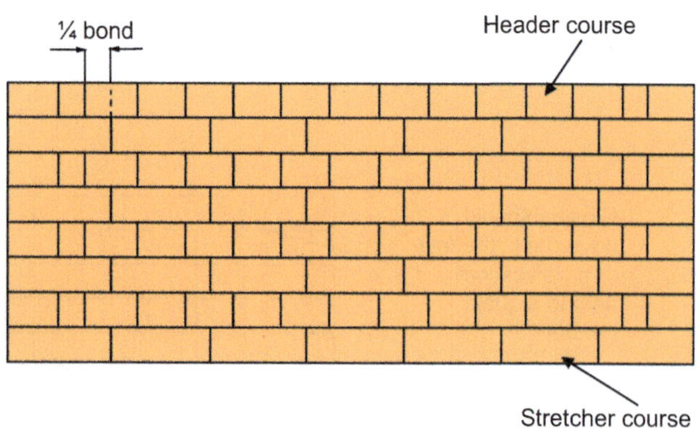

FIGURE 8.12
English bond wall

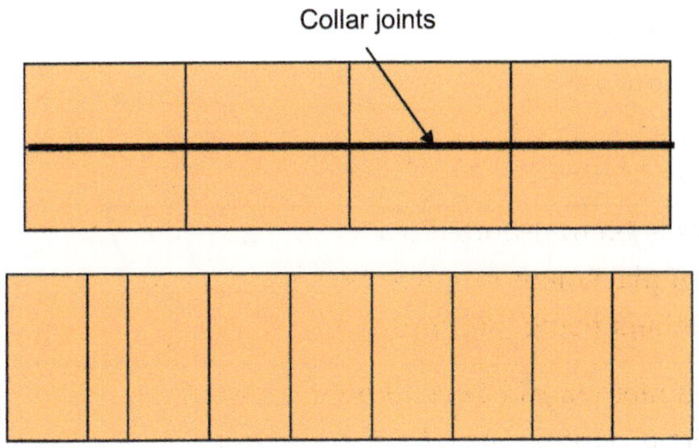

FIGURE 8.13
Collar joint

- Keep perpends uniform and plumb as large cross-joints can soon cause you to lose quarter-bond and can bring straight joints on to the face.
- Remove mortar from the back of the bricks against the collar joint as this could prevent the backing up from being laid level.

When backing up avoid the use of too much mortar near the collar joint, since the backing up could cause the face bricks to move out in front of the line.

Different skills are required when building one-brick walls.

Collar joints

The collar joint runs along the centre of the stretcher course when building a wall in English bond (Figure 8.13).

Stretcher course

Care should be taken when laying the stretcher course as too much mortar between the bricks will cause the face of the wall or back of the wall to bend (Figure 8.14).

The face of the wall should be laid first to the line provided.

When the mortar is laid for backing the wall it should be pulled back from the face brick so that when the backing brick is laid it does not push out the facing brick.

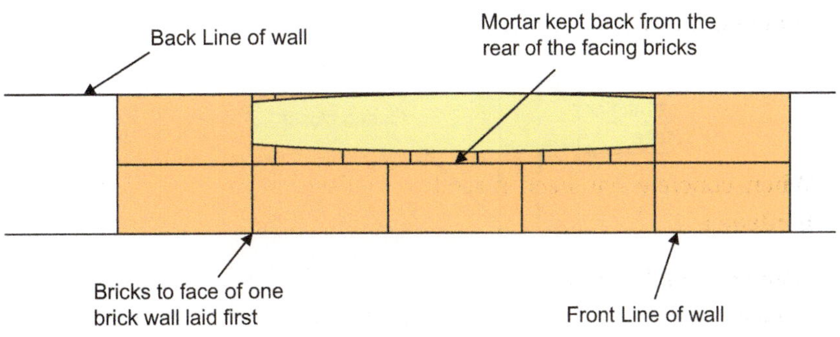

FIGURE 8.14
Laying the back stretcher course

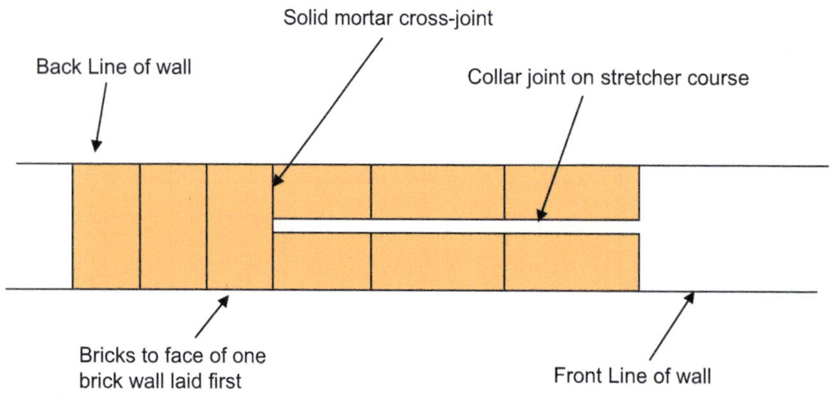

FIGURE 8.15
Laying the header course

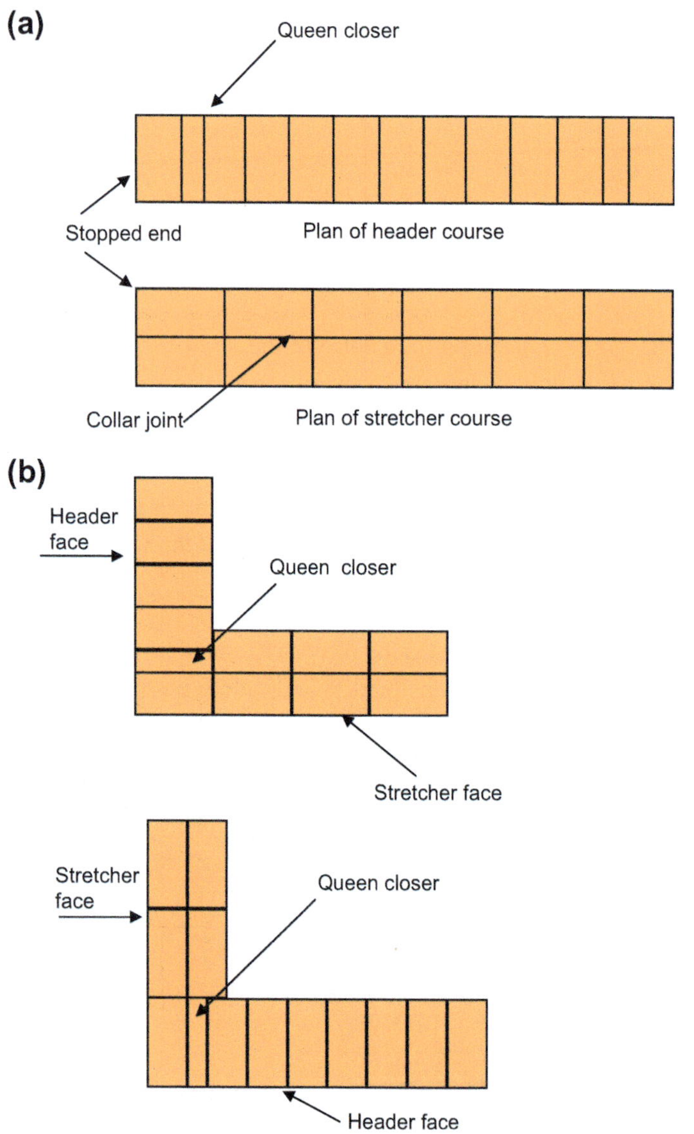

FIGURE 8.16
(a) Straight walls and (b) quoins in English bond

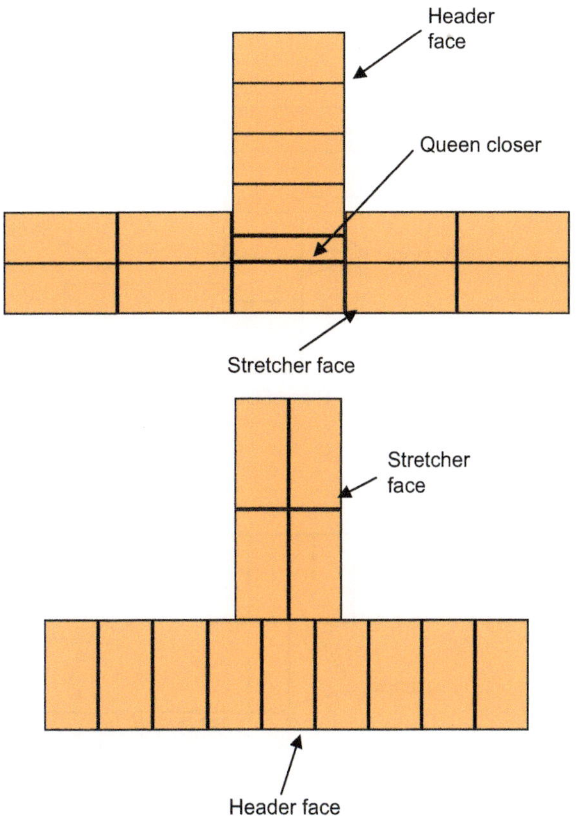

FIGURE 8.17
Tee junctions in English bond

Header course

When the header course is being laid special attention should be made to the long cross-joint (sectional joint) along the stretcher face of the brick (Figure 8.15).

The brick will have to be turned in the hand to expose the longer face of the brick.

The cross-joint should be applied in full, not top and bottom, to produce a solid joint.

Remember

There is no line to the back of the wall so it is essential to use the eye to line the bricks in with the course below.

The correct action is vital if full joints are to be achieved. Practice is necessary to achieve full joints.

It is usual to lay the back course from the front of the wall. This is another skill that has to be practised to obtain a straight face to the rear of the wall.

English bond details are shown in Figures 8.16–8.18.

Flemish bond

Flemish bond consists of courses of alternate headers and stretchers with the headers in one course placed centrally over the stretcher in the course below (Figure 8.19).

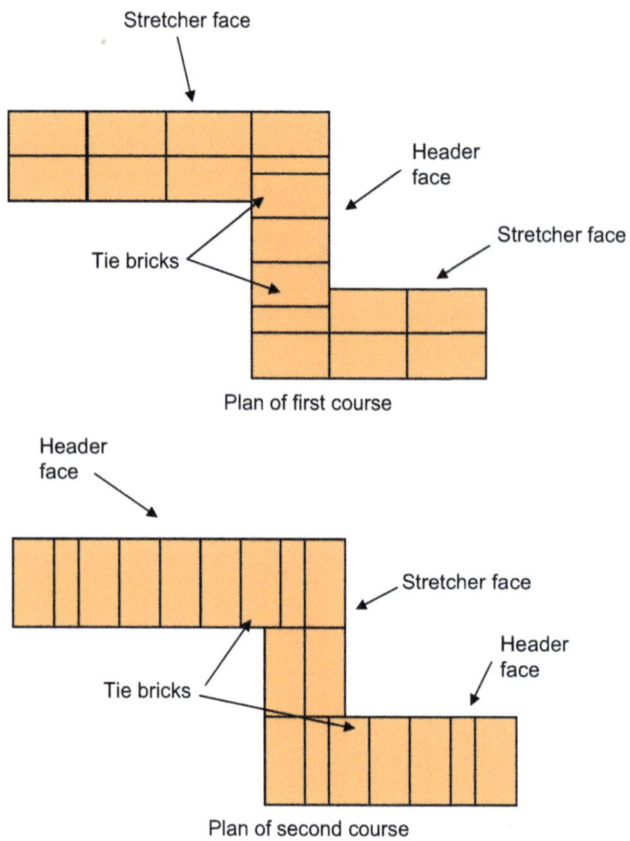

FIGURE 8.18
Double returns in English bond

A closer is placed next to the quoin header to form the correct quarter-bond.

Bonding quoins, tee junctions and double returns in Flemish bond are shown in Figures 8.20–8.22).

Setting-out facework in a wall without openings

Before setting-out the bond on the face side of any wall, it is wise to make a gauge rod (Figure 8.23). This should be made of straight, smooth timber 50 × 50 mm in cross- section, and approximately 3 m long. Along one face fine saw cuts carefully made at 225 mm intervals will allow stretcher bond to be set out consistently from end to end, leaving any broken bond near the middle of the wall.

Setting out facework with openings, broken bond and reverse bond will be dealt with in Level 2.

On another surface of the rod, saw cuts at 75 mm intervals serve to check regular vertical gauge when bricklaying commences and quoins are raised. Use of this dual-purpose storey rod, for checking horizontal and vertical gauge, avoids the potential risk of error when using a measuring tape.

When the brickwork of a building approaches ground level it is normal practice to change from common bricks to facing bricks.

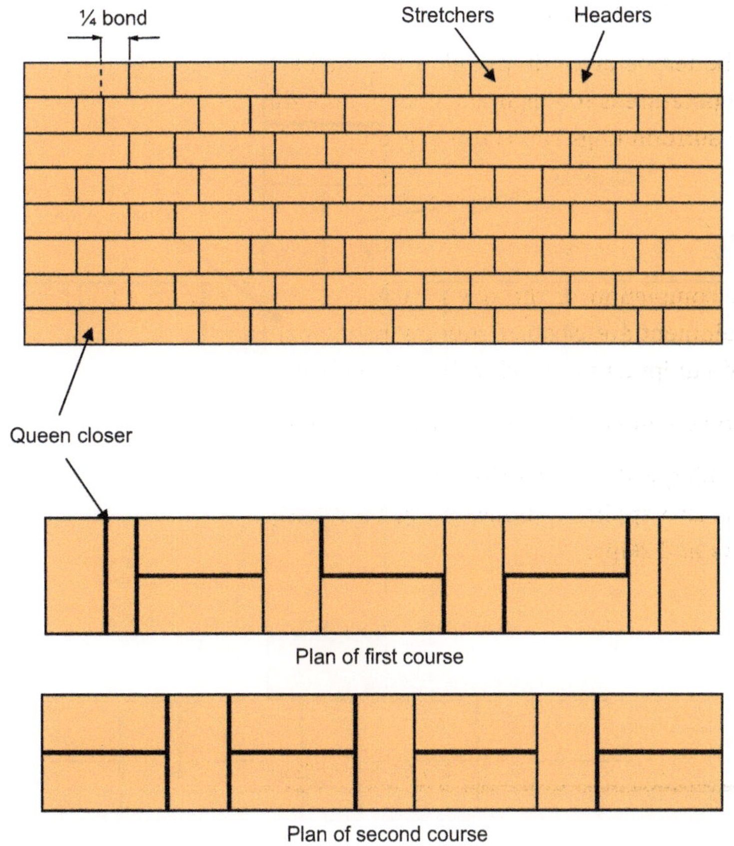

FIGURE 8.19
Straight walls in Flemish bond

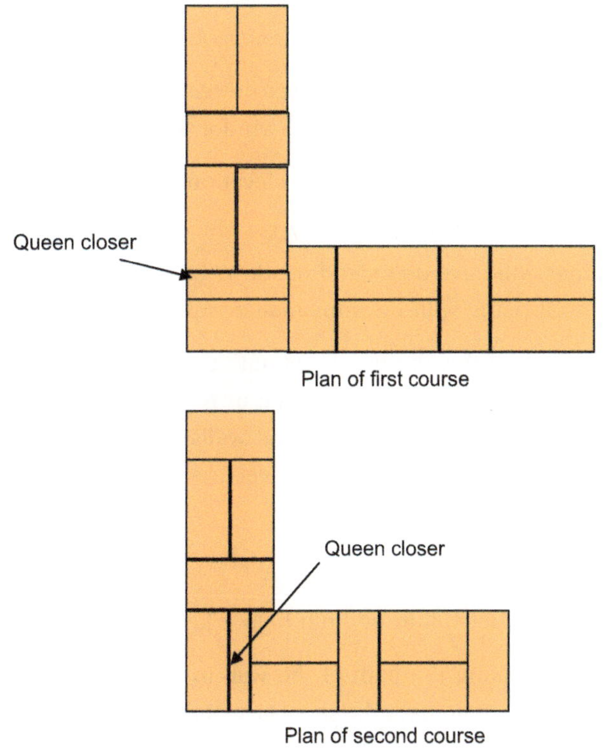

FIGURE 8.20
Quoins in Flemish bond

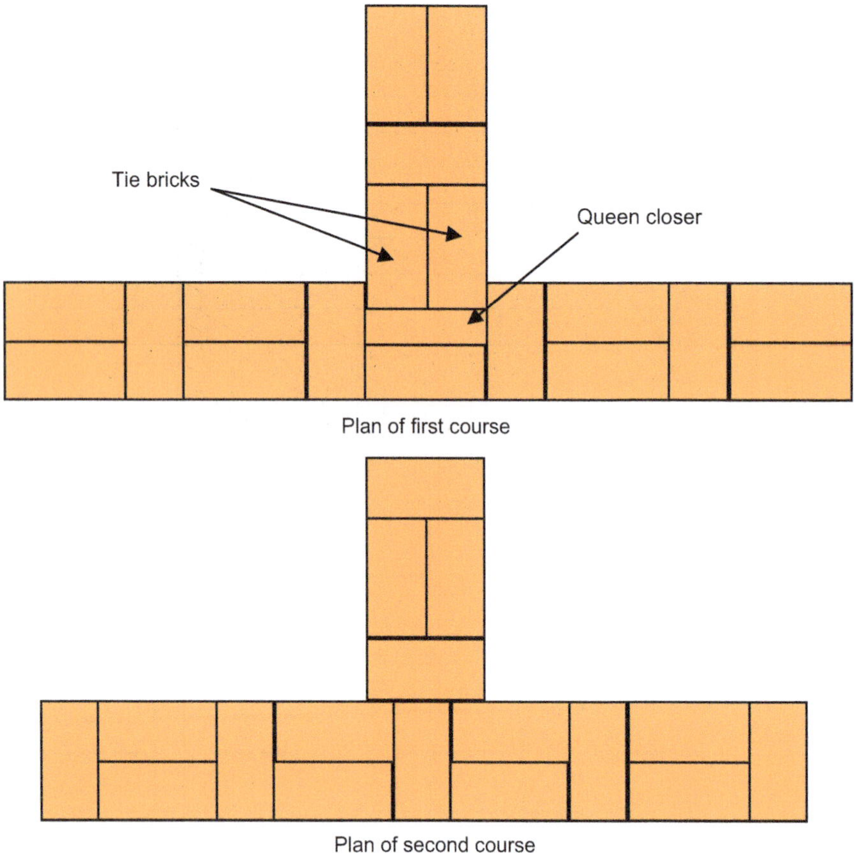

FIGURE 8.21
Tee junctions in Flemish bond

This change over normally takes place two courses below finished ground level to avoid commons being shown when the building is complete.

When the facings reach ground level preparation must be made in the bond arrangement for inserting door and window openings at a future height.

It is at this stage that the proposed elevation of the building must be carefully studied.

All buildings will have doors and windows built into them and this will lead to the long wall being split up into smaller walls.

It is essential that the bonding is set out at ground level to avoid the problems dealt with in the last module - broken or reverse bond when the windows are bedded into position later. Setting out walls with openings, broken bond and reverse bond will be dealt with at Level 2

Dry bonding

Always set out the wall to be built before laying any bricks.

Lay the bricks along the path of the wall in a dry state. This is known as dry bonding. Always try to have the same face of the brick at either end (Figure 8.24).

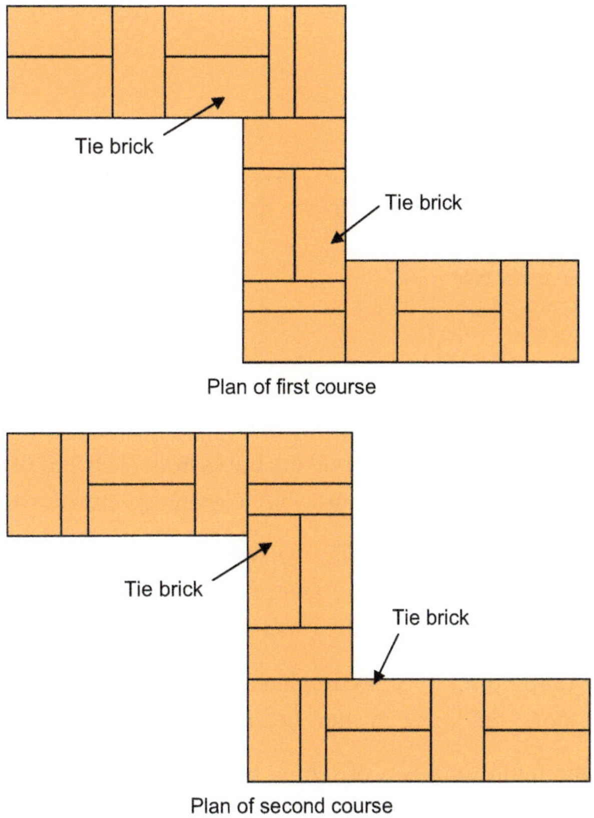

FIGURE 8.22
Double returns in Flemish bond

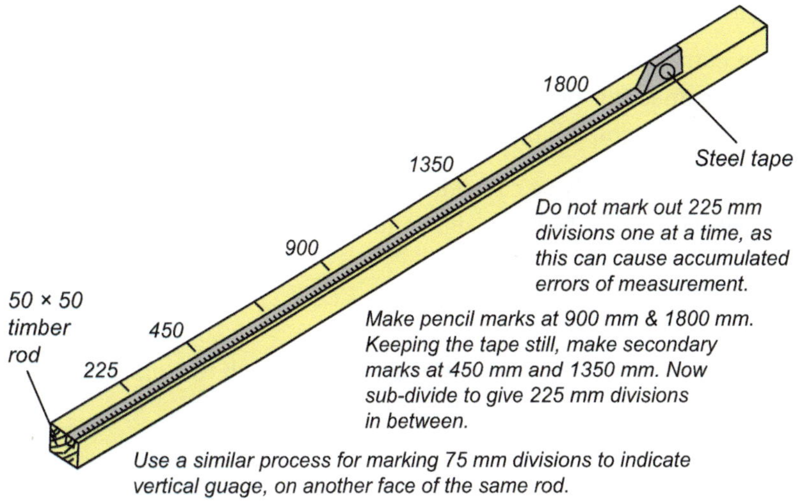

FIGURE 8.23
Marking out a gauge or storey rod

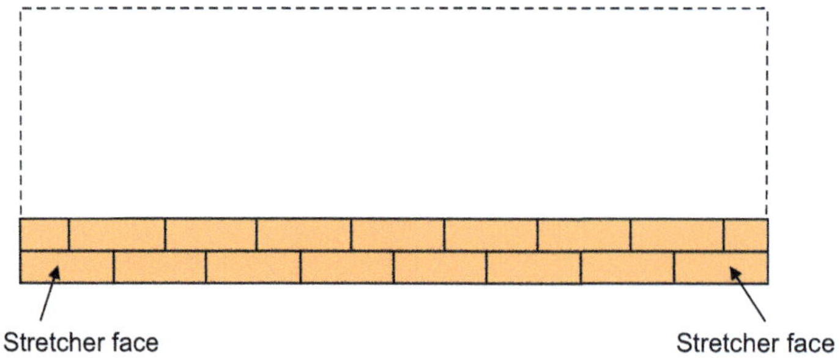

FIGURE 8.24
Dry bonding a wall

The normal practice is to dry bond the bricks from each end of the wall, placing the 'broken bond' under the door or window opening where it will be lost for a while.

Setting out facework is usually the responsibility of the supervisor who will, after consulting the architect, determine the detailed bond pattern and the location, if any, of broken or reverse bond.

Setting out the brickwork is completely different from setting out the building, which is done before excavation begins.

One of the main purposes of setting out facework is to create a matching and balanced appearance of bricks, particularly at the reveals on either side of the door and window openings and the ends of the wall.

It is essential that no straight joints occur.

Craft operations

Cutting bricks

The cutting of bricks is always difficult, especially in the case of hand-made and engineering bricks, the former creating excessive waste and the latter being almost impossible to cut with a hammer and bolster.

Bricks can normally be cut with a club hammer and bolster chisel (Figure 8.25). A selection of cut bricks is shown in Figure 8.26.

Bricklaying

The bricklayer has no conventional rules to which to work; rather, the rules to be used are the outcome of experience.

The apprentice will no doubt have other problems to handle besides those shown.

Time should be spent visualizing the job, considering the best methods of approach and using all the skills possible, whether the work is to be covered or highly decorative, for all to see.

As a learner, the apprentice should never allow quality to be sacrificed for speed, which will be attained by constant practice.

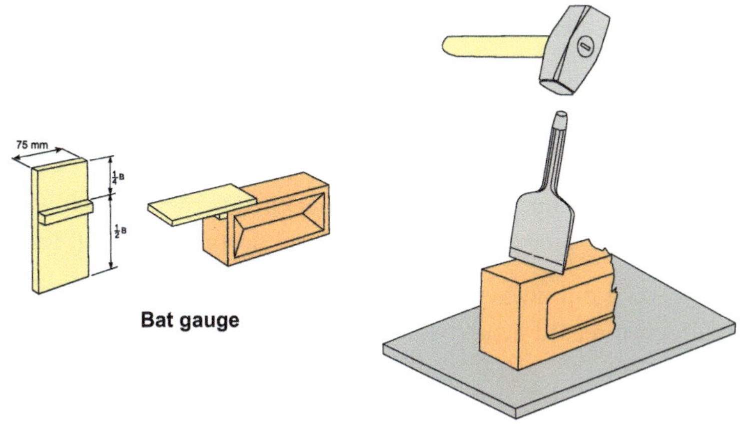

FIGURE 8.25
Cutting bricks with a club hammer and bolster chisel

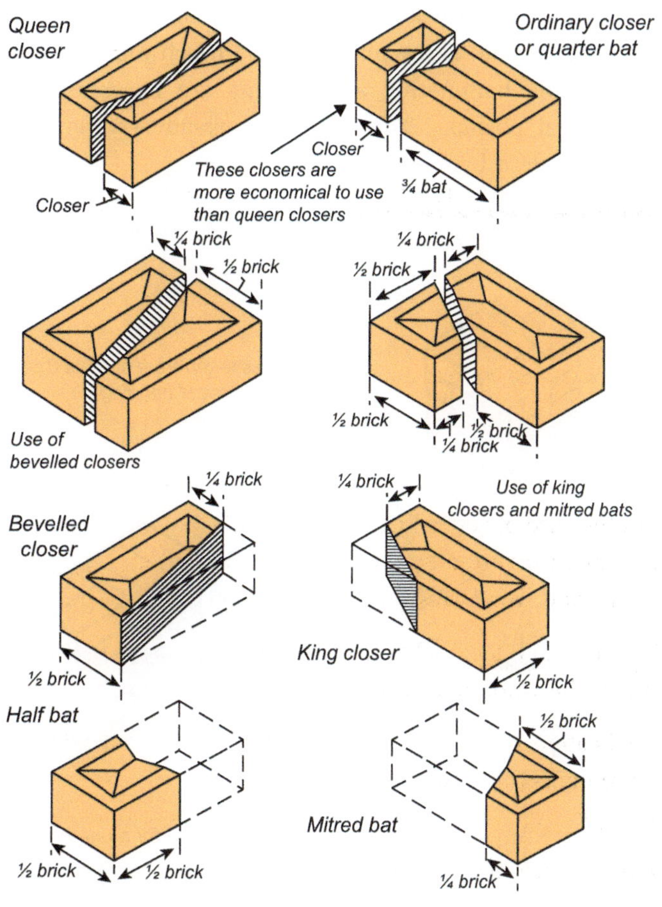

FIGURE 8.26
Definitions of cut bricks

The stability and appearance of the work should always be the master craftsperson's chief concern.

Good bricklaying entails the ability to master the art of spreading the mortar bed, dexterity in handling the brick to be laid and the possession of a keen eye. All of these can be acquired by practice.

Before laying bricks on any job, place the mortar or 'spot' board, with the bricks or blocks, in a convenient position. They must be within easy reach so that no unnecessary movement is involved when materials are required.

Block up the spot board on bricks, one at each corner, so that it is kept clean, and load out as shown for a one-brick wall or a cavity wall (Figure 8.27).

When loading out always take facing bricks from as many packs as possible, but at least three, to avoid banding of the bricks.

Handling the trowel and mortar

Do not grasp the trowel as if clenching the fist, but place the thumb on the ferrule and handle lightly, so that a flexible wrist action is possible (Figure 8.28). Pick up the mortar with an easy sweeping motion and spread it on the wall sufficiently thick to allow the brick to be placed by pressure of the hand. A common fault is the placing of too much mortar under the brick, so that considerable hammering and tapping are necessary before the brick reaches its final position. The bricklayer usually estimates the amount of mortar bed required by the feel of the brick.

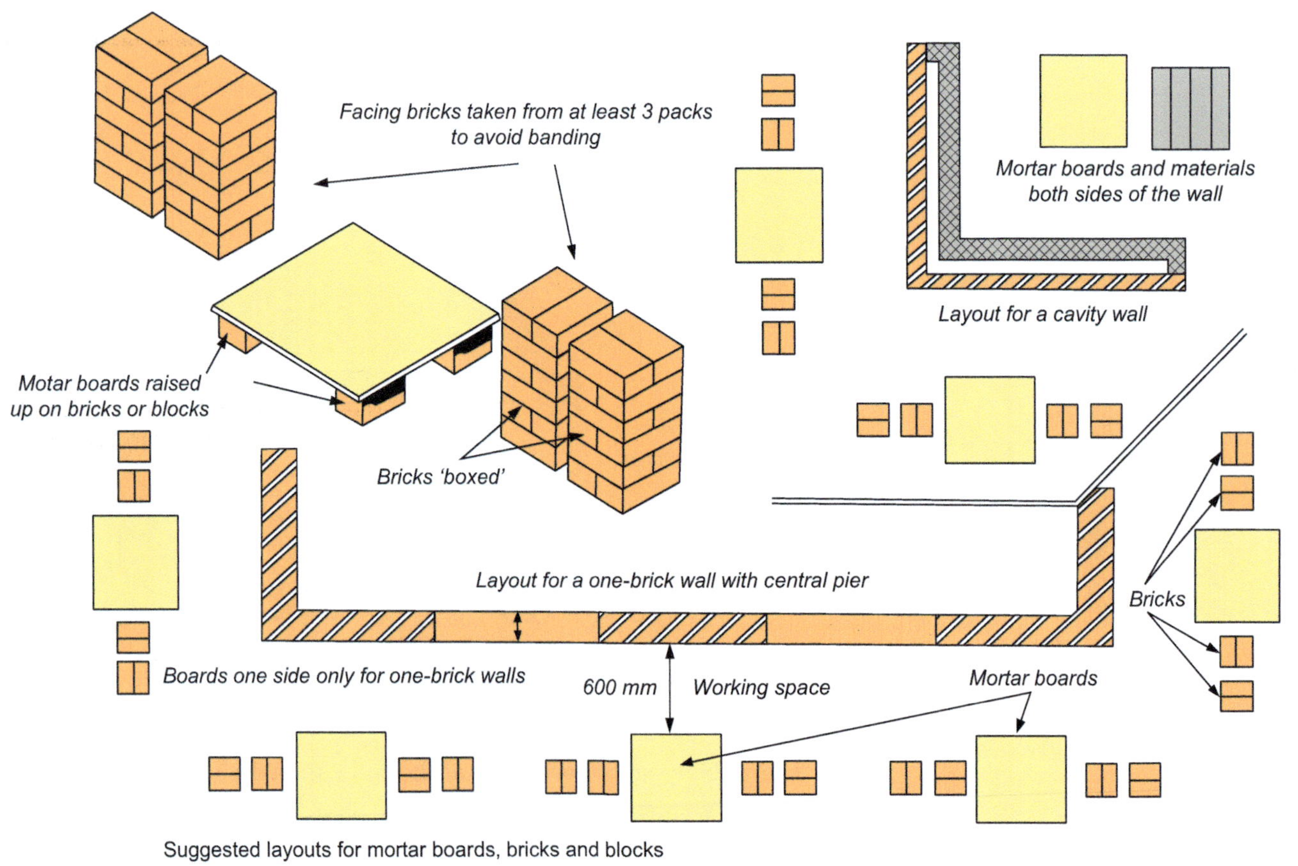

FIGURE 8.27
Layout of materials

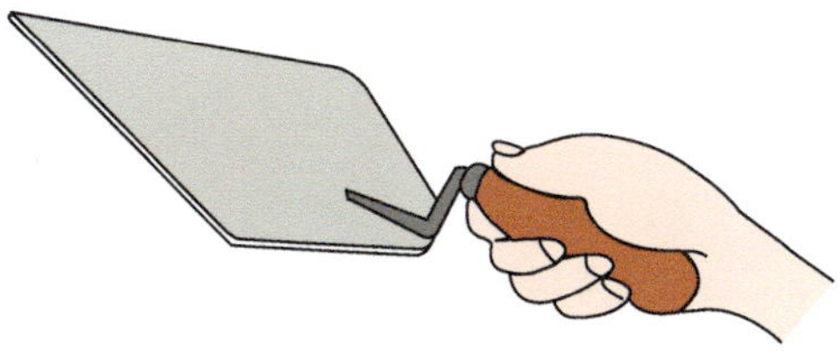

FIGURE 8.28
Method of holding the brick trowel

The following exercises should be practised until you are proficient.

1. Wet the spot board with clean water.
2. Set two shovels of freshly mixed mortar on to the spot board.
3. Stand in the correct working position for picking up the mortar, i.e. right- handed bricklayers forward and to the right facing the spot board, left-handed bricklayers reversed.
4. Cut away a quantity of mortar (Figure 8.29).
5. Using a sawing action draw the trowel of mortar across the spot board to form a roll.
6. Move the trowel back from the roll of mortar and turn it so that the blade is horizontal, 1 mm above the spot board and 50 mm diagonally from the roll.
7. With a sharp movement pick up the roll of mortar (Figure 8.30).
8. Set the trowel of mortar back with the rest on the spot board and repeat the action.

Spreading the bed joints

9. Once the mortar has been picked up on the trowel, hold the trowel of mortar over the edge of the spot board.
10. In a sweeping movement draw the trowel blade parallel to and along

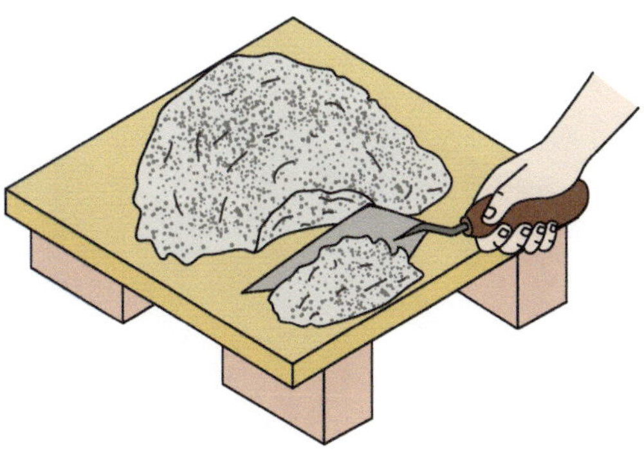

FIGURE 8.29
Method of cutting away a quantity of mortar on the mortar board

FIGURE 8.30
Method of picking up the mortar off the mortar board

the edge of the spot, simultaneously turning the trowel blade and spreading the mortar along the edge of the spot board (Figure 8.31).

11. With the trowel point, furrow the spread mortar along its length with a series of undulating trowel movements.
12. Cut off the surplus mortar along the edge of the spot board to produce a clean edge to the spread of mortar.
13. Clear mortar from the edge of the spot and repeat.

It is very important to practise these actions to become competent before moving on to the next exercise.

Handling bricks

If you are right handed then the left hand is used for handling the bricks. The trowel should not leave the hand.

A series of hand skills should be practised, such as picking up, turning to the correct face or bedding plane, placing on to the mortar bed and accurately aligning the bricks.

The correct way to hold the brick is across the width.

Remember

Protective footwear should b worn when handling bricks.

FIGURE 8.31
Method of spreading and furrowing the mortar along the edge of the mortar board

Applying the cross-joints

Before a brick is laid adjacent to another the cross-joint must be filled. Experienced bricklayers will often do this in situ using the last brick laid.

Trainees should mortar the cross-joints of the brick to be laid as an aid to accurate bedding of the brick and as practice in the co-ordination of hand movements.

1. Take a scoop of mortar on the trowel blade directly from the mortar on the spot board.
2. Flex the wrist of the trowel hand firmly to cause the mortar to spread across and adhere to the trowel blade.
3. Pick up a brick across its width. With the bedding plane towards the trowel hand, hold the brick and the trowel in front of the body and slightly apart (Figure 8.32).
4. Draw the trowel blade down across the width of the brick at its end so that a portion of the mortar on the blade is stuck on to the brick.
5. Lift the trowel again and turn it through 90 degrees. Draw the blade down across the face edge of the brick and its end. Again, a portion of the mortar should stick to the brick.
6. Move the trowel hand away from the body over the top of the brick. Push the trowel blade across the back edge of the brick. Again, a portion of mortar should stick to the brick. The brick should now be buttered on three arrises at its end and ready for bedding.

Laying bricks

Hand-made bricks are often slightly misshapen and some difficulty may be experienced in keeping a flat face and preventing 'hatching and grinning' (Figure 8.33a).

FIGURE 8.32
Holding the brick and applying the cross-joint

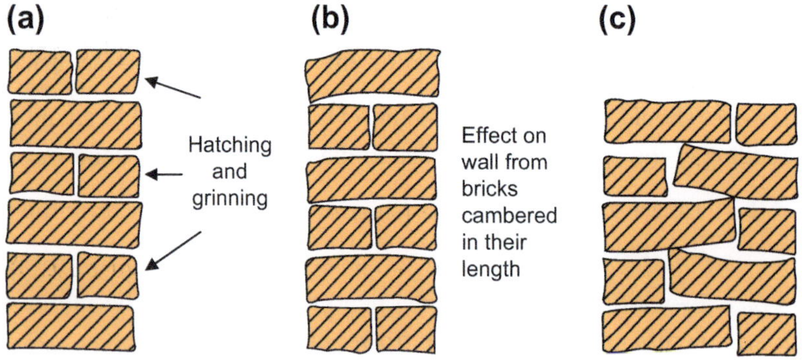

FIGURE 8.33
Dealing with the natural characteristics of hand-made and stock bricks

If bricks are cambered in their length, lay them as shown in Figure 8.33(b).

Never allow them to 'cock up' at the back as shown in Figure 8.33(c), as this makes the laying of the next course difficult and tends to place the wall out of level in its width.

The method shown sometimes necessitates laying the brick frog downwards and filling the frog with mortar before laying, to maintain the solidity of the wall. In ordinary circumstances, however, bricks should always be laid with the frog upwards.

When laying bricks of irregular size it is important to keep the top arris level and increase or decrease the bed joint (Figure 8.34).

When laying bricks to a line always ensure that they are laid up to but not touching the line. The gap should not be larger than the trowel thickness, but should be constant (Figure 8.35).

FIGURE 8.34
Laying irregularly shaped bricks

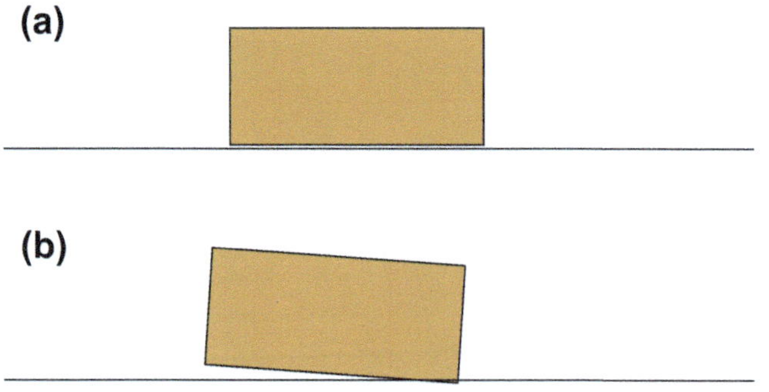

FIGURE 8.35
(a) Correctly laid brick with even gap from line; (b) incorrectly laid brick with one end over the line and the other too far back

Protection

In hot summer weather bricks, other than those of the engineering type, should be wetted to wash off surplus dust and to prevent undue absorption of moisture from the mortar bed.

In winter months this is not necessary as the atmosphere is usually sufficiently damp to achieve these purposes; however, brickwork should be protected overnight against frost. After the day's work is completed and before leaving the job, the last course of brickwork should be covered with hessian sacking or other suitable material which may be available.

Erecting a half-brick wall quoin

When building a wall over 1.125 m in length it is always advisable to use a line and pins, for this will ensure a neat job. Before bringing these into use, it is first necessary to build the corners, making sure that they are vertical or plumb, in alignment (Figure 8.36), and to gauge.

As each quoin brick is laid it should be plumbed with the aid of a spirit level and checked for gauge with a tape or gauge rod. One the quoins or stopped ends have been erected and checked the main part of the wall can be run in.

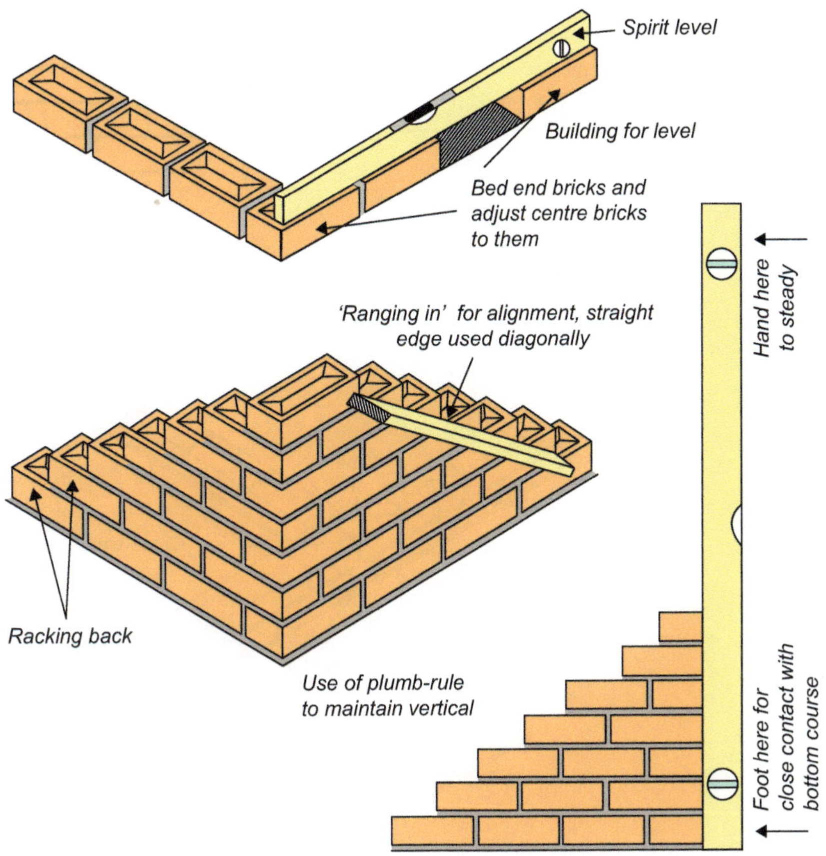

FIGURE 8.36
Raising quoins in half-brick walls

Lines, attached to pins or corner blocks, can be used to ensure level brickwork (Figure 8.37).

The bricks should be laid as explained in Figure 8.34 and 8.35(a).

Longer walls, when the line is likely to sag, should be built as shown in Figure 8.37.

Tee junctions and double angle returns

These are built using the same principle, but care should be taken with bonding the tee junction.

Quarter-bond is introduced by the use of three-quarter-bats (Figure 8.38).

Building a one-brick wall

To build a straight one-brick wall, first, erect approximately six courses of brickwork on the corners and then begin to use a line and pins.

Two ways of keeping the line taut are shown. Figure 8.39 shows a method commonly used: the placing of the line pin in a vertical joint. Figure 8.40, showing the use of corner blocks, illustrates the better method, especially where expensive facing bricks are being used.

FIGURE 8.37
Running in the main brickwork to line

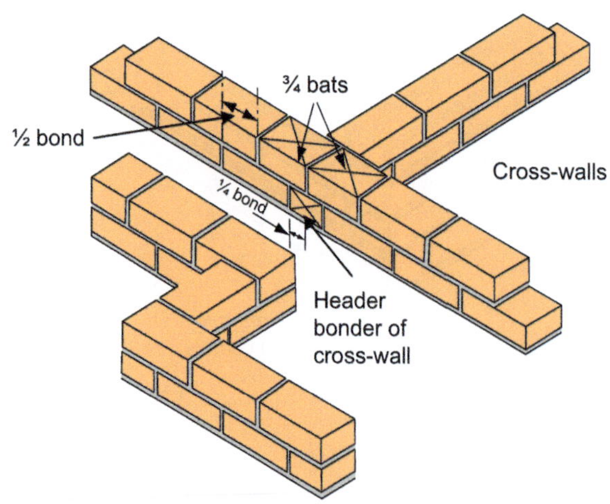

FIGURE 8.38
Junctions and double return angles

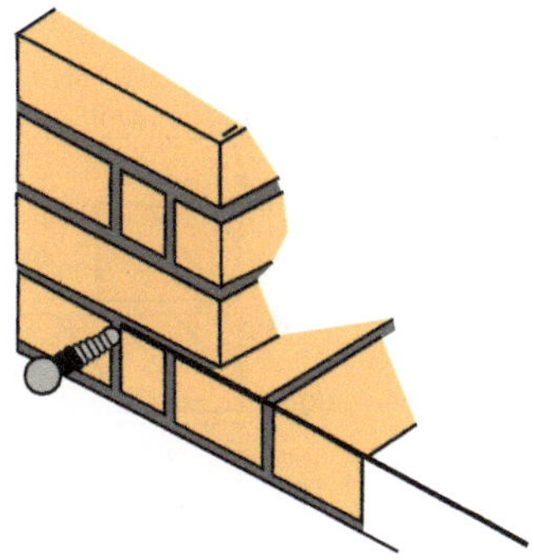

FIGURE 8.39
Using line pins

When laying bricks to a line, always ensure that a trace of daylight can be seen between the line and brick (Figure 8.35a). This prevents the laying of the bricks 'hard' to the line, which if continued, would eventually place the wall out of alignment to a considerable extent. To keep the wall flat and to prevent hatching and grinning, imagine that the bricks are being laid between two lines, one being the edge of the previous course laid to a line A (Figure 8.41) and the other to the present string line position at *B*.

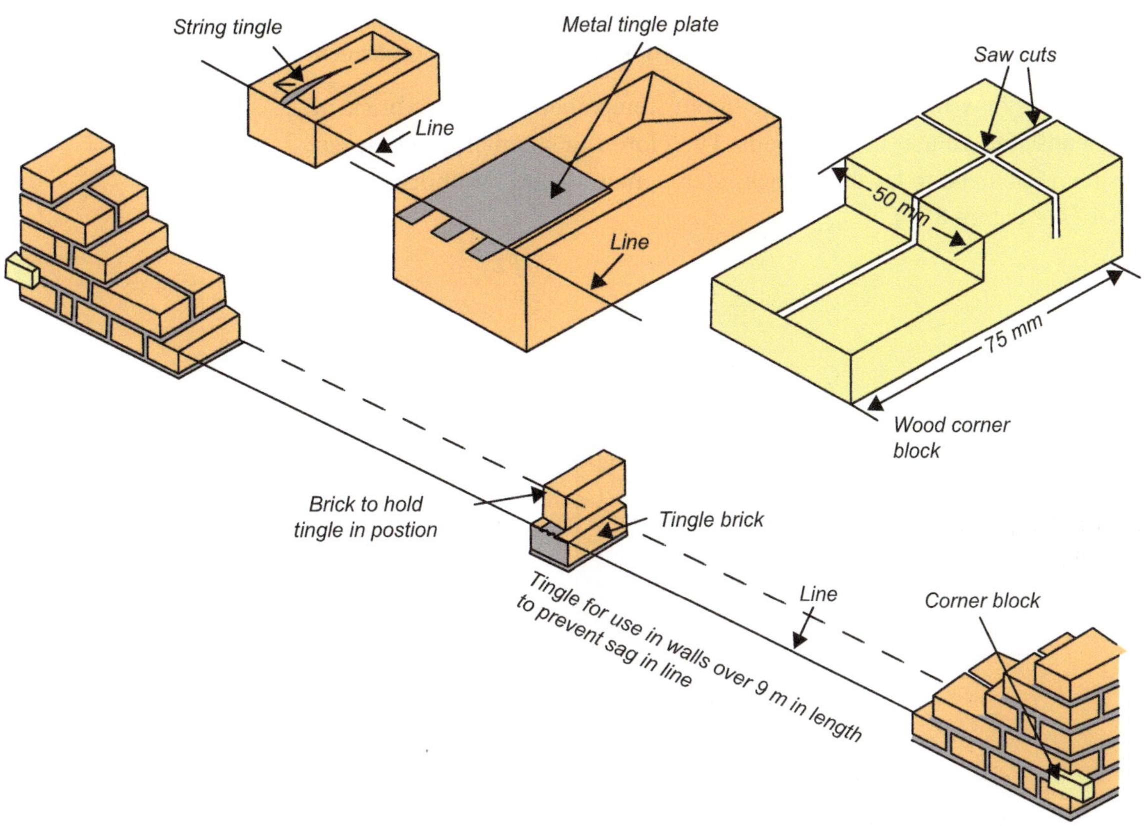

FIGURE 8.40
Using tingle plate and corner blocks

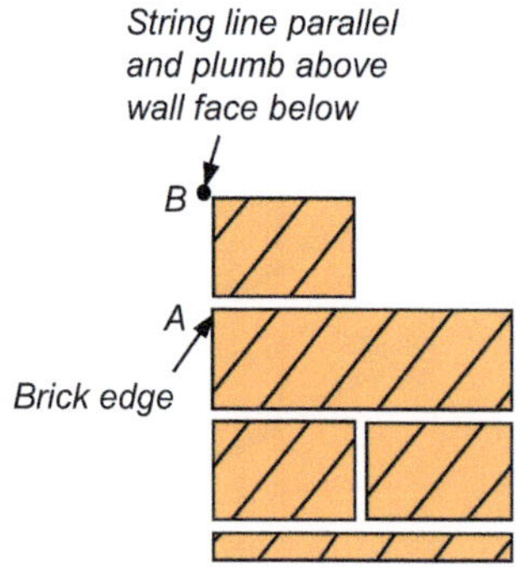

FIGURE 8.41
Laying to line

If a wall exceeds 9 m in length, it is necessary to use a 'tingle' to take the sag out of the line. The tingle brick should be as near the centre of the wall as possible, and must be sighted through from corner to corner every time the line is raised one course, to ensure that the string line is being supported at the correct level (see the bottom right-hand end of Figure 8.40). A tingle achieves its purpose up to a wall length of approximately 12 m.

Raising quoins on one-brick walls

The first course should be laid dry to ascertain the correct bonding and the position of the quoin bricks. Lay the quoin brick first to gauge and level. This brick is then used to gauge and plumb from, so it is important that it is correct.

When deciding how high to make the quoin, remember that every brick in length means one course in the quoin. Therefore, for a small quoin seven courses high you need to lay three bricks along one wall and four bricks along the other (see Figure 8.36).

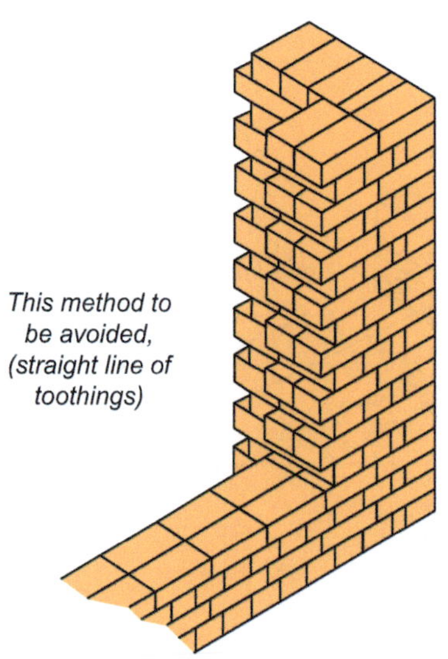

FIGURE 8.42
Bad practice

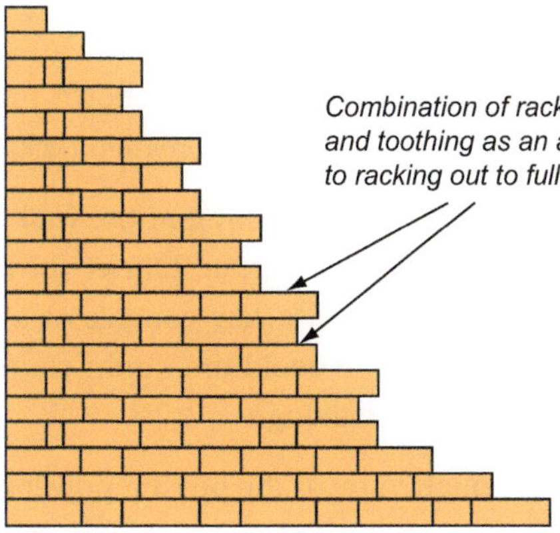

FIGURE 8.43
Better method

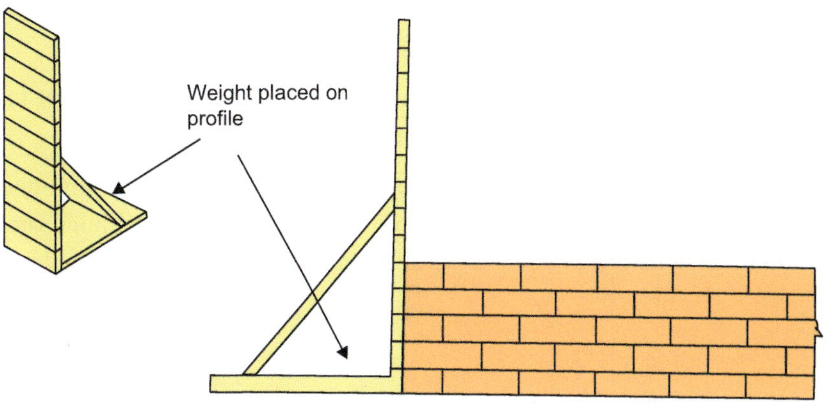

FIGURE 8.44
Wooden profile

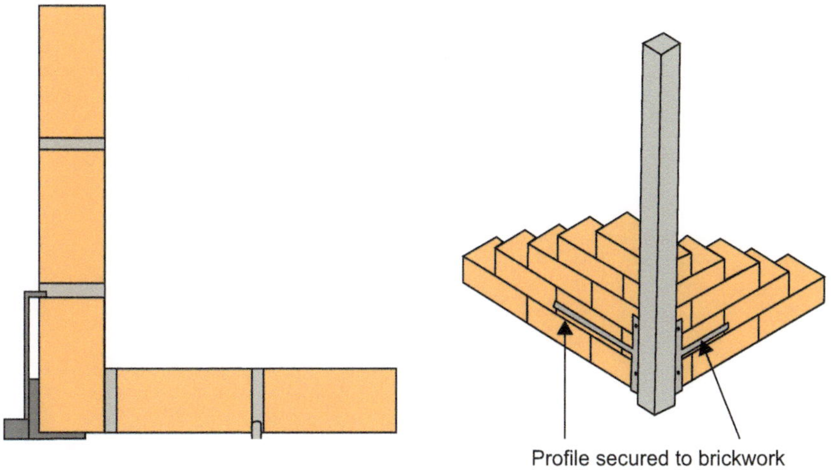

FIGURE 8.45
Patent corner profile

Corners should never be built with a straight line of toothings (Figure 8.42), as it is difficult to make good the toothings in all cases.

A combination of racking back and toothing is shown in Figure 8.43. This prevents a direct line of toothings and avoids excessive racking out of brickwork.

Corners must always be built before running in the wall so that there is always somewhere to fix line and pins or corner blocks throughout the day. It is preferable if corners are only raised a few courses at a time, say, only six or seven courses ahead of the line, to avoid the quoin courses getting out of face plane alignment with the main walling.

Quoins must be large enough to resist the pulling power of the line each time it is tightened up. They must also be of just sufficient height to allow accurate ranging- in with the spirit level, as shown in Figure 8.36.

Profiles

If it is necessary to build a large quoin, then a temporary timber profile (Figure 8.44) must be set up as indicated, to ensure that the racking-back courses will be truly in line with the overall wall face.

There are many patent profiles available which allow the bricklayer to run the line without setting up a quoin (Figure 8.45).

These should be erected first at both corners, checked for plumb and gauge, and lines attached. The wall is then run without the need to build the corners first. Versions are available as profiles for openings, cutting up gables, etc.

Multiple-choice questions

Self-assessment

This section of the book is designed to allow you to check your level of knowledge. The section consists of revision questions for this chapter. The questions are all multiple choice and have four possible answers. The answers are to be found at the end of the book.

The main type of multiple-choice question will be the four-option multiple-choice question. This will consist of a question or statement, known as the stem, followed by a choice of four different answers, called the responses. Only one of these responses is the correct answer; the others are incorrect and are known as distracters.

You should attempt to answer the questions by choosing either (a), (b), (c) or (d).

Example

The person employed by the local authority to ensure that the Building Regulations are observed is called the:

(a) clerk of works
(b) building control officer
(c) council inspector
(d) safety officer

The correct answer is the building control officer, and therefore (b) would be the correct response.

Laying bricks to line

Question 1 Identify the following item of equipment, which could be used when erecting a stopped end:

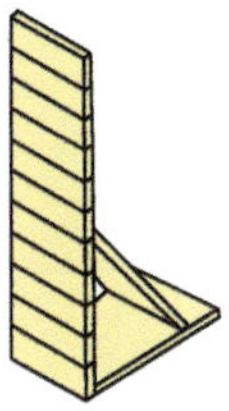

(a) gauge rod
(b) level
(c) straight edge
(d) profile

Question 2 How many bricks are required to carry out a test for dimensional deviation?

(a) 5
(b) 10
(c) 15
(d) 20

Question 3 What is the reason for bonding brickwork or blockwork?

(a) to balance the wall
(b) to strengthen the wall
(c) to remove straight joints
(d) to create a decorative effect

Question 4 Where should the queen closer be positioned in a one-brick wall in English bond?

(a) next to the quoin header
(b) next to the quoin stretcher
(c) at the end of the wall
(d) in the centre of the wall

Question 5 In the drawing of a one-brick wall which is the collar joint?

(a) 1
(b) 2
(c) 3
(d) 4

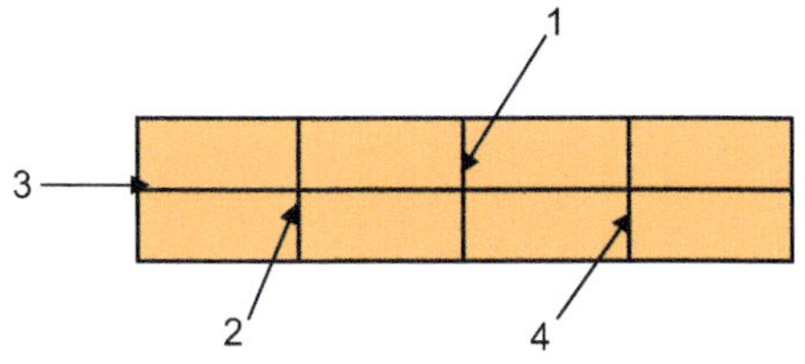

Question 6 Identify the brick shown in the drawing of a tee junction:

(a) queen closer
(b) tie brick
(c) stretcher
(d) header

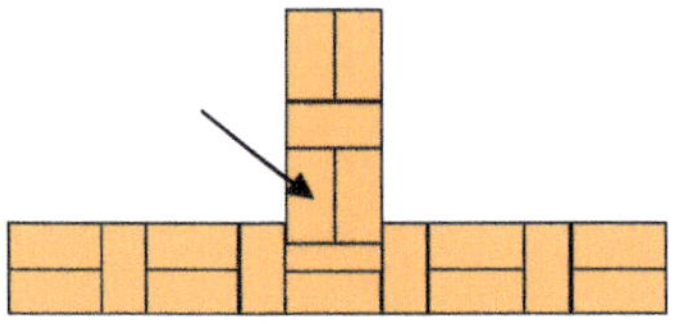

Question 7 Why is it good practice to dry bond a wall before laying?

(a) to ensure that any broken bond is in the centre of the wall
(b) to eliminate any cutting of the bricks
(c) to ensure that there is less waste
(d) to allow you to calculate the bricks required

Question 8 Identify the application shown:

(a) ranging in
(b) plumbing
(c) levelling
(d) gauging

CHAPTER 9

Laying Blockwork to Line

This chapter will cover the following City and Guilds units:

- Apprenticeship Module 5 S1.1, S1.2, S1.3
  S2.2
  K1.1, K1.2
- Diploma Unit 103 1.1, 1.2, 1.4
  2.1, 2.2, 2.4, 2.5, 2.6
  3.1, 3.2, 3.3, 3.5, 3.6
  4.1, 4.2, 4.3, 4.5, 4.6

This chapter is about:

- Interpreting instructions
- Adopting safe and healthy working practices
- Laying blocks to line

The following City and Guild performance criteria will be covered:

- Build single skin walls
- Set out returns and junctions in single skin walls

The following Diploma outcomes will be covered:

- Know how to set out blockwork to comply with workshop drawings
- Know how to build block walling using dense concrete blocks
- Know how to set out and build block walling with lightweight blocks

Types of block

A block is described as a walling unit exceeding the dimensions for bricks; its height should not exceed either its length or six times its width.

Blocks are produced from clay and concrete.

Concrete blocks

Concrete blocks are produced in a range of shapes and sizes.

The face side is usually 450 mm × 225 mm, the thickness varies from 37 mm to 225 mm and the weight from 6.3 kg to 15 kg.

They are produced in solid, hollow and multicut format. Multicuts enable a bolster cut to be made without wastage.

Special blocks

Special blocks such as the return block are usually designed to stiffen walls where bonding could cause weakness.

For closing cavities the reveal block could be used; another special is used to produce a splayed reveal.

Some manufacturers produce blocks with an insulant bonded to the outside face, while others produce hollow blocks with an insulant inserted in the voids.

Walling built with precast blocks may be divided into two main categories:

- load-bearing
- non-load-bearing.

Load-bearing blocks

These blocks are precast in moulds and compacted with the aid of vibration, or moulding machines involving the use of compressed air, or a combination of both.

These blocks are usually made of concrete comprised of Portland cement and a variety of aggregates, such as crushed stone, rock ballast or shingle.

Non-load-bearing blocks

These can also be precast in moulds, or can be produced in slab format and cut to size when set.

These blocks are usually made with cement and a variety of lightweight materials, such as fly ash or burnt coke.

Foundation blocks

These are manufactured in widths from 250 mm to 335 mm. These blocks are used below ground level and are designed to support cavity walls.

They may be dense or lightweight. The dense ones may require two people to bed them.

The current Building Regulations state that hollow blocks must have an aggregate volume of not less than 50 per cent of the total volume of the block calculated from its overall dimensions.

Hollow blocks must have a resistance to crushing of not less than 2.8 N/mm^2, if the blocks are to be used for the construction of a wall of a residential building having one or two storeys.

In all other circumstances blocks shall have a resistance to crushing of not less than 7 N/mm^2.

Block sizes

Precast concrete blocks are specified as type A, type B and type C. Their sizes are listed in Table 9.1.

Clay blocks

The manufacturing process is similar to that of clay bricks, using the extrusion/wirecut method.

Clay blocks are made from finely washed clay with certain special properties, which is forced through an extruding machine, in the process of which the blocks are cut off to length as the continuous length of clay emerges from the machine.

The thickness of the walls of the blocks is about 12 mm, allowing them to dry quickly and thoroughly. The green clay blocks are then burnt at a high temperature.

The blocks are usually 300 mm long and 225 mm high and the thicknesses range from 37 to 100 mm for partition walls.

They are available with a smooth face and dovetailed slots to provide a key for the plaster. See Table 9.2.

Table 9.1 Block sizes

Type of block	Length × height (mm)	Thickness (mm)
A	400 × 100 400 × 200 450 × 225	75, 90, 100, 140, 190, 140, 190 75, 90, 100, 140, 190, 225
B	400 × 100 400 × 200 450 × 200 450 × 225 450 × 300 600 × 200 600 × 225	75, 90, 140, 190 75, 90, 100, 140, 215
C	As above but intended for non-load-bearing walls	As above but intended for non-load-bearing walls

Table 9.2 Main types of block

Type of block	Uses	Material	
Lightweight varieties	Internal non-load-bearing walls and partitions generally	Breeze or clinker; waste coke or ash and cement; burnt clay	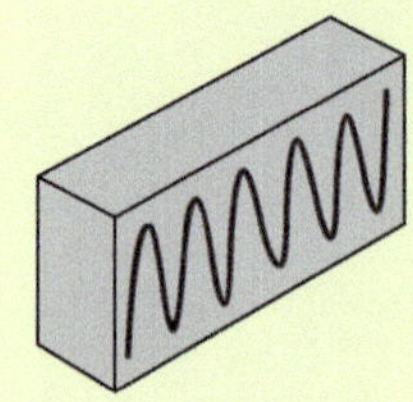
Dense heavy	Internal load-bearing walls and external walls	Usually concrete	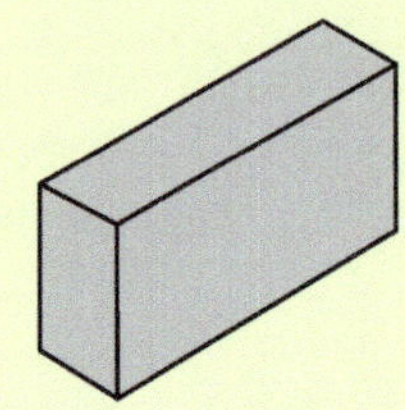
Hollow concrete	External walls, usually rendered	Concrete	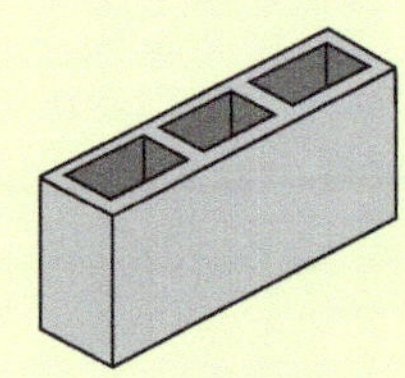
Cellular (light-weight)	Load-bearing internal walls	Burnt clay	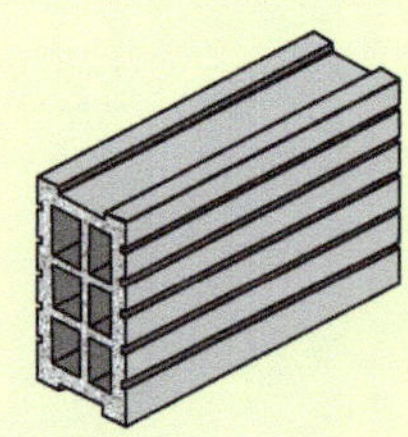

Clay blocks are not very easy to cut, so the manufacturer produces special half-units, enabling the bond to be formed without cutting any blocks.

Specials

Many block manufacturers provide specials cut to assist in the bonding on site and to prevent wastage from cutting (Figure 9.1).

Remember

Never use bricks and blocks in the same wall.

Setting out blockwork

Types and method of construction

Blockwork has become very popular not only as a quick method of producing inside walls and partitions but as a facing material in its own right.

Blocks are very economical in both handling and purchasing compared with bricks.

Apart from their size, the craft operations in erecting walls and corners are as for brick.

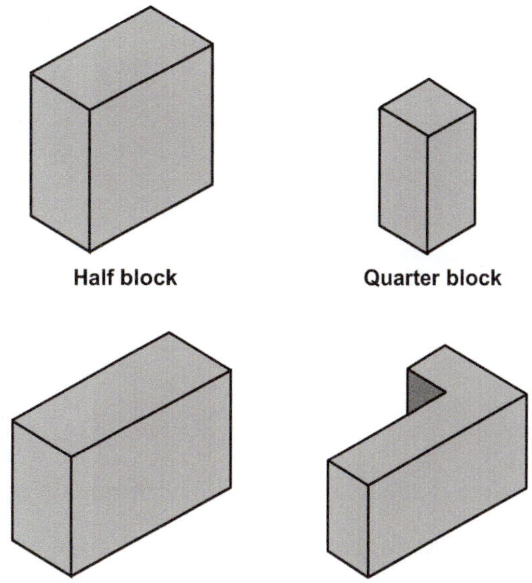

FIGURE 9.1
Special blocks

Remember

It is an offence under the Health and Safety at Work Act to use a masonry bench saw if you are under the age of 18.

Even when you are 18 and over the Act states that you must have been instructed in its safe use and carry an Abrasive Wheels Certificate.

However, there are several ways in which blockwork can be erected satisfactorily, depending on the thickness of the blockwork.

- When internal walls have been set out, fix profiles at angles and indents.
- To prevent buckling limit the height of wall to six courses per day.
- Support long walls to prevent buckling.

Setting out

It is important to dry bond the first course to avoid awkward cuts (Figure 9.2). Set out to a predetermined chalk line.

Many types of block are difficult to cut, although many specials are available such as quarter, half three-quarter and reveal/corner blocks (see Figure 9.1).

As mentioned in Chapter 5, a masonry saw is available to assist in cutting blocks, especially lightweight blocks, although many sites now have masonry bench saws.

Remember

Never mix different block types in the same wall.

Never use bricks, as they will cause a reduction in the insulation value of the cavity wall and could cause problems with drying shrinkage and pattern staining owing to different absorption of various materials.

Cutting blocks

When cutting blocks always use a club hammer and bolster chisel as this should result in neat, accurate cuts.

Select a block that has a good face and appears sound. Measure and mark the amount required (allowing for joints), place on a piece of 'softing' and cut with the hammer and bolster chisel.

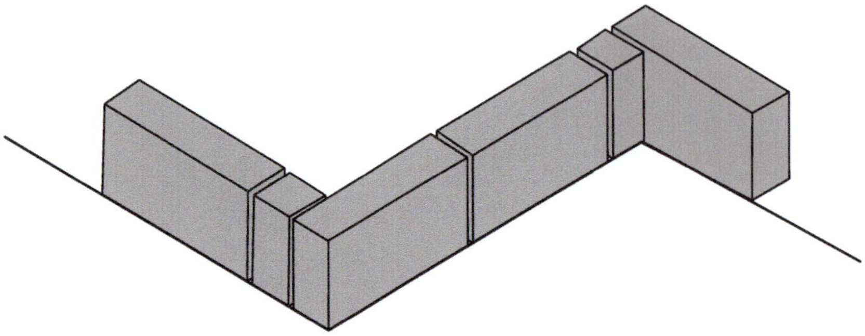

FIGURE 9.2
Dry bonding the first course

Cut along each end and face side until cut through (Figure 9.3).

'Softing' can be sand, rubber belting or an old piece of carpet which will take the 'jar' when cutting the block.

Final trimming of the cut block can be done with the comb hammer or scutch.

Unwanted block debris should be cleared away as soon as possible to ensure that the work area is kept clean.

Bonding

The bond used for blockwork should be half-bond where possible (Figure 9.4).

On no account should the bond be less than a quarter-block length.

Work to tight lines and check for plumb regularly. The maximum lift of blocks completed in one day should not exceed 1.5 m, which is normally six courses.

Quoins

Quoins are erected and racked back as for brickwork. A quarter-block is required next to the corner block to obtain half-bond (Figure 9.5).

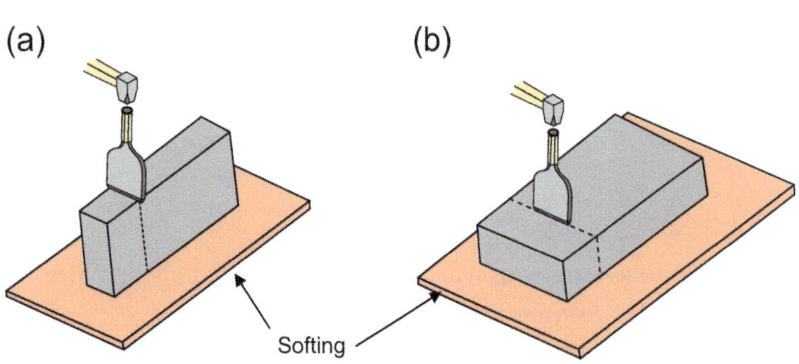

FIGURE 9.3
Cutting blocks: (a) block cut on edge; (b) block cut on face

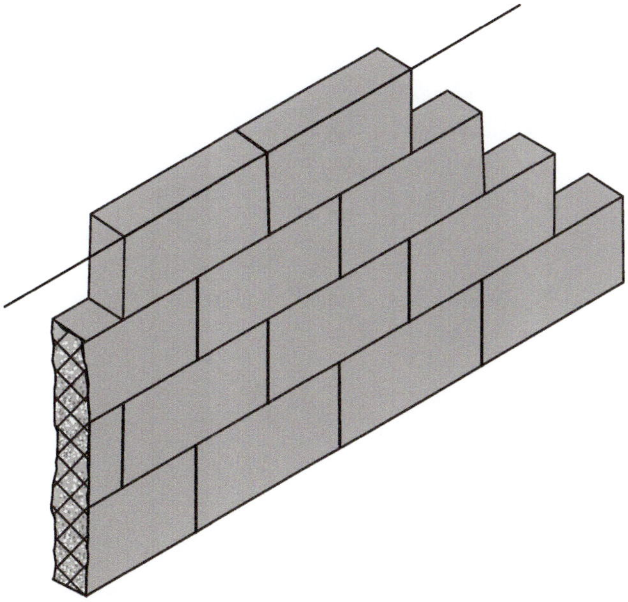

FIGURE 9.4
Half-bonding blockwork

Another method is to cut the first block to a three-quarter block to obtain the bond (Figure 9.6).

Special corner blocks could be used, but are an extra expense (Figure 9.7).

Plumbing blockwork

Plumbing blockwork is more difficult than plumbing brickwork as the wall rises more quickly and the mortar tends to squeeze out of the soft mortar bed joints, causing the wall to move out of plumb.

Always ensure that blocks are fully bedded with nominal 10 mm bed and cross-joints.

On no account tap the blocks sideways - this will only result in the bed joint opening up on one side. Always plumb the wall by tapping down on the high side of the block (Figure 9.8).

FIGURE 9.5
Quoin with quarter-blocks

FIGURE 9.6
Quoin with three-quarter-blocks

FIGURE 9.7
Quoin with return blocks

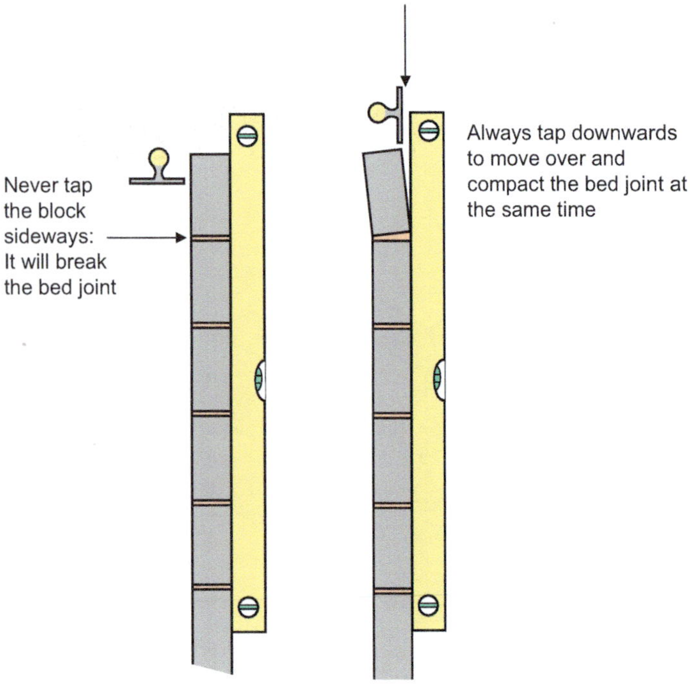

FIGURE 9.8
Plumbing blockwork

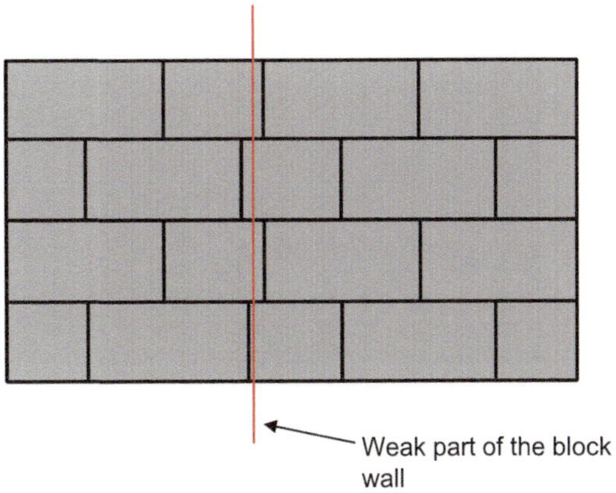

FIGURE 9.9
Broken bond

Broken bond

Broken bond may have to be adopted when wall lengths are not equal to block sizes. Block walls will usually require some method of broken bond.

It is important never to use cuts less than half a block as this can create a weak plane in the wall and if drying shrinkage occurs it may result in cracks appearing in the wall (Figure 9.9).

Blocks can be cut with a club hammer and bolster chisel, a masonry saw or a mechanical bench saw (Figure 9.3).

Try to make use of any special blocks that are available before starting to cut.

Always try to achieve a neat cut that retains the 10 mm cross-joint. Too large joints can again result in cracks appearing in the wall owing to drying shrinkage.

Broken bonds of less than half-lap can be avoided with using blocks (Figure 9.10).

Reverse bond

This is when the blocks at each end of the wall are different. The front elevation pictured in Figure 9.11 shows reverse bond being used.

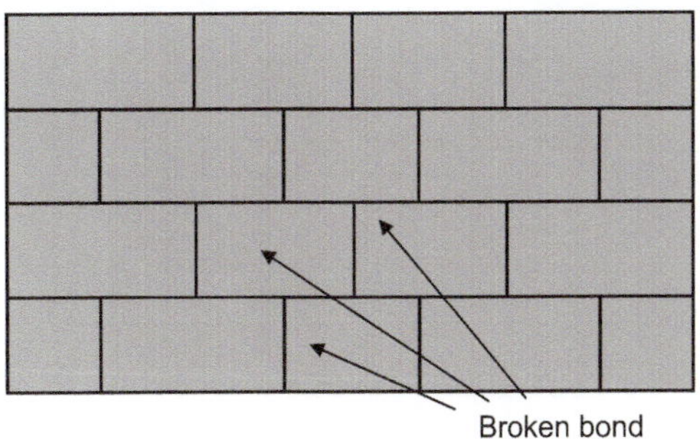

FIGURE 9.10
Broken bond

FIGURE 9.11
Reverse bond: end block different to opposite end block to avoid broken bond

When cut blocks are being used it is more economical to use three-quarter, half and quarter blocks provided by the manufacturer.

Whenever cut blocks are required they should always be of the same material.

Junction walls

It is often inconvenient to build junction walls at the same time as the main wall, so during construction you will need to make provision for work at a later date.

Indents should be left in the main wall to receive the junction wall later. These indents should be marked out correctly and kept plumb throughout the building of the block wall. The minimum lap at tee junctions should be quarter-lap.

An allowance of 20 mm is usually made over the width of the block to allow the blocks on the junction wall easy access into the indent (Figure 9.12).

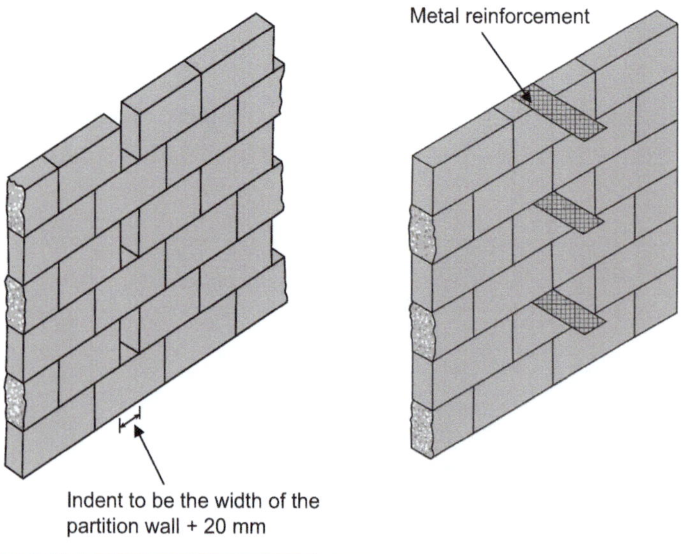

FIGURE 9.12
Junctions

Another acceptable method of providing a tie to junction walls is metal reinforcement (Figure 9.12). This is built into the main wall on every alternate bed joint.

This method can help to avoid numerous cuts having to be made at a junction of two block walls.

This method is recommended for bonding block walls together when two walls are constructed of different block types, especially when the blocks have different shrinkage properties.

Bonding to brick walls

Blocks may have to be bonded to brick walls. This can be done by leaving indents in the brickwork or by fixing proprietary wall connectors (Figure 9.13).

Pillars and piers

It may be necessary to support long walls by having attached piers designed into them.

There are several methods of bonding attached piers. Blocks could be laid flat, but this affects the bond on the face of the wall (Figure 9.14).

A better method is to have a pier the same size as the blocks, to avoid cutting, and to lay three together (Figure 9.15).

Blockwork quoins

Corners should be built as for brickwork. It is important to maintain half-bond.

The method most commonly used is to introduce a closer next to the first block; this will allow half-bond to be maintained (Figure 9.16).

It is important that blocks of the same material and strength are used.

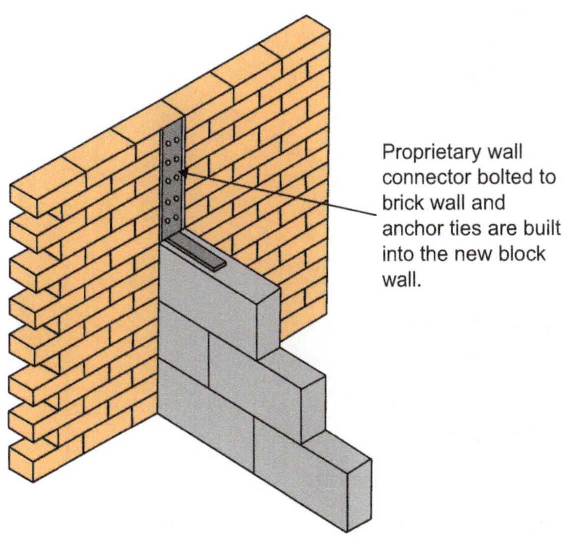

FIGURE 9.13
Proprietary wall connectors

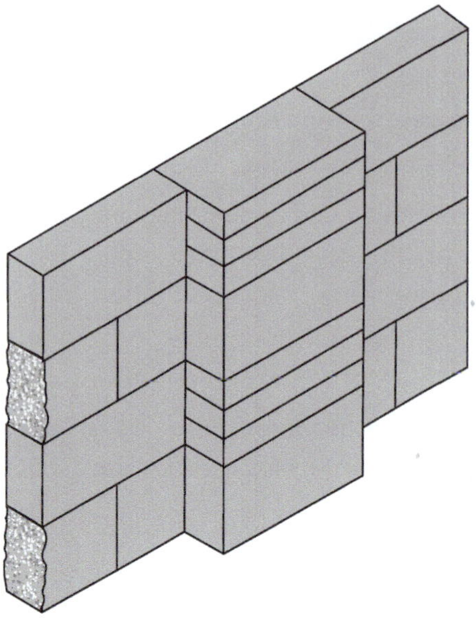

FIGURE 9.14
Attached pier with 100 mm blocks

Never use bricks to achieve bond.

Check course heights regularly with a vertical gauge rod.

In summer the blocks may be too dry to ensure adhesion with the mortar. It is possible to alter the consistency of the mortar and the mortar should be laid in shorter lengths.

NEVER dampen down the blocks.

As mentioned before, it is recommended that block walls forming partitions should only be built up to six courses in a day.

FIGURE 9.15
Attached pier with 150 mm blocks

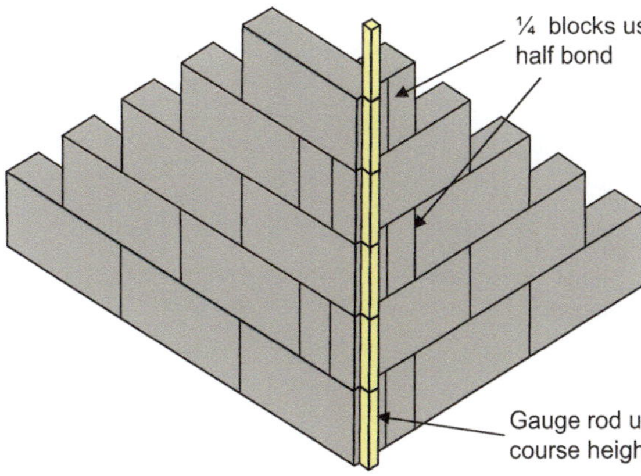

FIGURE 9.16
Quoin built in blockwork

This will allow for the wall to have set before building any higher.

It is also recommended that supports and profiles are pre-erected to assist in the building of the wall and to prevent buckling. When block walls are erected as part of a cavity wall they are supported with wall ties, so profiles are not necessary (Figure 9.17).

Lintel bearings

The lintel manufacturer's recommendation for minimum bearing should be followed. Lintels should bear on to a full-length block (Figure 9.18).

Block bonding

Partition walls can be block bonded to brick walls to gain maximum stability.

Three courses of bricks are equal to one course of blocks, so this is the pattern used to form the block bonding (Figure 9.19).

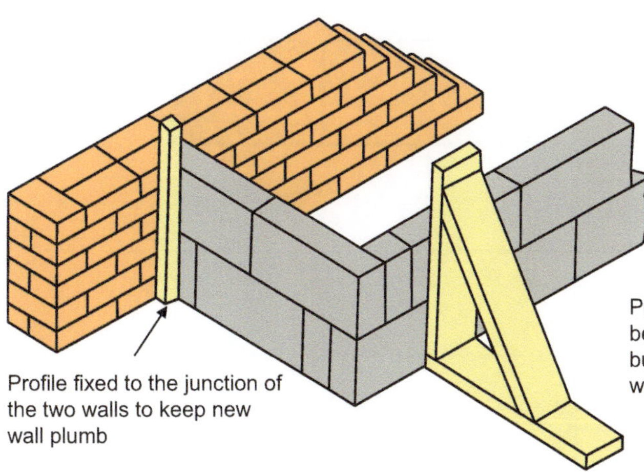

FIGURE 9.17
Use of profiles on internal blockwork walls

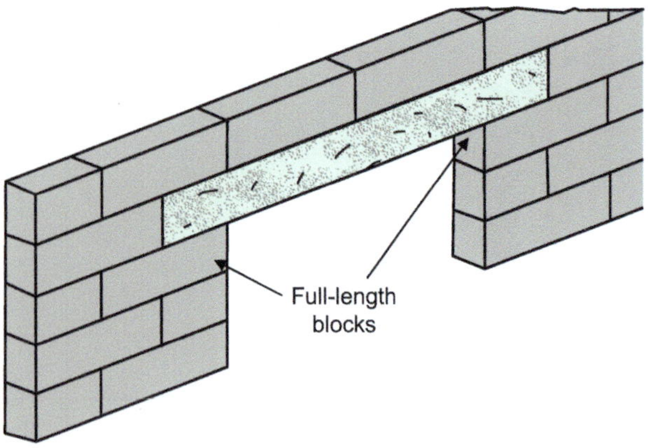

FIGURE 9.18
Bearings for lintels

Remember always to use the same density blocks in the same wall. Never mix bricks and blocks. Most manufacturers supply cut blocks to avoid excessive waste on construction sites. Figure 9.20 shows the various cuts available.

Partition walls

One of the most common uses of blocks is partition walls (Figure 9.21). These can be either a cavity or a solid wall, as required by the architect.

Remember that with thinner blocks for partition walls it is essential to provide temporary profiles to support the blockwork until it has set.

FIGURE 9.19
Block bonding

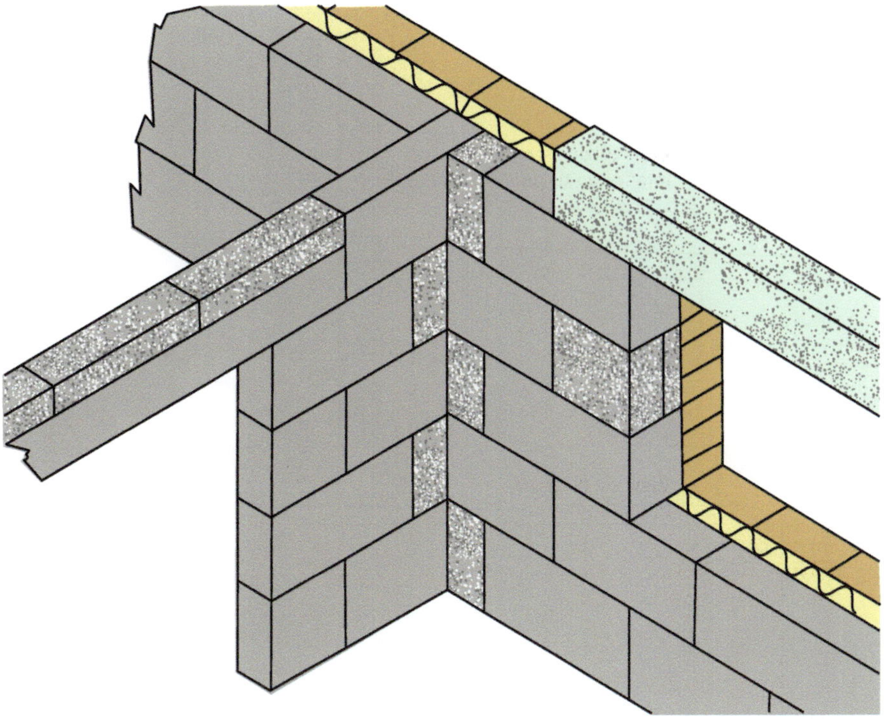

FIGURE 9.20
Use of cut blocks

Blocks below ground level

Certain quality blocks can be used below ground level (Figure 9.22). Always consult the manufacturer's instructions before using.

Laying blocks

The bricklayer has no conventional rules to which to work; rather, the rules to be used are the outcome of experience.

The apprentice will no doubt have other problems to handle besides those shown.

FIGURE 9.21
Partition walls: (a) cavity wall construction; (b) solid wall construction

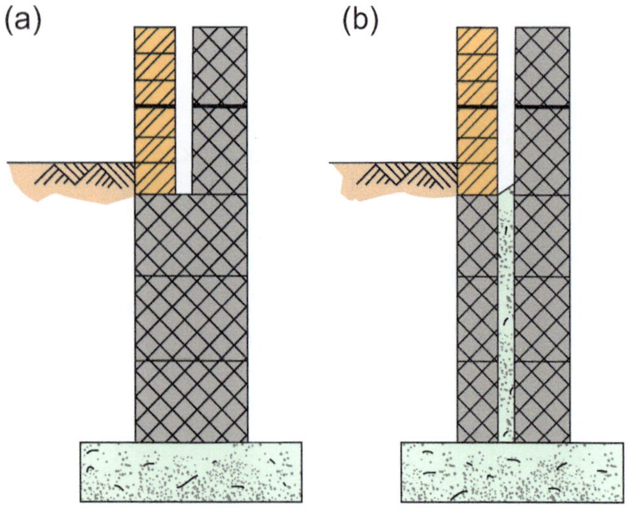

FIGURE 9.22
Blockwork below ground level: (a) solid foundation blocks (note: some concrete blocks require two people to lift them); (b) foundation blocks with weak concrete cavity fill below ground level

Time should be spent visualizing the job, considering the best methods of approach and using all the skills possible, whether the work is to be covered or highly decorative, for all to see.

As a learner, the apprentice should never allow quality to be sacrificed for speed, which will be attained by constant practice.

The stability and appearance of the work should always be the master craftsperson's chief concern.

Good blocklaying entails the ability to master the art of spreading the mortar bed, dexterity in handling the block to be laid, and the possession of a keen eye. All of these can be acquired by practice.

Before laying blocks on any job, place the mortar or 'spot' board, with the blocks, in a convenient position. They must be within easy reach so that no unnecessary movement is involved when materials are required.

Block up the mortar board on bricks, one at each corner, so that it is kept clean, and load out as shown for a one-block wall (Figure 9.23).

Do not grasp the trowel as if clenching the fist, but place the thumb on the ferrule and handle lightly, so that a flexible wrist action is possible (Figure 9.24). Pick up the mortar with an easy sweeping motion and spread it on the wall sufficiently thick to allow the block to be placed by pressure of the hand. A common fault is the placing of too much mortar under the block, so that considerable hammering and tapping are necessary before the block reaches its final position. The bricklayer usually estimates the amount of mortar bed required by the feel of the block.

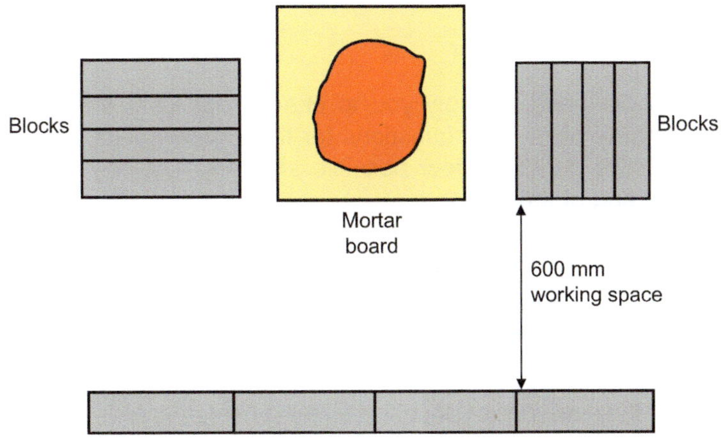

FIGURE 9.23
Suggested layout for blockwork

FIGURE 9.24
Method of holding the brick trowel

Multiple-choice questions

Self-assessment

This section of the book is designed to allow you to check your level of knowledge. The section consists of revision questions for this chapter. The questions are all multiple choice and have four possible answers. The answers are to be found at the end of the book.

The main type of multiple-choice question will be the four-option multiple-choice question. This will consist of a question or statement, known as the stem, followed by a choice of four different answers, called the responses. Only one of these responses is the correct answer; the others are incorrect and are known as distracters.

You should attempt to answer the questions by choosing either (a), (b), (c) or (d).

Example

The person employed by the local authority to ensure that the Building Regulations are observed is called the:

(a) clerk of works
(b) building control officer
(c) council inspector
(d) safety officer

The correct answer is the building control officer, and therefore (b) would be the correct response.

Laying blockwork

Question 1 Identify the following type of block:

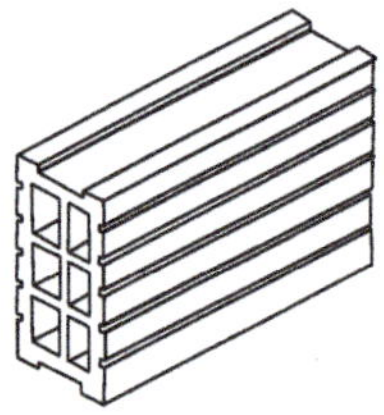

(a) concrete block
(b) clinker block
(c) fly ash block
(d) clay block

Question 2 What is the recommended height for blockwork each day?

(a) 3 courses
(b) 4 courses
(c) 5 courses
(d) 6 courses

Question 3 What is the recommended bond for blockwork?

(a) half-bond
(b) quarter-bond
(c) reverse bond
(d) broken bond

Question 4 Where would you find an indent?

(a) when setting up a corner
(b) when forming a junction
(c) when running to a line
(d) when dry bonding

Question 5 One course of standard blocks is equal to how many courses of bricks?

(a) 2 courses
(b) 3 courses
(c) 4 courses
(d) 5 courses

Question 6 What is the recommended distance from the wall to the materials?

(a) 300 mm
(b) 400 mm
(c) 500 mm
(d) 600 mm

Question 7 Why is it important to dry bond the wall before laying blocks?

(a) to find the cheapest method
(b) to show the labourer where to place the materials
(c) to avoid awkward cuts
(d) to be able to count the blocks required for the wall

Question 8 Identify what is shown in the drawing:

(a) a quoin built with return blocks
(b) a quoin built with three-quarter-blocks
(c) a quoin built with quarter-blocks
(d) a quoin built with half-blocks

CHAPTER 10

Cavity Walling

This chapter will cover the following City and Guilds units:

• Apprenticeship	Module 5	S1.1, S1.2, S1.3
	Module 5	K1.1, K1.2
• Diploma	Unit 105	1.1, 1.2, 1.3, 1.4, 1.5, 1.6, 1.8
		2.1, 2.2, 2.4, 2.5, 2.6, 2.7
		3.1, 3.3, 3.4
		4.1, 4.2, 4.3, 4.4, 4.6, 4.7

This chapter is about:

- Interpreting instructions
- Adopting safe and healthy working practices
- Laying bricks and blocks to line to form cavity walling

The following City and Guild performance criteria will be covered:

- Setting out cavity walls
- Materials used to build cavity walls
- Methods used to insulate cavity walls

The following Diploma outcomes will be covered:

- Know how to select the required quantity of resources to construct cavity walls
- Know how to construct cavity walls
- Know how to construct cavity walling return corners

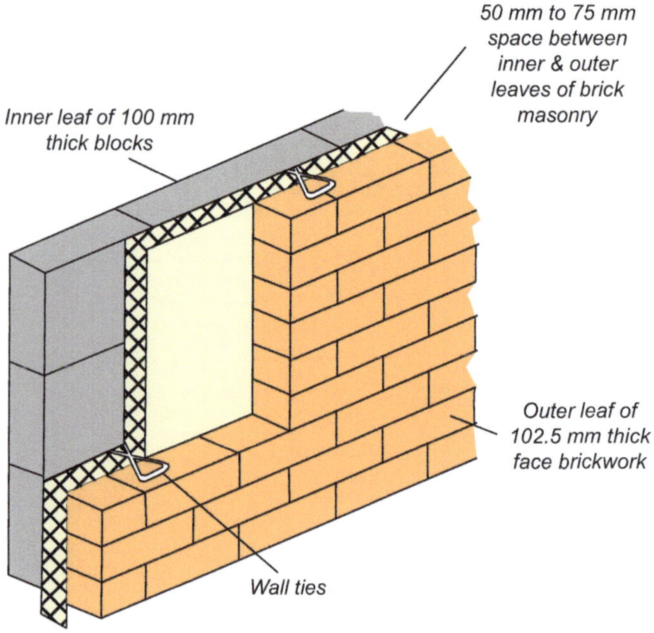

FIGURE 10.1
Typical insulated cavity wall

Definition of cavity walls

The standard form of construction for the external walls of brick buildings is called cavity walling. This means that the bricklayer builds the two separate 'leaves' or 'skins' of brick masonry (a general term indicating brickwork and or blockwork), with a 50–75 mm wide space between.

The outer skin is usually 102.5 mm thick face brickwork, but may be constructed from facing quality blocks. The inner skin is usually 100 mm thick common blocks that are later plastered to receive internal decoration (Figure 10.1).

Both skins of brick masonry are joined together with a regular pattern of corrosion- resistant ties, so that they behave as one single wall.

A continuous space termed a cavity, which is approximately 50 mm wide, is produced and the walls are supported by specially prepared ties of metal called wall ties.

The prevention of the passage of moisture from the outer to the inner walls is of the utmost importance in cavity wall construction, and to achieve this, damp-preventing materials must be inserted in the structure at special points, such as window and door reveals, and window and door heads.

Purpose of cavity walls

Cavity walls are very useful for damp and exposed positions to ensure a dry interior to the building. The reasons for this are:

- The cavity provides a break between the outer wall of the building, which may damp, and the inner wall.

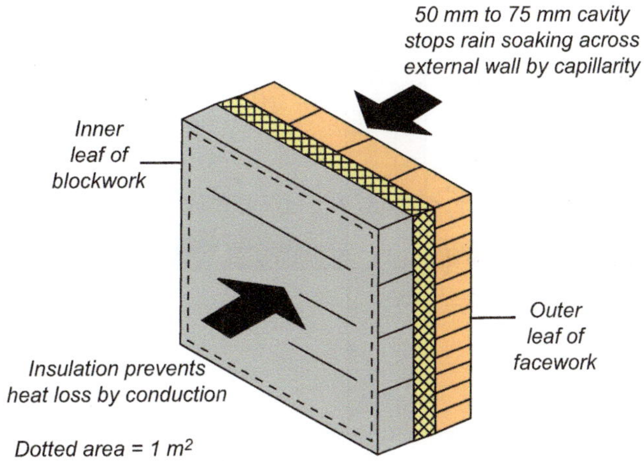

FIGURE 10.2
Function of typical cavity walling

- To achieve a better balanced temperature inside the building. The cavity keeps the inside cool in the summer and helps to retain heat in the winter.

Where proper care is exercised in the building of this type of wall, it will prevent the passage of moisture from the outer or protective wall to the inner wall.

Cavity wall construction began to be widely used from the 1920s, as a way of preventing dampness from soaking through the outer walls of buildings.

The 50–75 mm wide gap stopped rain penetrating from the outer surface to the plastered inner surface of external walls by capillary action (when water is drawn through hairline channels within a porous structure, e.g. a brick, by the action of surface tension). This was possible when outer walls were commonly 215 mm thick solid brickwork.

As a secondary advantage, this space between the inner and outer skins also provides thermal insulation for modern buildings, because heat energy mainly escapes by conduction through solid material. Air is a poor conductor of heat energy, therefore the rate of heat loss is very much slower than was the case when buildings had solid outer walls (Figure 10.2).

Building Regulations

The basic advantages of cavity wall construction, over solid 215 mm thick brickwork, for the outer walls of a building have been incorporated in the current Building Regulations.

In order to satisfy requirements of the current Building Regulations and enable planning permission to be obtained, cavity wall construction is usually specified whether the structure is low rise or multistorey.

Figure 10.3 indicates how the basic requirements of the Building Regulations are satisfied, where standard strip foundations are specified with a solid

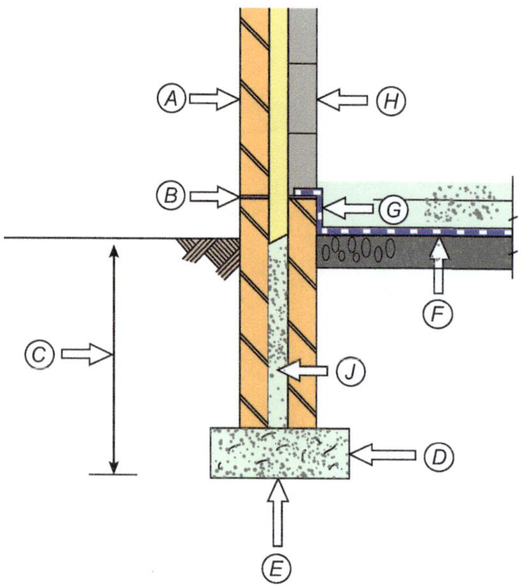

FIGURE 10.3
Cavity walling on standard strip foundation

ground floor slab. Figure 10.4 shows a trench-fill foundation, associated with a suspended ground floor construction of prestressed, precast concrete floor beams supporting standard size concrete blocks.

The following labels are used in Figures 10.3 and 10.4:

- A: external cavity wall, providing resistance to through-penetration of rain
- B: horizontal damp-proof course (DPC) in both leaves, not less than 150 mm above ground level, to prevent dampness rising from the soil

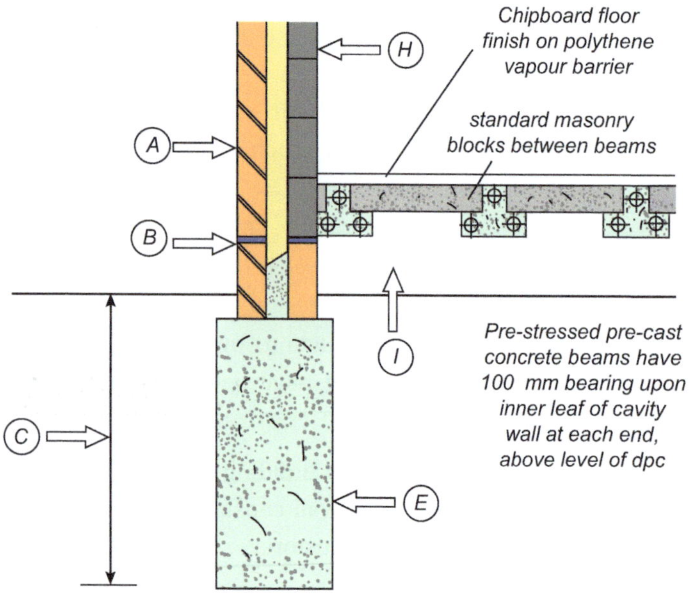

FIGURE 10.4
Cavity walling on trench fill foundation with suspended concrete floor

C: a minimum distance of 1 m between ground level and the underside of the concrete foundation, as a protection against frost heave in winter and drying shrinkage of clay subsoils in summer

D: a minimum 150 mm thickness of foundation concrete to transfer the building load adequately on to the natural foundation of the subsoil

E: sulphate-resisting cement in the foundation concrete and substructure brickwork, necessary where soluble sulphates are present in subsoil water

F: continuous damp-proof membrane (DPM) across the ground floor areas to prevent dampness rising from the soil; ground floor *u*-value of 1.13; insulation can be placed underneath, on top, or as a sandwich layer

G: DPM and DPC, lapped and joined within the floor thickness around the perimeter of rooms

H: external walls built of lightweight blocks and other thermal-insulating material to give a U value of 0.18 (in other words, heat energy must not escape through the outer walls at a rate greater than 0.18 W/m^2/hour, per degree difference in temperature internally and externally)

I: ventilated air space separating the suspended floor from damp soil

J: weak cavity fill to prevent collapse of walls owing to pressure from the ground.

Note

Please check with the current Building Regulations for the latest amendments.

Function of cavity walls

Figure 10.5 shows that it is the inner leaf of cavity walling that largely supports the load from floors and roof in a low-rise building of load-bearing wall

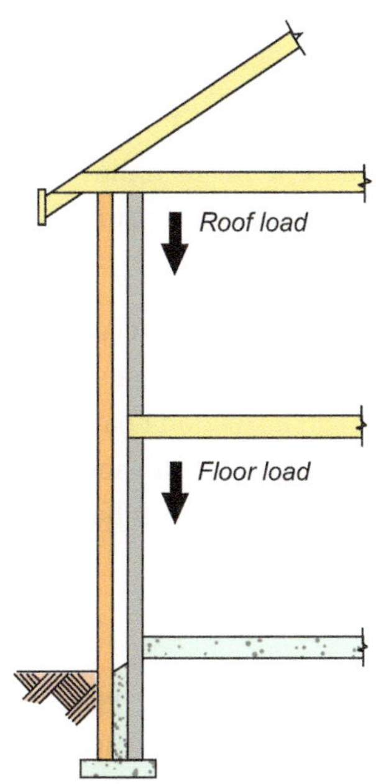

FIGURE 10.5
Loads on cavity walls

construction. Common building blocks have totally replaced common bricks for this inner leaf, owing to improved thermal insulation values and bricklayer output (one standard size 100 mm thick block is equal to six bricks).

The purpose of the outer skin of facework is to give the building a weather-resistant and pleasant appearance, by selection from the wide range of colours and surface textures of bricks and facing blocks available to the designer.

Although cavity walling is constructed with properly finished, solidly filled mortar joints, it is expected that this outer skin will let rainwater soak through as far as the cavity. This is because mortar, bricks and blocks are porous to varying degrees. This through-penetration will be highest on those elevations of a building exposed to prevailing wet winds.

Where cavity walling is used as cladding to a high-rise building the degree of exposure to wind-driven rain increases with height.

Thermal insulation

When cavity walling was first used, it was common practice to install air bricks at the top and bottom, at intervals around the whole perimeter of the building, to ventilate this 50–75 mm wide space to remove damp air. Since the 1950s, however, cavity walls have become sealed, with insulating material built in as work proceeds, to improve the thermal insulation value of the cavity space.

Two ways that a bricklayer may be told to install thermal insulation in cavity walling are shown in Figures 10.6 and 10.7

In a fully filled system, flexible fibre 'batts' of insulation completely occupy the cavity space. In a partially filled system, stiffer 'boards' of insulating material half-fill the cavity, but a 25–35 mm air space is retained as well. Both batts and boards are supplied in purpose-made sizes to fit neatly between layers of wall ties and are approximately 900 mm in length.

All insulation batts should be fixed staggering the vertical joints, and butting the vertical and horizontal joints as tightly as possible. Insulation batts can be cut with a sharp knife, but always ensure that the cut is square and forms a perfectly tight joint.

Insulation batts for full-fill insulation are available in various thicknesses and can be made from mineral fibre, which is soft and flexible. Insulation for partial cavity fill is made from more rigid insulation boards such as expanded polystyrene bead board. The boards in partial cavity fill should only be fixed with special wall ties with plastic clips to hold the insulation back against the internal blockwork.

Wall ties

Headers and courses of headers are used in solid brick walling to tie the wall from back to front. In cavity wall constructions headers cannot be used for this purpose, because they would allow dampness to cross the cavity by capillary action. Therefore, a range of proprietary ties is made for this job of tying

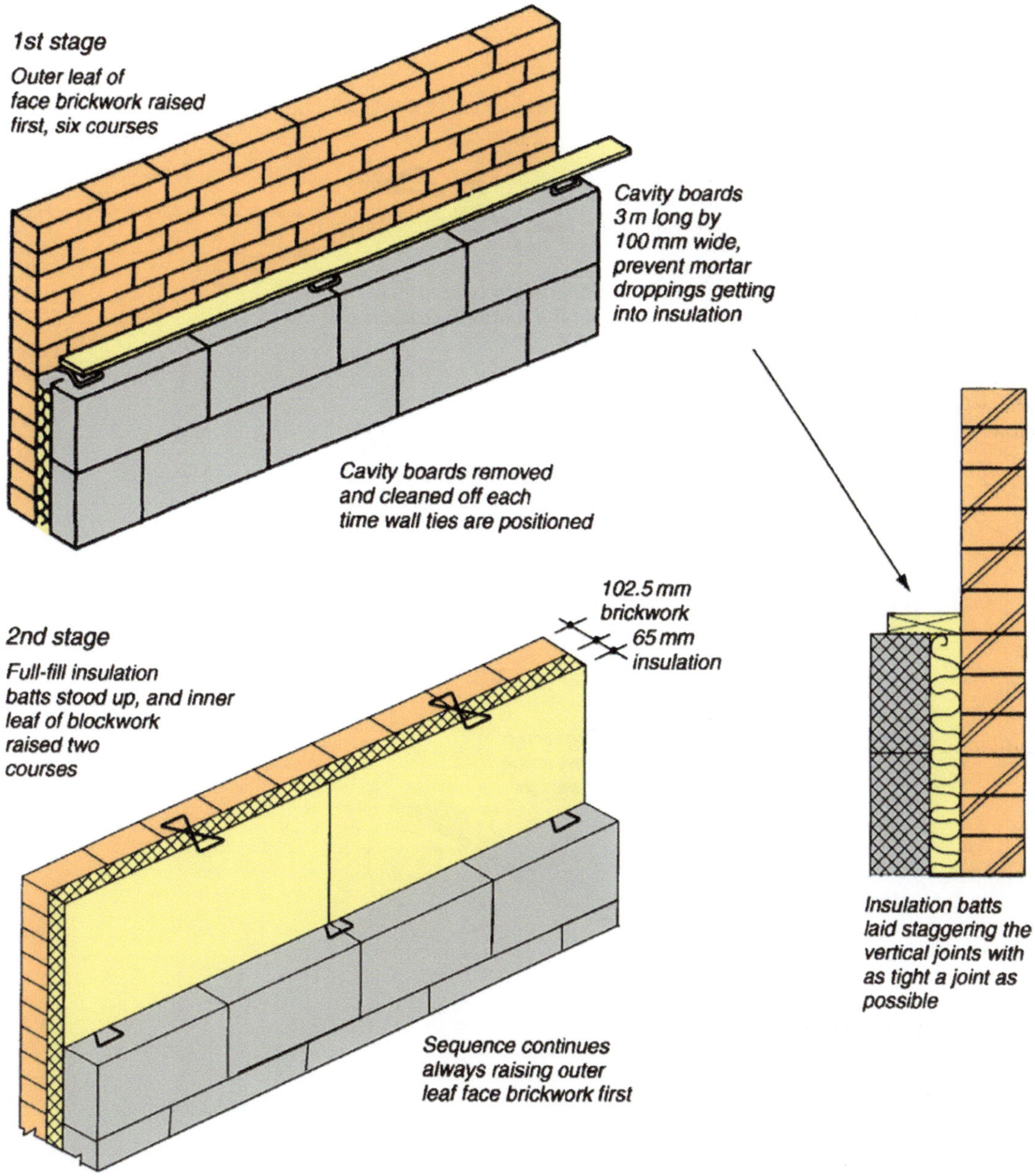

FIGURE 10.6
Fully filled cavity insulation system

together inner and outer skins of brick masonry, as 'substitute headers' so that both leaves behave as one wall. These wall ties are made from stainless steel, galvanized steel or polypropylene, so they do not provide a passage for moisture.

Ties are shaped so as to form a drip at their centre. Examples are shown in Figure 10.8. If moisture penetrates the outer wall and passes along the wall tie it will meet this drip, fall into the cavity and seep away through 'weep holes' placed at the base of the outer wall below the horizontal DPC (Figure 10.9).

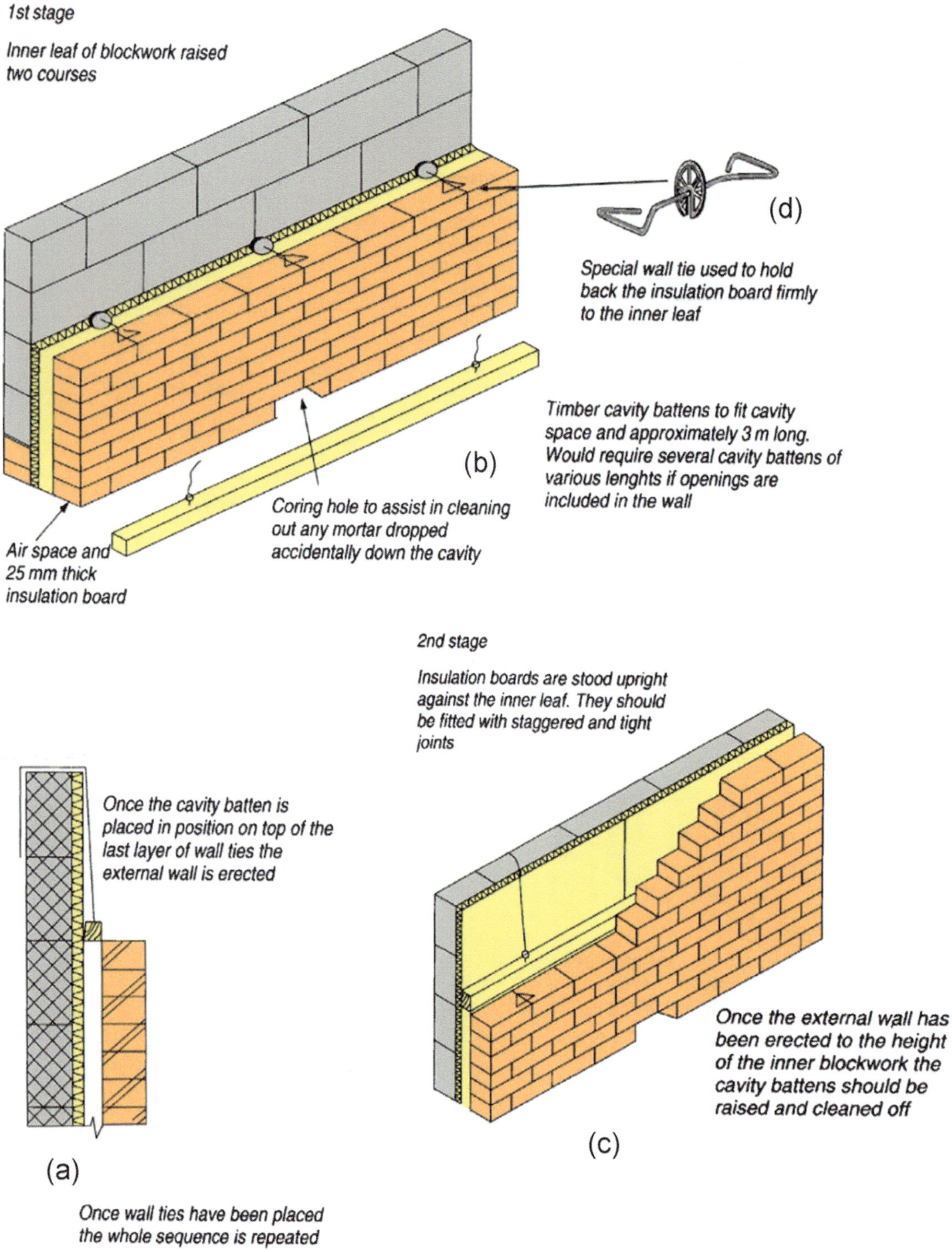

FIGURE 10.7
Partially filled cavity insulation system

Weep holes are formed by leaving open a mortar cross-joint, or inserting special weep holes, at intervals along the wall face.

The standard maximum spacing for wall ties is at intervals of 900 mm horizontally and every sixth course vertically. Each horizontal layer should be

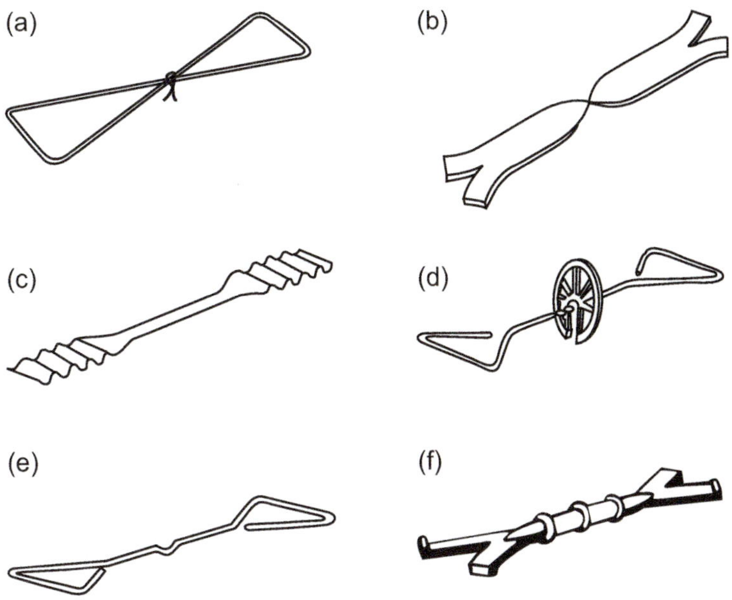

FIGURE 10.8
Types of wall tie: (a) butterfly; (b) galvanized fishtail; (c) stainless steel tie; (d) double triangle with plastic clips; (e) double triangle; (f) plastic

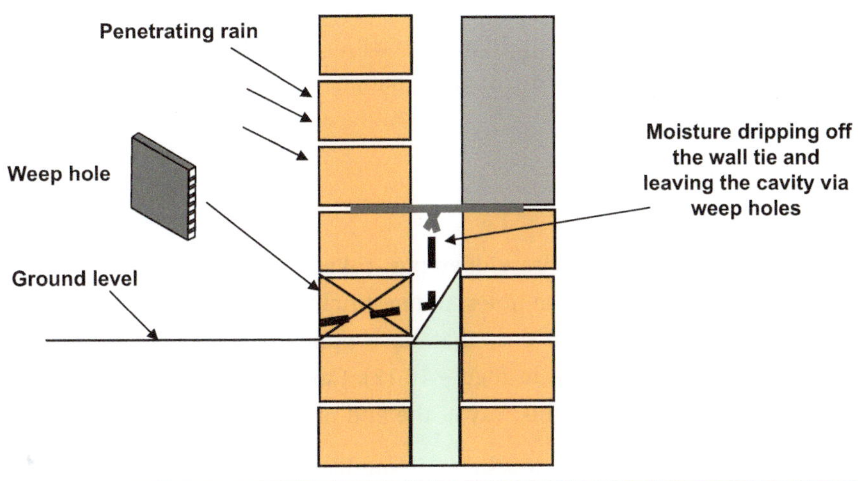

FIGURE 10.9
Use of drip on wall tie

offset. For purposes of estimating quantities of wall ties required, this works out at approximately 2.5 per square metre.

For cavity walling to be effective, wall ties, insulation and cavity gutters must be kept free of mortar droppings as work proceeds. If, owing to carelessness and poor supervision, cavities are not kept clean, then dampness will be able to cross the cavity through porous mortar droppings.

Cavity battens or boards, approximately 3 m long, raised and cleaned off every six courses as work proceeds, are the best way of preventing mortar droppings

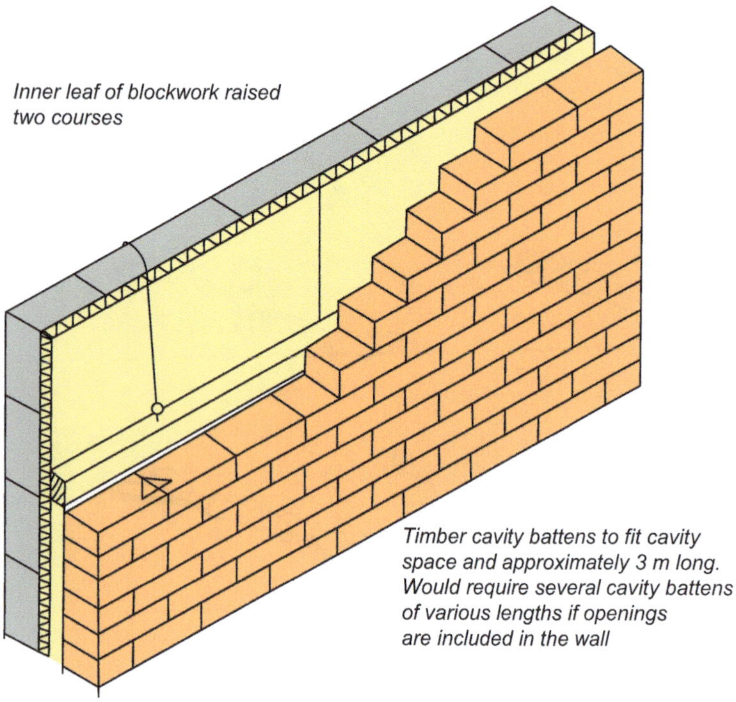

FIGURE 10.10
Cavity battens with partial fill insulation

falling into cavity walling (Figure 10.10). Alternatively, where fully filled cavity insulation is specified, plain battens without lifting wires are used (Figure 10.11).

Coring holes

It is usual to leave temporary openings, called coring holes, in the outer leaf, over all cavity trays. These holes, of one-brick size, are for the removal of any mortar droppings, tools, etc., that have gone past the cavity battens (see Figure 10.7b and point A in Figure 10.12). Cavity ties and trays should be inspected and cleaned in this way at the end of every day's work.

Where possible, however, it is better if one-block-sized coring holes are left out of the inner leaf (see point B in Figure 10.12). This is recommended as it provides a bigger temporary opening. It also avoids the risk of the slight difference in mortar colour that would highlight the coring holes in the finished facework. The coring holes are sealed up when the external scaffolding is removed.

Temporary bedding bricks or blocks in sand at point A or B provide support for those above, and make for easy removal to form the coring holes.

Coring holes should only be regarded as a back-up procedure, and not a substitute for cavity battens or boards.

See also Figure 10.7(b), which shows a coring hole at ground level.

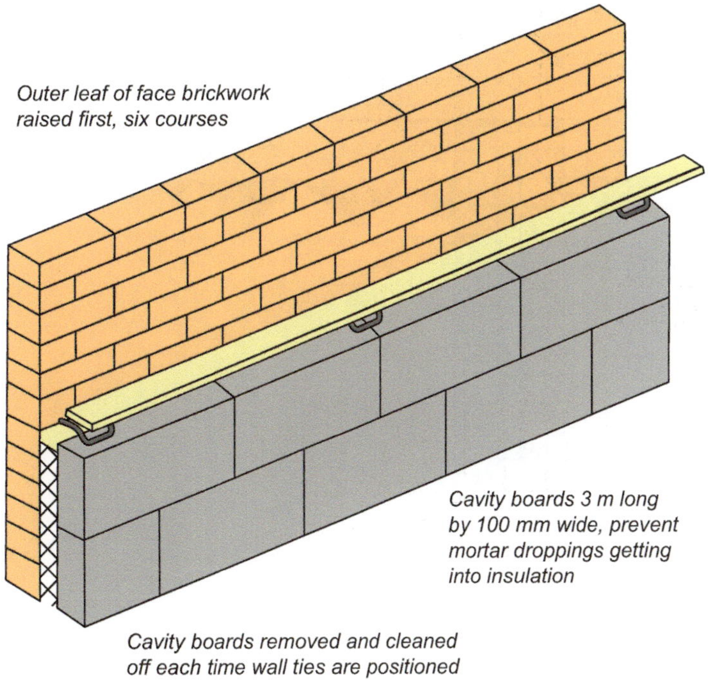

FIGURE 10.11
Cavity battens with full insulation

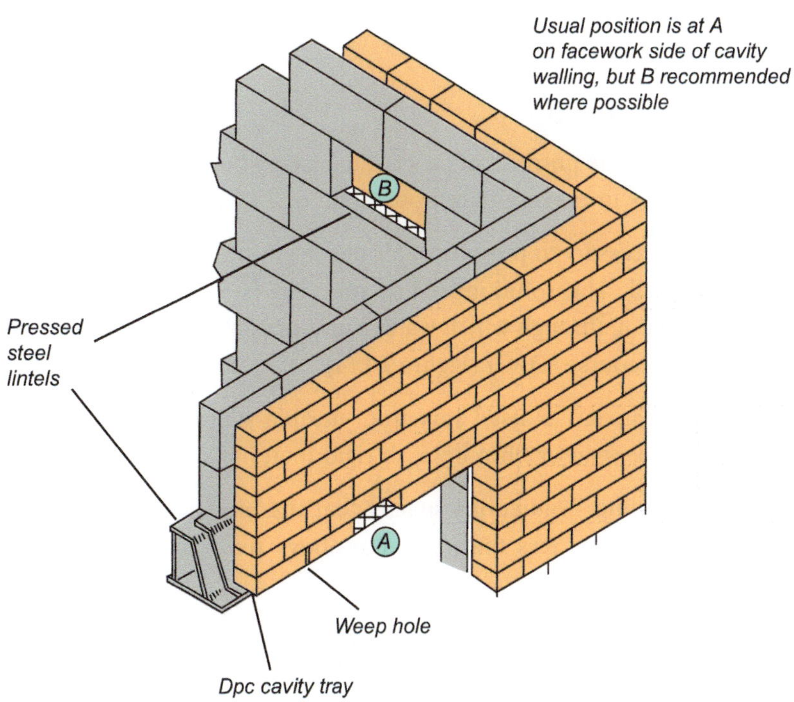

FIGURE 10.12
Location of coring holes

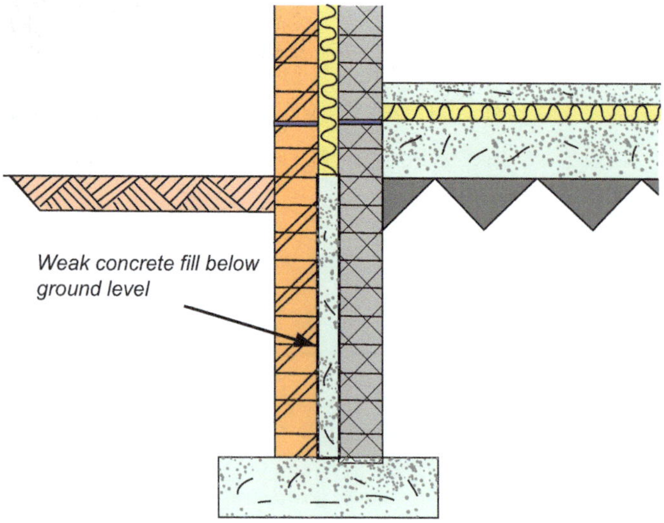

FIGURE 10.13
Cavity walls below ground level

Cavity walls below ground level

As an aid to stabilizing a cavity wall below ground level the cavity is filled with weak concrete up to ground level.

The top of the concrete filling is sloped towards the outer face wall to ensure that any water dripping down the cavity escapes through the weep holes.

The insulated cavity should extend at least 150 mm below the lowest horizontal DPC (Figure 10.13).

Note

Always try to keep the cavity clean.

Resistance to damp

We have seen that the purpose of providing a cavity between the two leaves of an external wall is to prevent damp penetration.

It is, however, necessary to prevent water in the ground from rising up the two leaves of the wall. The solution to this problem is to insert a layer of impermeable material, through which water cannot penetrate. The layer is known as a damp-proof course (DPC) (Figure 10.14).

The damp-proof materials used vary from flexible sheet material made from bitumen with a hessian or fibre base, to rigid material such as slate.

The current Building Regulations specify that the external wall must have a DPC at a height of not less than 150 mm above the finished surface of the adjoining ground.

Working below ground level

Working below ground level is very different from working above ground.

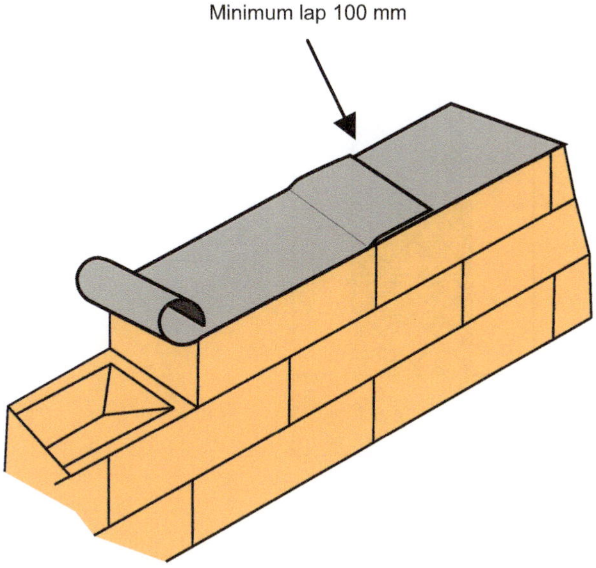

FIGURE 10.14
Horizontal damp-proof course

The materials have to be set out above ground on the banks of the excavation, taking care not to place too much pressure on the trench sides (Figure 10.15).

The space in the trench will be very tight; it may be limited to only 150 mm. If the walls being built are cavity then the facing wall will be easier to build because there is more room. The inner wall will be very tight to construct.

Sometimes the bricklayer may decide to work from the ground level and bend down into the trench.

Foundation blocks

The use of foundation blocks is now very common, which eliminates the need to fill the cavity with weak concrete fill (Figure 10.16).

Remember

Take special care when working below ground level. Ensure that the appropriate protective clothing and equipment are available

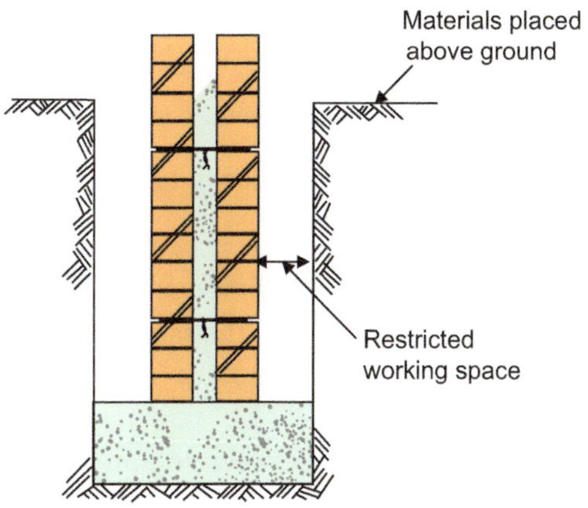

FIGURE 10.15
Cavity wall below ground level using brickwork

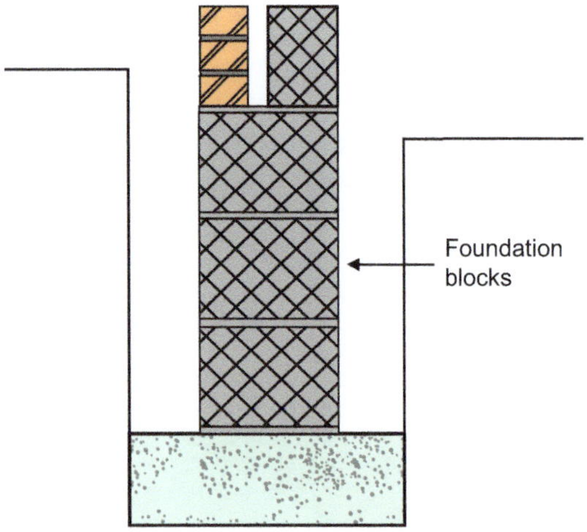

FIGURE 10.16
Wall below ground level using foundation blocks

Stepped foundations

When building foundation brickwork on stepped foundations extra care should be taken.

Always start at the lowest level. The first course should be set out as for ordinary strip foundations, but care should be taken when ranging in the first course owing to the difference in level.

The concrete foundation should be marked out as for ordinary level foundations but care should be taken to ensure each step is in multiple brick courses.

Before starting to lay bricks it is important to set out the bond dry to avoid broken bond or reverse bond.

The following figures show a basic rule for setting out stepped foundation brickwork.

Rule: If the difference between the first course and the top course is an *odd* number then the first brick should be opposite to the top brick (Figure 10.17a):

- Header is on the top course.
- The number of courses between top and bottom is odd
- Therefore, the first course needs to be opposite to the top course = stretcher.

Rule: If the difference between the first course and the top course is an *even* number then the first brick should be the same as the top brick (Figure 10.17b):

- Header is on the top course.
- The number of courses between top and bottom is even.
- Therefore, the first course needs to be the same as the top course = header.

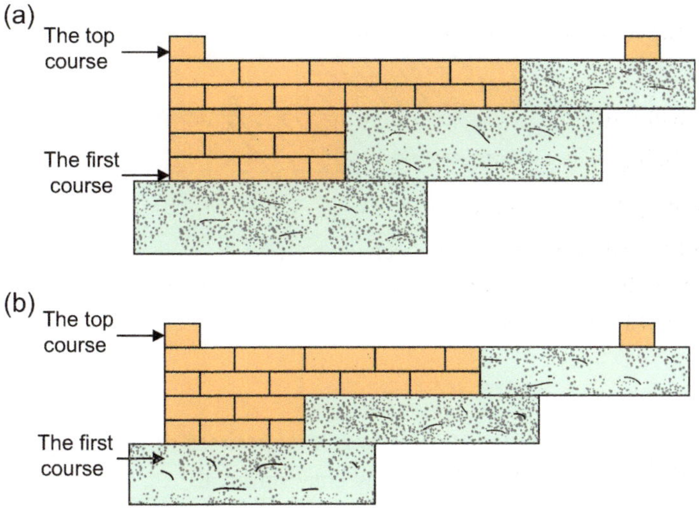

FIGURE 10.17
Calculating the correct number of brick courses

When the wall positions have been determined the corner can be built and the wall completed (Figure 10.18).

Once the steps have been levelled the remainder of the foundation brickwork is the same as for normal level foundations.

Provision for services

Sand courses

Occasionally provision will have to be made for services passing through the external walls of the building.

Typical examples are openings for drainage, gas, electricity, etc.

Quite often the exact position is not known, so sand courses are used to allow bricks to be removed at a later date to allow the services to pass through the wall.

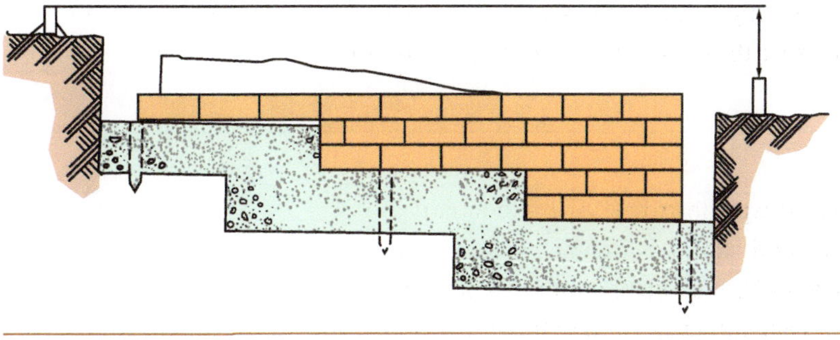

FIGURE 10.18
Completed steps

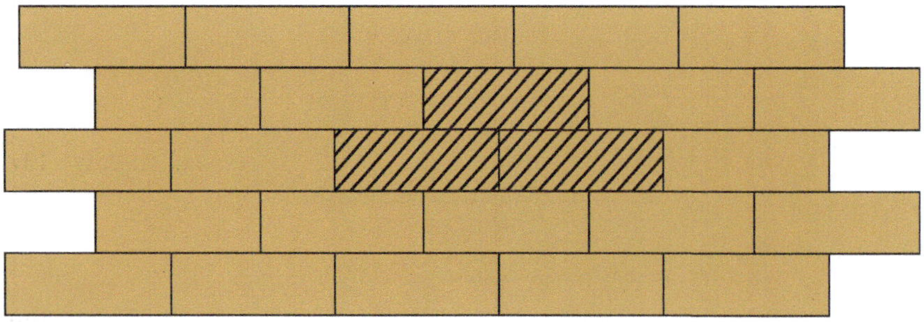

FIGURE 10.19
Sand courses

An example is shown in Figure 10.19, where two courses of bricks have been bedded in sand instead of mortar to allow easy removal.

Concrete beams

When larger openings have to be left out or the architect has designed the opening with support over it to prevent settlement, a concrete beam can be inserted (Figure 10.20).

Welsh arch

For smaller openings a Welsh arch can be used. These are also often used above ground when the appearance is important.

The Welsh arch consists of three cut bricks (Figure 10.21).

Raft foundations

Depending on the design of the raft foundation, very little work will be required up to the DPC. With a simple raft foundation two courses of brickwork are required (Figure 10.22). Some raft foundations go deeper into the ground and therefore require more foundation brickwork.

Once all the work has been finished up to DPC level the excavation will require backfilling and any surplus subsoil will have to be removed from the site.

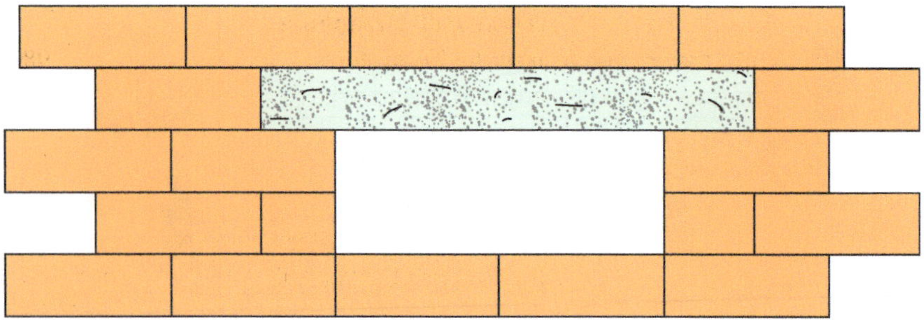

FIGURE 10.20
Concrete lintel

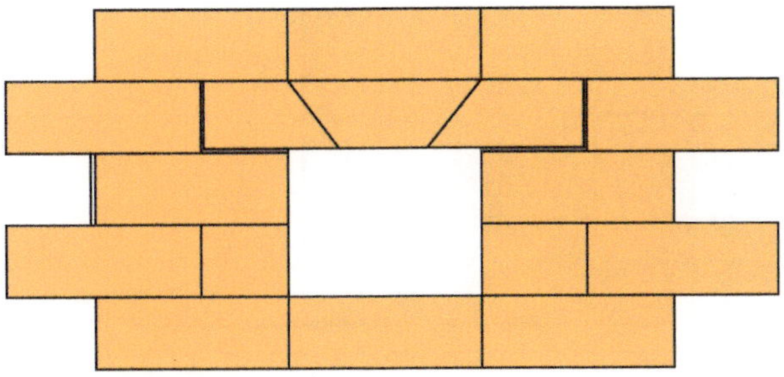

FIGURE 10.21
Welsh arch

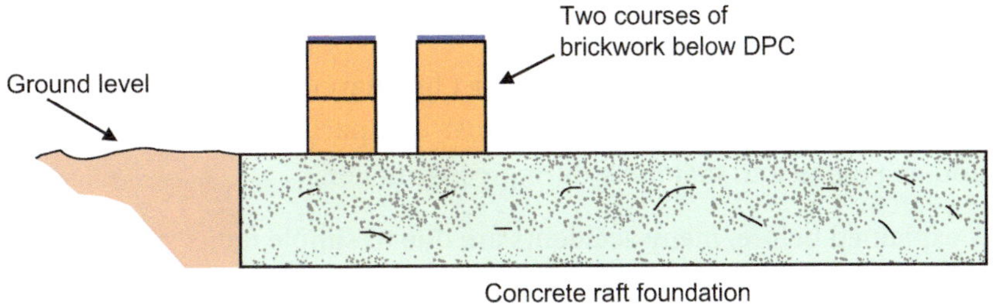

FIGURE 10.22
Cavity wall on raft construction

Protection of work

It is important that the work is protected at the end of each day and when there is bad weather during the day.

All work should be protected from the dangers of frost. Frost can cause the mortar to crumble and fall out, therefore resulting in the need to repoint the wall, causing extra expense. One of the easiest and simplest methods of protection is hessian sheets. These can be draped over the brickwork to protect from frost.

Excess rainwater can have adverse effects on cement and lime mortars. Joints will start to run, causing staining to the face of the components. Polythene sheeting will have to be used to protect freshly laid brickwork and blockwork from driving rain.

Materials such as bricks and blocks should arrive on site on pallets in a protective layer. This will keep them dry until they are required.

Hot summer weather can also have adverse effects on mortar and concrete, causing drying-out cracks.

Weather protection

Components can be protected from adverse weather by sheeting down the work as soon as it is completed. Examples are shown in Figure 10.23.

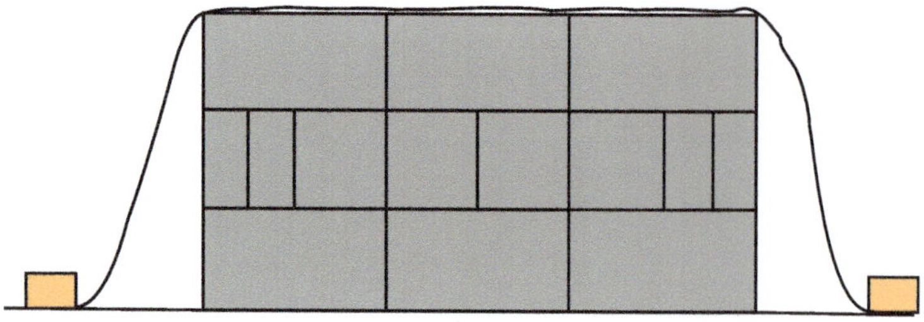

FIGURE 10.23
Protecting a finished wall: sheeting can be used to wrap over the finished wall

Working at heights

General scaffold information

Working platforms have been described in Chapter 2.

A scaffold is a temporary staging to assist bricklayers and other tradespeople to construct a building. The scaffold must be spacious and strong enough to support people and materials during construction.

As explained before, many accidents are due to simple faults such as misuse of tools, untied ladders or a missing toeboard.

Unfortunately, many accidents are due to ignoring the Construction (Health, Safety and Welfare) Regulations, which specify basic scaffolding requirements.

The three basic requirements for scaffolds are:

- They should be suitable for the purpose.
- They should be safe.
- They should comply with the Regulations..

General safety

No scaffold should be erected or be substantially added to or altered or be dismantled except under the immediate supervision of a *competent person*; and so far as possible by *competent workers* possessing adequate experience of such work.

When work cannot be safely carried out from ground level or from part of a building or other permanent structure, there shall be provided either scaffolds or where appropriate ladders or other means of support, all of which shall be suitable for the purpose.

It is important that you are familiar with a wide range of access equipment.

Access

It is usual to access a bricklayer's scaffold from a ladder, and this must be positioned so that this can be done easily and safely.

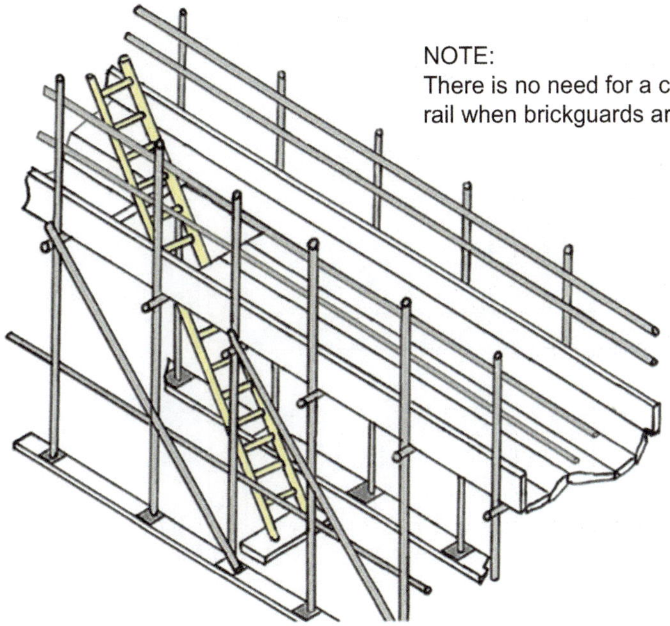

FIGURE 10.24
Access to working platform

Remember

The Work at Heights Regulations 2005 require that ladders should only be considered where a risk assessment has shown that the use of other more suitable work equipment is not appropriate because of the low risk, and the short duration of the task or considerations of where the work is located.

In the case of access to a scaffold, the ladder should be securely attached to the scaffold, preferably inside the structure (Figure 10.24).
Any surface on which the ladder rests should be stable and of sufficient strength to support the ladder so that its rungs remain horizontal and it can support any loading placed on it.

The final rung of the ladder from which the operative steps onto the platform should ideally be just above the surface of the platform.

The ladder, which should be secured at both the top and bottom, should extend at least 1.05 m (approximately five rungs) above the platform.

Bricklayers require a scaffold wide enough to allow:

- the stacking of materials
- working space for the bricklayer
- room for the passage of people and materials (Figure 10.25).

Guardrails and toeboards

Access platforms more than 2 m high must have guardrails and toeboards. The top guardrail should be at least 950 mm above the platform with intermediate guardrails fixed so that any gap between the rail and any other means of protection is not greater than 470 mm.

The risk of falling materials causing injury should be minimized by keeping platforms clear of loose materials.

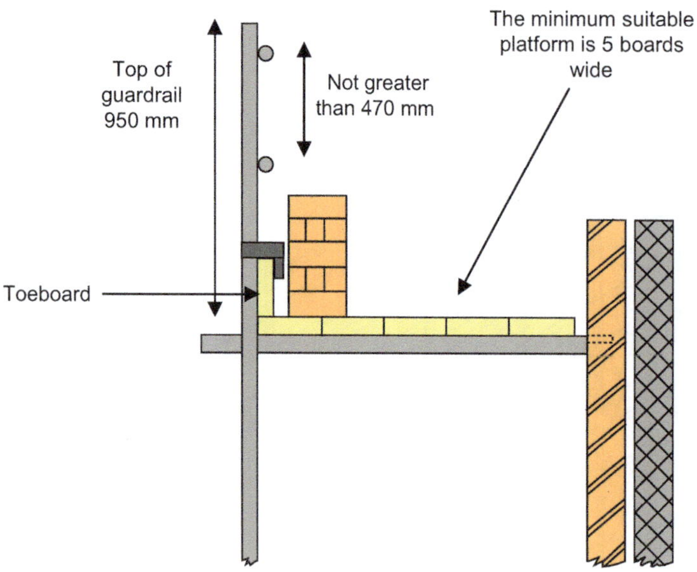

FIGURE 10.25
Bricklayer's working platform

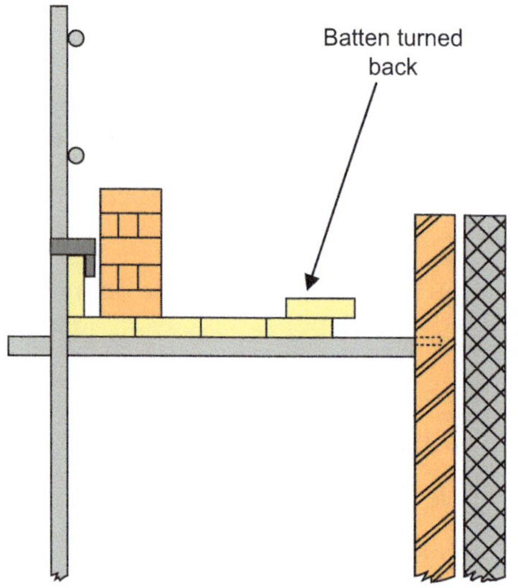

FIGURE 10.26
Inner batten turned back at the end of the working day

In addition, materials or other objects must be prevented from rolling, or being kicked, off the edges of working platforms. This can be achieved by fixing toeboards, solid barriers, brick guards or similar at open edges.

At the end of the working day it is important to turn back the batten nearest the new brick wall to prevent any splashing if it rains (Figure 10.26).

Multiple-choice questions

Self-assessment

This section of the book is designed to allow you to check your level of knowledge. The section consists of revision questions for this chapter. The questions are all multiple choice and have four possible answers. The answers are to be found at the end of the book.

The main type of multiple-choice question will be the four-option multiple-choice question. This will consist of a question or statement, known as the stem, followed by a choice of four different answers, called the responses. Only one of these responses is the correct answer; the others are incorrect and are known as distracters.

You should attempt to answer the questions by choosing either (a), (b), (c) or (d).

Example

The person employed by the local authority to ensure that the Building Regulations are observed is called the:

(a) clerk of works
(b) building control officer
(c) council inspector
(d) safety officer

The correct answer is the building control officer, and therefore (b) would be the correct response.

Cavity walling

Question 1 What is the minimum width of a cavity?

(a) 50 mm
(b) 67 mm
(c) 100 mm
(d) 150 mm

Question 2 The cavity wall should begin not less than how far below the horizontal DPC?

(a) 300 mm
(b) 225 mm
(c) 150 mm
(d) 75 mm

Question 3 Identify the type of wall tie shown:

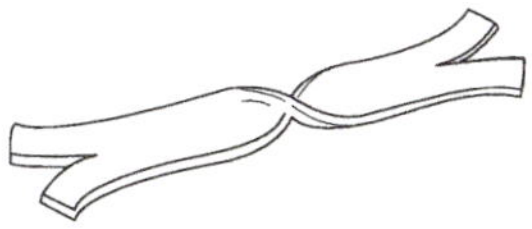

(a) butterfly wall tie
(b) plastic wall tie
(c) double triangle tie
(d) galvanized fishtail wall tie

Question 4 Cavity battens are used to:

(a) maintain the correct cavity width
(b) collect any mortar droppings in the cavity
(c) enable the wall to be built more quickly
(d) set out the correct bond

Question 5 The maximum horizontal distance for positioning wall ties is:

(a) 600 mm
(b) 700 mm
(c) 900 mm
(d) 1000 mm

Question 6 Which of the following is the correct method for fixing partial insulation slabs in a cavity wall?

(a) use special nails
(b) use special screws
(c) use special glue
(d) use a special wall tie

Question 7 What is the use of a coring hole?

(a) to allow the cavity to be cleaned out
(b) to allow services through the wall
(c) to provide decoration
(d) to save on bricks

Question 8 What is the purpose of sand courses?

(a) to provide decoration
(b) to allow bricks to be removed easily at a later date
(c) to prevent settlement of the wall
(d) to allow for expansion

CHAPTER 11

Joint Finishes

This chapter will cover the following City and Guilds units:

• Apprenticeship	Module 5	S3.3, K3.4
	Module 12	K3.3
• Diploma	Unit 103	4.4
	Unit 104	4.4
	Unit 105	4.5

This chapter is about:

- Interpreting instructions
- Adopting safe and healthy working practices
- Selecting materials, components and equipment
- Jointing and pointing brick and block structures

The following City and Guild performance criteria will be covered:

- Apply joint finishes to walls

The following Diploma outcomes will be covered:

- There is no comparable Diploma Level 1 Unit but all three practical units require pointing and jointing methods

Joint finishing

The surface-finishing treatment of new facework may be a jointing or pointing operation for bricklayers, and has an important effect on the finished appearance of brickwork.

When looking at stretcher-bonded facework, 18 per cent of what you see is mortar colour.

After the bricks have been bedded it is necessary to treat the exposed joints in some way to prevent the weather getting into the structure, and to provide a decorative appearance.

Jointing

Jointing is the craft term applied when joints are finished with the same mortar as is being used for the bricklaying, while the work proceeds.

Joint finishing is usually left until a convenient moment. Bricklaying will generally stop before break times and the end of the day to leave time for jointing up.

It is important to allow sufficient time for finishing the joints correctly but the need to do so at the right times throughout the day is also extremely important.

Pointing

Pointing is the term used to describe the surface finish applied to the cross-joints and bed joints of a brick wall when raked out to a depth of approximately 12 mm, and filled with a mortar of different colour, texture and sometimes density from that used for laying the bricks.

Appearance

Good jointing can improve poor brickwork, but bad jointing can spoil good brickwork.

Careful and skilful jointing can minimize the effect of small deficiencies in bricks and bricklaying, but careless jointing can make them look worse.

Jointing up is a critical part of building facework and is not something to be rushed at the end of the day.

It considerably affects the permanent appearance of facework, as almost one-fifth of the total surface consists of mortar joints.

Mortar mixes

Where facework is to be jointed as work proceeds, the bricklaying mortar of course provides the joint finishing colour. Ironing-in bricklaying mortar made from a fine-grained building sand, for example, will leave a smoother surface than if coarser local sand is used.

If fine-grained building sand is used to produce pointing mortar, then a 'weather struck and cut' finish will polish up better and may be cut or trimmed more cleanly with the frenchman than when the sand is coarser.

Cement-rich or strong pointing mortar should be reserved for very dense class A engineering bricks only, while $1{:}^1/_2{:}4^1/_2$ is suited to class B bricks.

A slightly stronger mix than 1:1:6 is in order for the majority of bricks with compressive strengths between 20 and 40 N/mm^2, e.g. 1:1:5. The reduction in sand improves the fattiness of the mortar, so that it sticks to the pointing trowel. If the fattiness of a pointing mortar needs further improvement, it is better to increase the proportion of lime rather than the cement.

Very careful and consistent batching is necessary, with strict control of mix proportions by volume using gauge boxes or buckets each time, if mortar is always to finish up the same colour and strength.

Timing

The timing is probably the most important aspect of jointing-up, particularly when making a neat flush joint without smudging the facework.

The right time to joint-up is determined by both the suction rate of the bricks and the weather conditions.

At one extreme, bricks of low water absorption that are very wet will have a low suction rate. The bricks will tend to 'float' and the mortar will dry very slowly, especially during wet or cold weather.

At the other extreme, high water absorption bricks that are very dry will have a high suction rate and the mortar will dry out very quickly. This can also affect the bond and in some cases bricklayers will dampen the bricks before they are laid.

During summer months it is necessary to joint-up every course in a length of walling. In winter months many course can be laid before the mortar is dry enough to joint.

The mortar should be soft enough for the jointing tool to leave a smooth surface and to press the mortar into contact with the brick arrises in order to maximize rain resistance.

Trying to finish a mortar joint that is too dry and pressing too hard with the jointing tool can 'blacken' the face of the joint. Trying to joint-up too soon spreads the mortar and leaves a rough joint surface.

To achieve the correct finish the bricks should be laid to perfect gauge. The jointing tool must remain in contact with the brick arrises above and below the bed joints and each side of the cross-joints, otherwise 'tramlines' will be left.

Cross-joints should always be finished first, whatever the type of joint finish.

Types of joint finish

Weather struck joint

The weather struck joint (Figure 11.1a) is a popular method of finishing joints on external facework. It gives protection against rain penetration as the slope of the joints runs water off the joint.

The upper edge of the joint is struck back with a pointing trowel. The bottom of the joint can be stuck by using a straight edge and trowel. This makes irregular bricks appear straighter than they are.

Struck joint

The struck joint (Figure 11.2) is normally used on internal fair-faced work, especially where the brick or blockwork is to receive an applied decorative finish of paint, etc.

It should not be used for external work as water could collect on the upper arrises of the bricks.

The lower edge of the joint is struck with a pointing joint.

Flush joint

In a flush joint (Figure 11.3), the mortar is compressed into the joint and finished flush with the face of the brickwork. It may be used for internal or external work.

The finish is obtained by rubbing over the joints lightly with a piece of cloth.

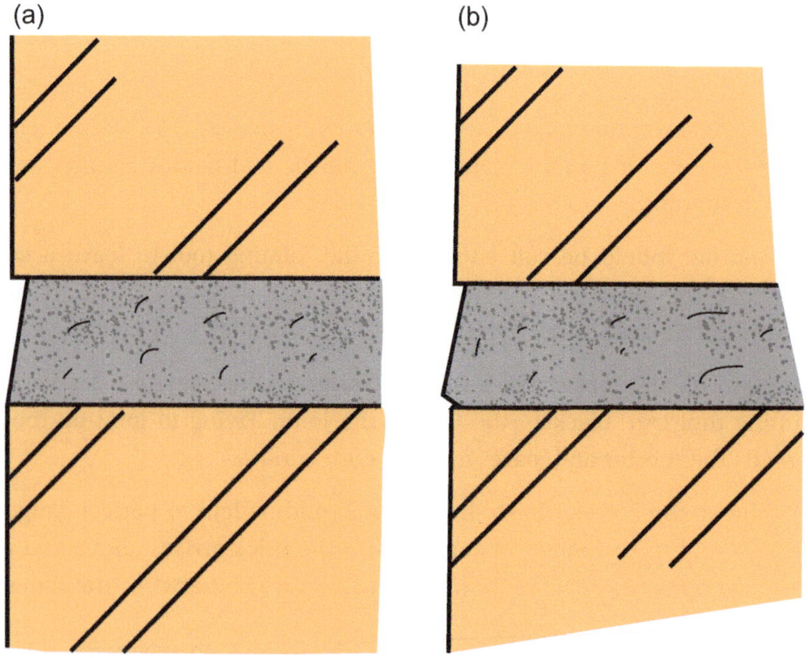

FIGURE 11.1
(a) Weather struck jointing; (b) weather cut and struck

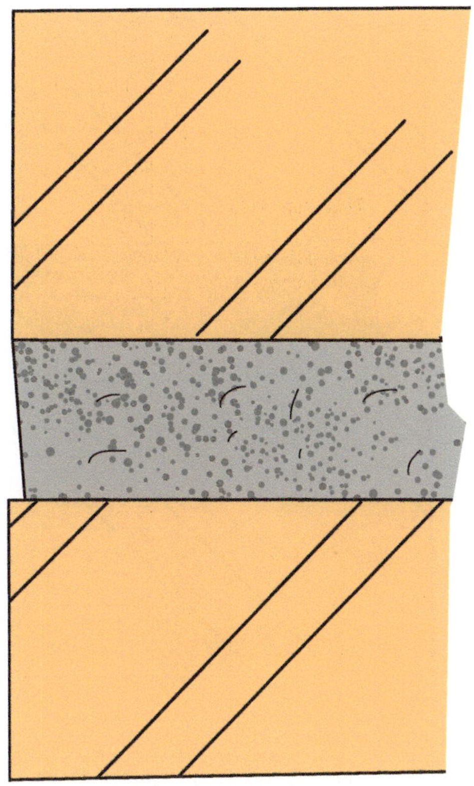

FIGURE 11.2
Struck jointing

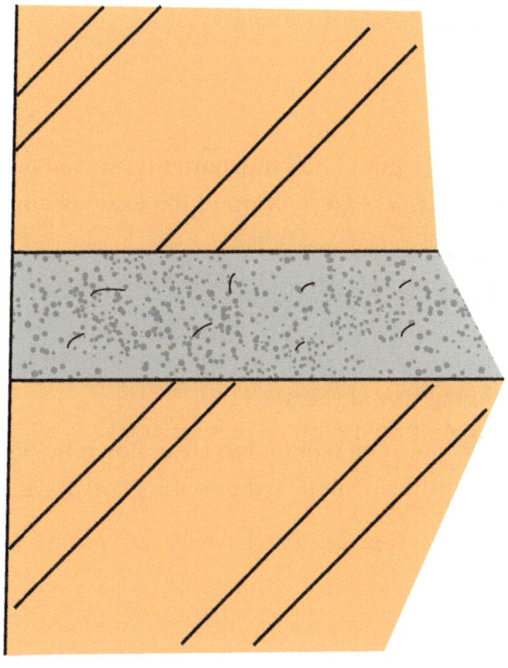

FIGURE 11.3
Flush jointing

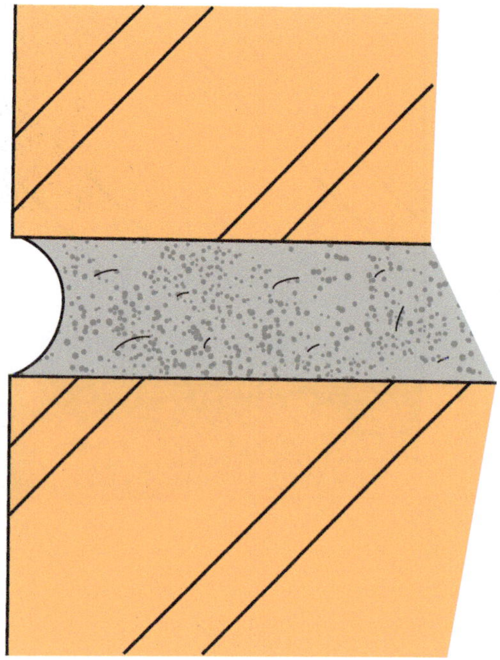

FIGURE 11.4
Keyed jointing

Keyed joint – rounded or tooled

A concave finish is obtained by rubbing a suitably shaped tool over the joint (Figure 11.4). The simple tool is often a piece of bucket handle or rounded mild steel piping.

This joint is mostly used for external work.

Recessed joint

In a recessed joint (Figure 11.5), the mortar is pressed back firmly into the joint with a metal jointer or a piece of wood the exact width of the joint.

Remember

Care is needed when finishing joints at external and internal angles.

Finishing joints in internal angles must emphasize the tie bricks or bonding at these points, finishing alternately left to right, and not with a straight joint.

Take care to continue jointing under projections, copings and soffits of arches.

Tools and equipment

The equipment required to carry out the treatment of joints includes the hand hawk and pointing trowel (Figure 11.6).

A small brush will also be required to clean down the face of the components after the work has been completed and the joints have set.

Brushing brickwork before the joints have set can spoil the appearance of facework.

Pointing

Today most face brickwork is 'jointed', which means that the joints are finished as the work proceeds and should require no further attention at the end of the

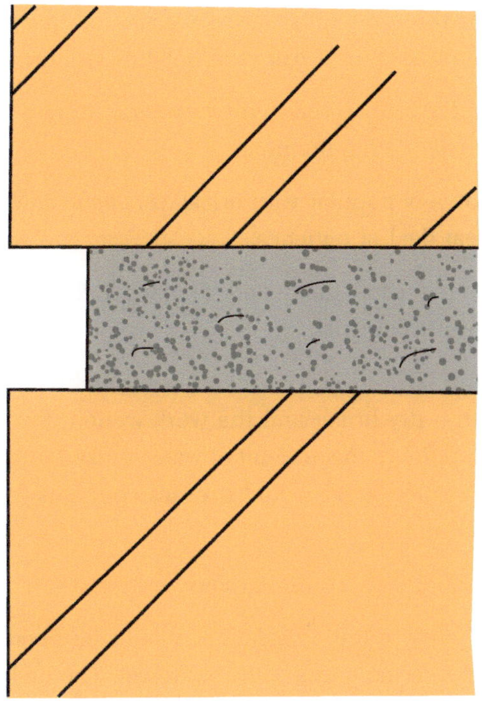

FIGURE 11.5
Recessed jointing

day. Occasionally, however, architects will specify that the joints shall be 'pointed' in order to achieve a particular effect.

When new work is to be pointed all joints are raked 12–15 mm deep on the day the wall is built ready to receive a different mortar, or at a later date.

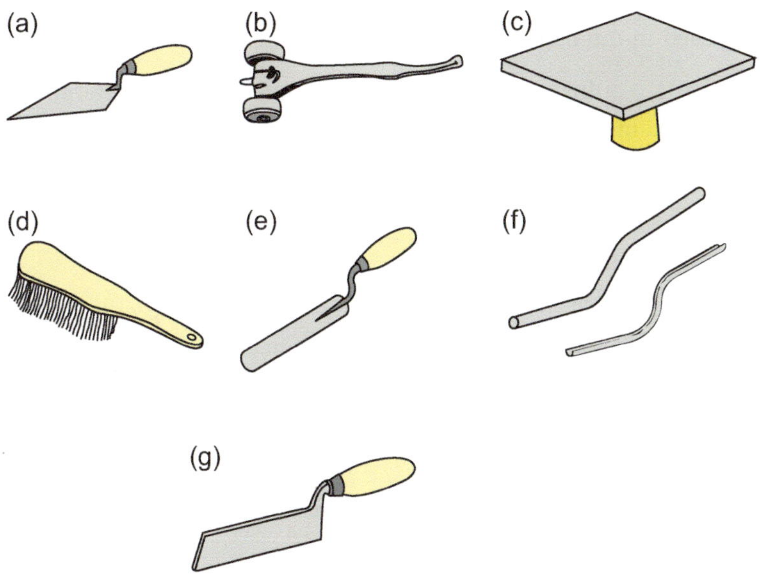

FIGURE 11.6
Tools and equipment used for jointing and pointing: (a) pointing trowel; (b) chariot jointer; (c) hand hawk; (d) brush; (e) finger trowel; (f) bucket handles; (g) recessed jointer

An architect may have requested pointing to ensure that the finished texture and colour of the joint finish are constant; or the architect may require a differently coloured mortar from the bedding mortar to form the joint finish.

Any of the styles can be selected when repointing a wall, depending on the architect's or client's requirements.

Pointing is not very popular with bricklayers as it slows the process down and requires a great deal of patience.

Careless pointing can spoil good brickwork, but good pointing can considerably improve facework.

Before pointing begins, all loose mortar and debris should be removed from the joints with a dry brush and the work wetted down to a damp condition. Wetting down reduces the amount of water sucked into the brickwork from the mortar, which if too great would prevent complete hydration, resulting in a weak, crumbly mortar.

Repointing should start at the top and work down.

Again, as for jointing the cross-joints should be filled in first. The mortar is pressed firmly into the joints with the trowel. The bed joints are filled in next and also pressed firmly into the joints.

The mortar should inset approximately 1 mm at the top and 'cut projecting' the lower edge by the same amount. This will leave a sloping weathered surface which will allow rain water to fall away quickly and therefore provide better rain resistance than recessed or flush joints.

Repointing

Repointing may be required when the joints of a wall have eroded after a considerable amount of time, owing to exposed weather conditions or poor materials used in the original mortar.

If repointing is required the original mortar joints need to be removed to a depth of approximately 12 mm.

The whole area should be dusted down to remove any powder from the joints and then dampened down to provide a key for the new mortar. See Figure 11.7 for the tools and equipment required.

The existing joints will have to be removed with a plugging chisel and lump hammer. There are also many modern devices that can be set to rake out joints to various depths.

Prior work

Before old brickwork is repointed it is essential that the cause of the deterioration is established.

It is usually the result of slow erosion over many years, but if it is due to sulphate attack on the mortar then the cause should be sought out and corrected before any repointing is carried out.

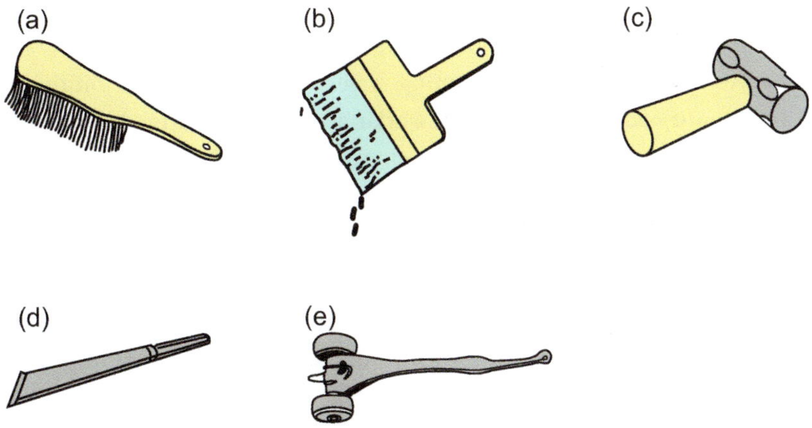

FIGURE 11.7
Tools and equipment for repointing: (a) dry brush; (b) wet brush; (c) lump hammer; (d) plugging chisel; (e) chariot jointer

Access

The sequence of operations is virtually the same as for new brickwork, except that a working platform will have to be erected. When jointing and pointing new work the scaffold would already have been erected.

The most common method is for a tower scaffold to be used (Figure 11.8), but it all depends on the height of the wall to be repointed.

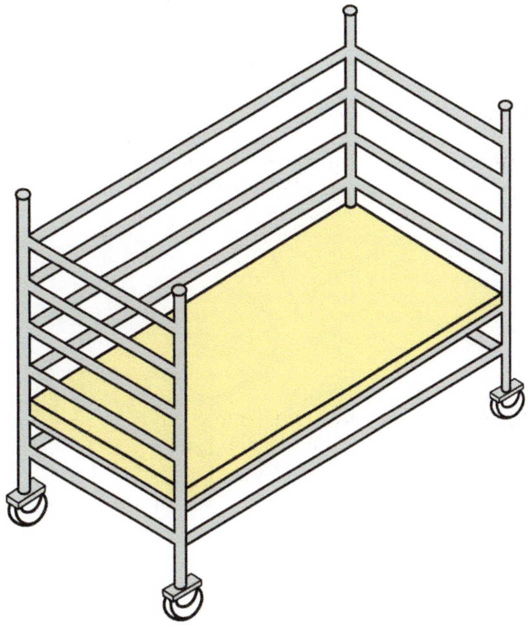

FIGURE 11.8
Small tower scaffold

Techniques

Jointing

Whichever finish is to be used, timing is very important when jointing-up as the bricks are laid throughout the working day. The mortar between bricks should be allowed to stiffen up just enough due to brick suction, so that the jointing tool can pass smoothly and cleanly. Too soon and the mortar smears and does not leave a smooth profile. Left too long before jointing, heavy pressure on the jointer leaves black metal marks on the dried mortar face.

With the single exception of tuck pointing, whether jointing or pointing, always do the cross-joints first, followed by bed joints, each time you stop bricklaying to joint-up.

Brushing-off with a soft bristle hand brush, to remove any loose crumbs of mortar, should be left until the end of the day. Brush lightly and on no account leave bristle marks in the mortar face. It is better to leave brushing until the following morning than to risk marking the joints. Take particular care when jointing face brickwork at those points shown in Figure 11.9.

Pointing

This craft operation, carried out some weeks or months after the wall has been built, requires patience and is a skill that takes time to develop. The joint finish commonly specified for pointing brickwork is weather struck and cut (see Figure 11.1b).

1. Always start at the very top of the walling to be pointed.
2. Remove any obvious hardened crumbs of mortar clinging to the wall face when the joints were raked out some weeks or months ago.

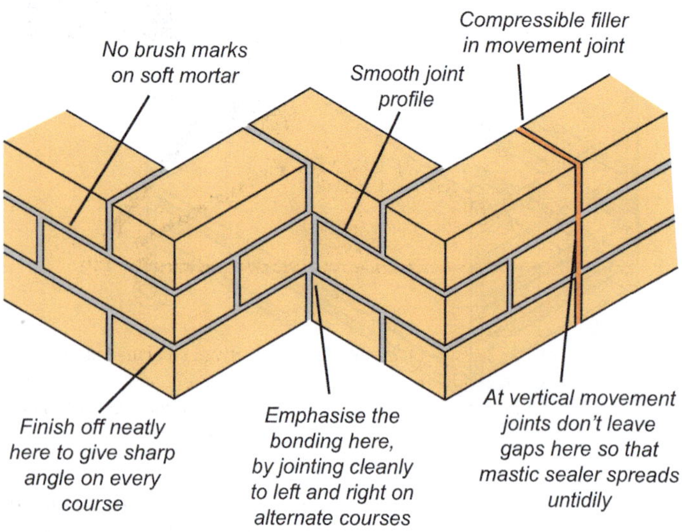

FIGURE 11.9
Fine points of jointing and pointing

3. Brush the whole lift of brickwork using a stiff-bristle hand brush.
4. Wet the wall face generously if the bricks are very absorbent, less generously if the bricks have a lower suction rate.
5. Load the hand hawk with mortar flattened out to approximately 10 mm thickness.
6. Using the small pointing trowel or 'dotter', pick up joint-sized pieces of mortar from the hawk and press carefully and firmly into cross-joints, but see also item 14 below.
7. Completely fill each cross-joint with a second application if necessary, polish the mortar surface and indent on the left-hand side.
8. After completing approximately $^1/_2$ m^2 of cross-joints, cut or trim the right-hand side of all these joints in the manner shown in Figure 11.10, so that they all look the same width on face.
9. Using a longer pointing trowel, pick up joint-sized pieces of mortar from the hawk and start pointing one bed joint, pulling the loaded trowel up to the last piece of mortar applied each time.
10. After filling a 500 mm length of bed joint, polish it with the pointing trowel and indent the top.
11. When half a dozen bed joints have been pointed in this way, cut or trim the bottom edge of each one (Figure 11.11) using a frenchman or the tip of a pointing trowel together with a feather edge pointing rule. The amount of mortar to be left projecting from the wall face after trimming or cutting joints should not exceed the thickness of the trowel blade.
12. Sensibly adjust areas of pointing to suit the drying conditions of bricks and weather, so that joints cut cleanly. Too early and the mortar will not fall away cleanly when trimmed. Too late and the joint edges will crumble and not leave a clean straight line when cut.

FIGURE 11.10
Weather struck and cut pointing: cutting the right-hand side of cross-joints

FIGURE 11.11
Weather struck and cut pointing: trimming the bottom of bed joints with straight edge and frenchman

13. Brush very lightly at the end of the day, or preferably on the following day if there is the slightest risk of marking the sharp cut edges of the pointing.
14. If a wall has alternate bands of class A and absorbent facing bricks, after wetting the whole wall, point absorbent facings first. When the wall has dried off, return and point the class A bands.

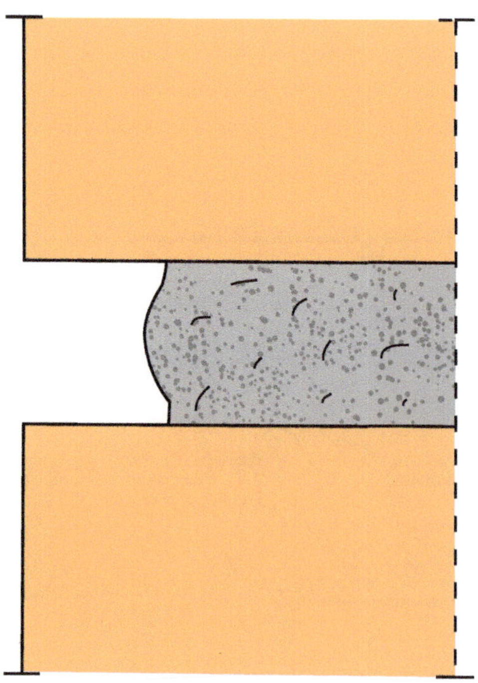

FIGURE 11.12
Correct way of raking out joints in preparation for repointing

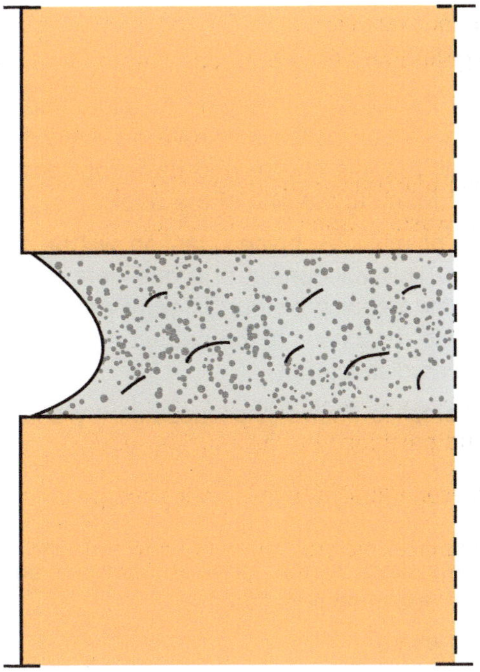

FIGURE 11.13
Incorrect way of raking out, which will lead to pointing failure due to poor adhesion

Summary

The great merit in surface finishing mortar as a jointing process is that the joint profile is an integral part of the mortar bed and there is no possibility of failure through insufficient adhesion between the main mortar bed and the surface finish.

Failure of pointing, i.e. its separation from the main mortar bed and consequent falling away, is caused by careless raking out and lack of suitable preparation. To overcome possible failure, joints in new brickwork should be raked out to a depth of at least 12 mm, as shown in Figure 11.12, and not as shown in Figure 11.13.

Multiple-choice questions

Self-assessment

This section of the book is designed to allow you to check your level of knowledge. The section consists of revision questions for this chapter. The questions are all multiple choice and have four possible answers. The answers are to be found at the end of the book.

The main type of multiple-choice question will be the four-option multiple-choice question. This will consist of a question or statement, known as the stem, followed by a choice of four different answers, called the responses. Only one of these responses is the correct answer; the others are incorrect and are known as distracters.

You should attempt to answer the questions by choosing either (a), (b), (c) or (d).

Example

The person employed by the local authority to ensure that the Building Regulations are observed is called the:

(a) clerk of works
(b) building control officer
(c) council inspector
(d) safety officer

The correct answer is the building control officer, and therefore (b) would be the correct response.

Jointing and pointing

Question 1 When the joint finish is completed as the work proceeds it is known as:

(a) pointing
(b) jointing
(c) finishing
(d) flushing

Question 2 Identify the following joint finish:

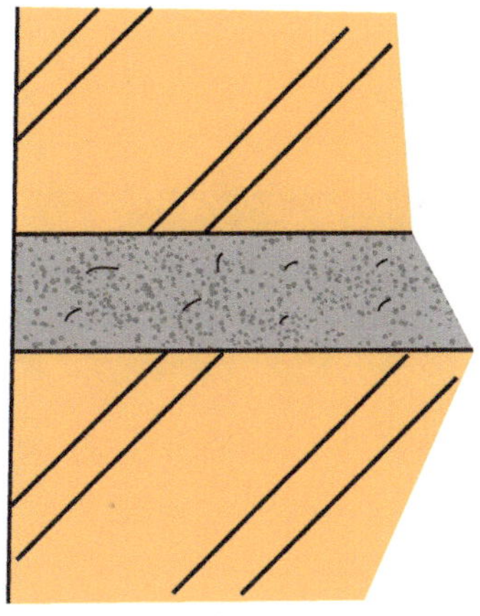

(a) flush joint
(b) half round joint
(c) recessed joint
(d) weather struck joint

Question 3 Where is the correct start position when repointing a wall?

(a) around doors and windows
(b) highest point
(c) lowest point
(d) anywhere

Question 4 When the joint finish is applied after the whole area has been completed it is known as:

(a) pointing
(b) jointing
(c) finishing
(d) flushing

Question 5 What is the correct depth for raking out mortar joints before repointing a brick wall?

(a) 16 mm
(b) 14 mm
(c) 12 mm
(d) 10 mm

Question 6 Identify the tool used for finishing joints:

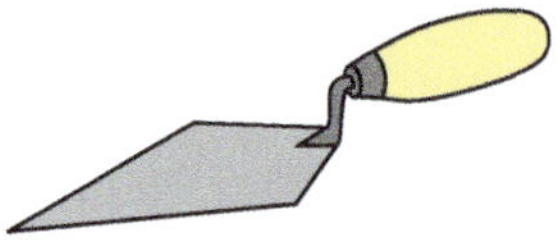

(a) jointing iron
(b) frenchman
(c) pointing iron
(d) pointing trowel

Question 7 Identify the following joint finish:

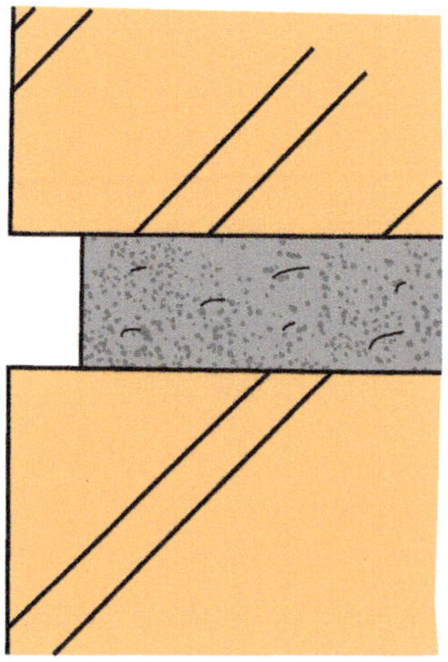

(a) flush joint
(b) half round joint
(c) recessed joint
(d) weather struck joint

Question 8 The main purpose of a joint finish is to:

(a) make the brickwork more decorative
(b) save on mortar
(c) compact the mortar
(d) make the brickwork stronger

CHAPTER 12

Answers to Multiple-Choice Questions

Chapter 1. The Construction Industry (Residential)

1 (b); 2 (a); 3 (c); 4 (d); 5 (b); 6 (b); 7 (a); 8 (c)

Chapter 2. Health and Safety in the Construction Industry

1 (b); 2 (a); 3 (a); 4 (c); 5 (d); 6 (c); 7 (d); 8 (a)

Chapter 3. Communication

1 (c); 2 (a); 3 (b); 4 (d); 5 (c); 6 (d); 7 (a); 8 (a)

Chapter 4. Construction Technology

1 (a); 2 (d); 3 (a); 4 (b); 5 (b); 6 (b); 7 (a); 8 (c)

Chapter 5. Tools

1 (c); 2 (b); 3 (d); 4 (a); 5 (a); 6 (b); 7 (b); 8 (b)

Chapter 6. Preparing and Mixing Concrete and Mortar

1 (b); 2 (b); 3 (c); 4 (d); 5 (a); 6 (c); 7 (a); 8 (b)

Chapter 7. Setting Out Basic Masonry Structures

1 (a); 2 (d); 3 (b); 4 (c); 5 (a); 6 (c); 7 (b); 8 (d)

Chapter 8. Laying Bricks to Line

1 (d); 2 (b); 3 (b); 4 (a); 5 (c); 6 (b); 7 (a); 8 (a)

Chapter 9. Laying Blockwork to Line

1 (d); 2 (d); 3 (a); 4 (b); 5 (b); 6 (d); 7 (c); 8 (a)

Chapter 10. Cavity Walling

1 (a); 2 (c); 3 (d); 4 (b); 5 (c); 6 (d); 7 (a); 8 (b)

Chapter 11. Joint Finishes

1 (b); 2 (a); 3 (b); 4 (a); 5 (c); 6 (d); 7 (c); 8 (c)

Index

Abrasive wheels 50
Accidents 59-65
  book 70
  causes of 60
  emergency 68
  reporting 69-70
Adjustable steel trestle 84
Aggregates 149, 196-198
Agrément certificate 193
Anchor straps 168
Apprentice 133
Approval 140
Applying cross joints 299
Architect 131
Arris 300
Asbestos regulations 44
Asphalt DPC 164
Atmospheric depth 333
Attached piers
  bonding 320

Backactor 145
Back-to-back houses 20
Banding of bricks 160
Baroque style 14
Barrows 232
Bench marks
  temporary 143
  ordnance 143-144
Bevelled closers 295
Bill of quantities 248
Blocks
  bearing 322
  below ground level 324
  bonding 315
  broken bond 317
  clay blocks 310
  concrete 310
  foundation blocks 324
  laying 325
  reverse bond 318
  setting out 313
  sizes 311
  special blocks 310-313
  types 310-312
Block bonding 322
Blockwork
  attached piers 320
Blockwork indents 318
Blockwork pillars 320
Bonding
  principles 283
  purpose 282
  rules of 284
  types 286-293
Bonds *see separate names*
Boning rods 267-268
Brick profiles 305
Bricklaying technique 298
Bricklaying trowels
  types 133
  handling 297
Bricks 113
  actual size 113
  co-ordinating size 279
  dimensional 279
  on site test 281
  proportions 278
  tolerances 280
Bricks and mortar quantities 227-228
British Standards 192
British Standards Code
  of Practice 193
Broken bond 317
Builders 2
Building styles 4
  Adam 16
  Cruck house 6
  Baroque 14
  Edwardian 21-22
  Elizabethan 11
  Georgian 14
  Gothic style 7
  Modern 22
  Norman house 6
  Palladian 15
  Regency 16
  Renaissance 10
  Roman 5
  Saxon 5
  Stuart 11
  Tudor 9-10
  Victorian 18-20
Builder's square 251
Building team 128
  types 26
Building line 261
Building Regulations

cavity walls 331
general 192
Building sand 196
Building control officer 131
Bulking of sand 216

Calculations
areas 118
blockwork 115
brickwork 114
circles 112
concrete 115, 224
cubic 112
drainage 116
linear 112
mortar 120
number 112
percentages 120
perimeters 119
ratios 121
setting out 260
superficial 113
triangles 116
types 112
Careers 133
Carrying 78
awkward shapes 79
materials 81
Cavity wall
battens 336
below ground 340
boards 335
cleanliness 336
concrete fill 333, 340
definition 330
foundations 340–342
functions 330–332
heat loss 331
insulation 334–336
lintels 322
loads 333
openings 339
purpose 330
wall ties 337
weep holes 337
CDM Regulations 36, 58
CE marks 64
Cement
general 199
types 200
storage 150
Chemical injuries 195
Chisels
bolster 177
cold 178
comb 178
plugging 178
Civil engineer 131
Classification of dwellings 4
Clerk of works 131
Clients 126–129
Closers 295
Cob walling 12–13
Collar joint 287
Colouring agent 202
Communal buildings 25
Communication
methods 121
negative 125
oral 121
positive 125
site diary 123
taking notes 124
telephone 123
written 122
Companies
medium 2
large 2
limited 7
small 2
Competent person 83
Compound 147
Concrete 209
compaction 210, 238
constituents 210–233
curing 240
finishing 239
manufacture 233
mixing - hand 211–213, 229
mixing - machine 214
placing 210, 238
pumps 237
quantities 220–225
recommended mixes 211
reinforced 235
requirements 145–146
testing 216–220
transporting 236
type of mix 211–233
workability 212
yield 221
Construction industry 2, 3
Construction process 3
Contractors 2
Contracts manager 130
Coring holes 337–339
Corner blocks 184, 303
Compound 149
Compressor 50
COSHH 36, 42
Cowley level 269–270
CPR 55
Cross joints 299
Cross walls 302
Cruck house 6
CSCS scheme 77
Cube test 218–220
Cut bricks 295
Cutting
blocks 314
bricks 294

Damp proof
course 162
types 162–164
Datum 143, 265
Decorated gables 10
Detached building 26
Diagonals 263
Double insulated drill 51

Double returns 290–293
DPC
  bedding 165
  engineering bricks 165
  laps 164–166
  materials 164
  types 164
Drawings
  abbreviations 103–105
  assembly 107
  block plan 106–109
  colour 105–106
  details 107–109
  graphical symbols 104–106
  hatchings 102
  location plan 105–106
  range drawings 107–108
  scales 100–101
  setting out 247
  simple 97
  site plan 108
  types 105
  understanding 108
  working 247
Dry bonding
  blocks 314
  bricks 294
Dry silo mortars 208
Dumpers 232
Dwelling units 25

Electric fuses 53
Electricity at work regulations 36, 51
Electricity on site 32
Electric shock 54
Elizabethan 10
Emergency
  action 50–68
  procedures 68
  resuscitation 55–56
Engineering bricks 163
English bond 286
  double returns 290
  quoins 288, 305
  straight 287
  tee junctions 289
E-platforms 125
Estimator 129
European Standards 193
Excavator 142
Excavations 146

Feather edge 184
Fire
  emergency procedures 71
  factors 71
  fire-fighting equipment 72–73
  regulations 36, 47
Firms 2
First aid 69
First aid box 69
Flats
  high rise 6
  low rise 6
  medium rise 6
Flemish bond 291
  double returns 293
  quoins 291
  straight walls 291
  tee junctions 292
Flush jointing 354
Formwork 233
Forward loading shovel 232
Foundations 145
  rafts 345
  stepped 343
  strip 146, 332
Frenchman 184, 362
Frog 113
Frontage line 261–263
Fully filled cavity
  insulation 335

Ganger 132
Gauge
  horizontal 285
  rods 114, 293
  vertical 285
Gauge boxes 206
General operatives 132
General supervisor 130
Generator 31
Georgian 14
Goal posts 52
Gothic 7
Government department 127
Ground floors
  solid concrete 332
  suspended concrete 332
Ground-work activities 141
  foundations 142
  site clearance 142
  site preparation 142
  trench excavation 142

Half bond 283
Half brick quoin 301
Half round joints 356
Hammers
  brick 176
  club/lump 176
  comb 176
  scutch 176
Hand hawk 179
Hand-made bricks 300
Handling bricks 296
Handling materials 78
Hardwick Hall 11
Hatching and grinning 300
Hazards 61
  spotting 61
Header course 288
Health and hygiene 75
Health and Safety at Work Act 36
Health and Safety Executive 41
Health and Safety Inspector 41
Health and safety risks 43
Heritage 28
High rise 6
Hop ups 83

Industrial buildings 25
Information
  interpreting 97
  relaying 121
  signs 42–45, 64–66
  sources 97
  storing 97

Jacobean 11
Joint finishing 352
  appearance 353
  mortar mixes 352
  reasons for 352
  types 352–356
Jointing 352
  definition 352
  techniques 360
  timing 352
  tools 357
Junction walls 318

Keyed jointing 356
King closer 295
Knapped flint walling 12

Ladders
  lifting 91
  safe use of 90
  types 89
  tying ladders 91
Laying bricks 299
Layout of materials brick walls 296
  block walls 325
Level pegs 267
Levels
  checking 182
  Cowley level 269–270
  laser 271
  site square 271–272
  transferring 265
  water level 268
Lightweight staging 86
Lifting operations (LOLER) 36, 55
  equipment 81
  materials 80
Lime 199
Link detached 27
Lintels
  bearings 322
  bedding 322
  steel 339
Listed buildings 28
  Grade I 30
  Grade II^{x} 30
  Grade II 30
Local authority 28
Location plan 105–106

Main contractors 130
Maintenance work 2
Mandatory signs 36, 58
Manufacturers technical information
  97, 191
Masonry saws
  bench 50
  hand 181

Materials
  handling 148
  layout 149–150
  protecting 151–153
  setting out 160
  transporting 161
Mathematical tiles 16
Mechanical tools and equipment 50
Mitred bats 295
Mixing equipment 230–231
Mechanical equipment hazards 82
Mechanical injuries 194
Mobile plant 49, 232
Mobile towers 92
Mortar 202
  batching 225
  calculations 226
  constituents 225
  designation 205
  dry silo mortars 207
  mixing 229
  mortar mixes 203
  premixed 231
  reinforced 209
  requirements 204
  selection 203
  spreading 297
  testing 207
  transporting 232
  types 202–203
  typical uses 208
Multiple-choice answers 367

New Towns Act 24
New work 2
Noise at Work Regulations 36, 45
Norman house 6

One-brick walls 287
Ordinary Portland cement 199
Ordnance bench mark 144
Overhanging upper storey 8

Palladian 15
Partial fill cavity 336
Partnership 126
Patent profiles 305
PAT testing 54–55
Perpends 285
Personal hygiene 77
Personal safety 50, 51
Planning officer 131
Plasticiser 201
Platforms 84
Plumbing
  blockwork 316
  brickwork 301
Pointing 356
  definition 357–358
  rules 358
  technique 360
  tools 359
PPEs 47, 62
Prefab 23
Private individual 126
Private quantity surveyor 129

Private sector 126
Profile
  bricklaying 305
  foundation 265
Prohibition signs 64
Proprietary wall connectors 319
Protection of work 345
Public sector 126
Public undertaking 127
PUWER 36, 48

Qualifications 135
Quantities of materials 111
Quarter bond 286
Quoins
  blocks 321-322
  bricks 301

Racking back 301
Range of activities 127
Ranging in 301
Ranging lines 250
RCD 53
Ready-to-use mortars 206
Recessed jointing 357
Recreational units 25
Refurbishment 3
Regency 16
Release agent 236
Renaissance 10
Re-pointing 358
Residential units 25
Restoration 3
Retarder 201
Reverse bond 318
Reversing a spirit level 182, 254
RIDDOR 36, 47
Right angle
  3:4:5 method 252
Risk assessments 73-75
Roman Empire 5
Roofs
  connectors 167
  eaves detail 169
  flat roofs 169
  terminology 166
  trusses 167
  types 167
  wall plates 167
Rules of bonding 284
Running laps in DPC 164

Safety
  documentation 41-46
  duties 39
  in the workplace 41-62
  legislation 35, 192
  officer 132
  policy 39
  practices 190
  stress injuries 193
Sand courses 343
Saxon house 5
Scaffolding
  access 359
  boards 85
  guard rails 348
  hop ups 83
  low towers 93
  system 92
  toe boards 348
  tower 91
Scale rules 99
Schedule 99, 111, 248
Security
  individual 66, 257
  site 67, 257
Semi-detached units 27
Services provision 343-344
Setting out 259-265
  builder's square 251
  building line 266
  calculations 255
  centre line 141
  diagonals 263
  drawings 247
  equipment 249-251
  face brickwork 290
  kerb line 261-264
  profiles 264-265
  steel pegs 145
Silt test 196
Site agent 132
Site clearance 142, 254
Site location 105
Site preparation 142
Site safety
  signs 64
  supervisor 39
  survey 40
Site square 271-272
Slate DPC 162-164
Sliding bevel 180
Slump test 216-218
Solid ground floor 332
Source of work 126-129
Specifications 97-109
Spirit level 181
  checking 182, 254
Split heads 84-85
Spot boards 296
Spreading mortar 297
Standard strip foundation 332
Static plant 50
Statutory Authority 127
Step ladders
  aluminium 88
  wooden 87
Stepped foundations 343
Storage 146
  flammable liquids 155
  materials 148
  protection 151
  workshop/yard 148
Storey rod 184-185, 293
Straight edge 184, 251
Straight joints 282-283
Stretcher bond 301
Stretcher course 288
Strip foundation 332
Struck joint 354
Stuart 11

Stucco 18
Subcontractor 133–134
Supervisor 132
System scaffold 93

Taking off 111
Team working 241
Tee-junctions
  English 289
  Flemish 292
Temporary bench mark 143
Terraced units 27
Thermal insulation 334–336
Three:four:five method 266
Tingle plate 303
Tools 174–184
  brick gauge 163
  brick trowels 175–176
  chisels 177–178
  corner blocks 185
  dividers 179
  hand brush 179
  hand hawk 179
  jointing irons 175
  lines and pins 176, 303
  sliding bevel 180
  square 179
  trammel heads 180
  tapes 178, 249
  tool kit 176
Toothings 305
Top soil 140
Tower scaffold 92
Town and Country
  Planning 192
Trades people 132
Transferring levels 265
Transformers 52
Traveller 268
Trestle platforms 84
Tudor 9
Types of buildings 25

Vandalism 68
Vibration at work 36, 57
Victorian 18

Waste disposal 158
Water 201
Water:cement ratio 216
Water level 268
Wall plate 168
Wall ties 337
Warning signs 66
Weather struck jointing 354
Weep holes
  inserts 337
Welfare facilities 58
Welsh arch 345
Window tax 14
Work at Height Regulations 36, 46
Working below ground 241
Working drawings 247
Working platforms 82
Working space 296
Workshops 60

For Product Safety Concerns and Information please contact our EU representative GPSR@taylorandfrancis.com Taylor & Francis Verlag GmbH, Kaufingerstraße 24, 80331 München, Germany

Batch number: 10731606

Printed by Printforce, the Netherlands